AF702650

"This groundbreaking book is an indispensable resource for anyone concerned about the impacts of mineral extraction on Indigenous Peoples and local communities worldwide. With meticulous research and a commitment to amplifying marginalized voices, the authors offer a comprehensive analysis of the complex dynamics surrounding mining encounters. By emphasizing comparative perspectives and the holistic understanding of social, environmental, economic, and cultural factors, this work sheds light on both the opportunities and challenges presented by extractive industries. Through collaboration with Indigenous organizations and partners across multiple countries, the MinErAL network has produced invaluable insights that contribute to the global dialogue on sustainable development and Indigenous rights. A must-read for policymakers, researchers, and advocates alike."

—**Rosa Galvez,** *Canadian Senator representing Québec (Bedford); formerly a professor and head of the Civil and Water Engineering Department at Université Laval in Québec City, Canada*

"This collection is an excellent and necessary resource for those interested in the relations between mining and Indigenous Peoples. At a time when Indigenous Peoples are increasingly facing pressures from extractive industry, this book convincingly shows the need for a comparative approach and a holistic understanding of the impacts of these activities."

—**Chris Southcott,** *professor in the Department of Sociology, Lakehead University, Canada*

Mining and Indigenous Livelihoods

This book maps the encounters between Indigenous Peoples and local communities with mining companies in various postcolonial contexts.

Combining comparative and multidisciplinary analysis, the contributors to this volume shine a light on how the mining industry might adapt its practices to the political and legal contexts where they operate. Understanding these processes and how communities respond to these encounters is critical to documenting where and how encounters with mining may benefit or negatively impact Indigenous Peoples. The experiences and reflections shared by Indigenous and non-Indigenous contributors will enhance our understanding of evolving practices and of the different strategies and discourses developed by Indigenous Peoples to deal with mining projects. By mobilizing in-depth fieldwork in five regions—Australia, Canada, Sweden, New Caledonia, and Brazil—this body of work highlights voices often marginalized in mining development studies, including those of Indigenous Peoples and women.

This book will be of great interest to students and scholars of mining and the extractive industries, sustainable development, natural resource management, and Indigenous Peoples.

Thierry Rodon is a professor in the Department of Political Science at Université Laval, Canada, and holds the INQ Research Chair in Northern Sustainable Development.

Sophie Thériault is a full professor in the University of Ottawa's Faculty of Law (Civil Law Section), Canada, where she served as vice-dean, Academic (2019–2023), and as vice-dean, Graduate Studies (2015–2017).

Arn Keeling is a settler-scholar and professor of geography at Memorial University of Newfoundland and Labrador in St. John's, Canada.

Séverine Bouard is a human geographer (PhD) at IAC, New Caledonia.

Andrew Taylor is an associate professor of demography at the Northern Institute of Charles Darwin University in Australia.

Routledge Studies of the Extractive Industries and Sustainable Development

The Shaping of Greenland's Resource Spaces
Environment, Territory, Geo-Security
Mark Nuttall

Indigenous Responses to Mining in Post-Conflict Colombia
Violence, Repression and Peaceful Resistance
Diana Carolina Arbeláez Ruiz

Mining, Mobility, and Social Change in the Global South
Regional Perspectives
Edited by Gerardo Castillo Guzmán, Matthew Himley, and David Brereton

Local Responses to Mine Closure in South Africa
Dependencies and Social Disruption
Edited by Sethulego Matebesi, Lochner Marais, and Verna Nel

Tree Plantation Extractivism in Chile
Territories, Fundamental Human Needs, and Resistance
Alejandro Mora-Motta

Mining and Indigenous Livelihoods
Rights, Revenues, and Resistance
Edited by Thierry Rodon, Sophie Thériault, Arn Keeling, Séverine Bouard, and Andrew Taylor

For more information about this series, please visit: www.routledge.com/Routledge-Studies-of-the-Extractive-Industries-and-Sustainable-Development/book-series/REISD

Mining and Indigenous Livelihoods

Rights, Revenues, and Resistance

Edited by Thierry Rodon, Sophie Thériault, Arn Keeling, Séverine Bouard, and Andrew Taylor

Routledge
Taylor & Francis Group
LONDON AND NEW YORK

earthscan
from Routledge

First published 2025
by Routledge
4 Park Square, Milton Park, Abingdon, Oxon OX14 4RN

and by Routledge
605 Third Avenue, New York, NY 10158

Routledge is an imprint of the Taylor & Francis Group, an informa business

© 2025 selection and editorial matter, Thierry Rodon, Sophie Thériault, Arn Keeling, Séverine Bouard, and Andrew Taylor; individual chapters, the contributors

The right of Thierry Rodon, Sophie Thériault, Arn Keeling, Séverine Bouard, and Andrew Taylor to be identified as the authors of the editorial material, and of the authors for their individual chapters, has been asserted in accordance with sections 77 and 78 of the Copyright, Designs and Patents Act 1988.

The Open Access version of this book, available at www.taylorfrancis.com, has been made available under a Creative Commons Attribution-Non Commercial-No Derivatives (CC-BY-NC-ND) 4.0 International license.

Any third party material in this book is not included in the OA Creative Commons license, unless indicated otherwise in a credit line to the material. Please direct any permissions enquiries to the original rightsholder.

Funded by Université Laval.

Trademark notice: Product or corporate names may be trademarks or registered trademarks, and are used only for identification and explanation without intent to infringe.

British Library Cataloguing-in-Publication Data
A catalogue record for this book is available from the British Library

ISBN: 978-1-032-51628-8 (hbk)
ISBN: 978-1-032-52394-1 (pbk)
ISBN: 978-1-003-40643-3 (ebk)

DOI: 10.4324/9781003406433

Typeset in Times New Roman
by Apex CoVantage, LLC

Contents

Contributors

Glenn Banks is a geographer in the School of People, Environment and Planning at Massey University, New Zealand.

Mathilde Baritaud is a research assistant at the University of French Polynesia (Puna'auia, Tahiti), affiliated to the CETOP, specialized in the study of tourism in Pacific Ocean.

Justine Becker is a former research assistant in the School of Labour Studies at McMaster University, Hamilton, Canada.

Zoé Boirin-Fargues is a PhD law candidate at the University of Ottawa, Canada, in a joint program with Université Paris 1 Panthéon-Sorbonne.

Séverine Bouard is a human geographer (PhD) at l'Institut Agronomique néo-Calédonien (IAC), New Caledonia.

Sabrina Bourgeois is a PhD candidate at the Political Science Department, Université Laval, Canada.

Jodi Cowdery is an Aboriginal PhD candidate at the Northern Institute at Charles Darwin University, Australia.

Christine Demmer is an anthropologist and researcher at Centre Norbert Elias (CNRS) in Marseille, France.

Julie Fortin, PhD, is a consultation analyst at the Impact Assessment Agency of Canada.

Sarah Holcombe is an associate professor and principal research fellow at the University of Queensland, Australia.

Leah S. Horowitz is an associate professor in environmental studies and American Indian and Indigenous studies at the Nelson Institute for Environmental Studies at the University of Wisconsin-Madison, USA.

Arn Keeling is a settler-scholar and professor of geography at Memorial University of Newfoundland and Labrador in St. John's, Canada.

Pierre-Yves Le Meur is an anthropologist and senior researcher at the French Research Institute for Sustainable Development (IRD, SENS Research Unit) in Montpellier, France.

Francis Lévesque is a professor of anthropology and history at the School of Indigenous Studies of Université du Québec en Abitibi-Témiscamingue (UQAT), Canada.

Katie Mazer is a is a postdoctoral fellow at the School of Labour Studies at McMaster University in Hamilton, Canada.

Suzanne Mills is an associate professor in the School of Labour Studies and the School of Earth, Environment and Society at McMaster University in Hamilton, Canada.

Ella Myette is a researcher with a master's in geography from McGill University in Montreal, Canada.

Ciaran O'Faircheallaigh is professor emeritus of politics and public policy at Griffith University in Brisbane, Australia.

Thierry Rodon is a professor in the Department of Political Science at Université Laval, Canada, and holds the INQ Research Chair in Northern Sustainable Development.

Stephan Schott is a professor in the School of Public Policy and Administration at Carleton University in Ottawa, Canada.

Jean-Marc Séguin worked as a director for mining affairs for Makivik until September 2024. He now serves as the liaison agent for the Qiniqtiq Landholding Corporation of Kangiqsualujjuaq.

Kristina Sehlin MacNeil is an associate professor of Sámi studies and the deputy director for Várdduo—Centre for Sámi Research at Umeå University in Umeå, Sweden.

Andrew Taylor is an associate professor of demography at the Northern Institute of Charles Darwin University in Australia.

Sophie Thériault is a full professor in the University of Ottawa's Faculty of Law (Civil Law Section) in Ottawa, Canada.

Jean-Louis Thydjepache is a Kanak and a community relations officer at Koniambo Nickel SAS in New Caledonia.

Guillaume Vadot is a sociologist associated with Institut des mondes africains, Paris, and an ecological restoration professional.

Sandy Worden is an applied social researcher and communication management specialist at the University of Queensland, Australia.

Ana Catarina Zema is a senior researcher at Armazém Memória, Brazil.

Acknowledgments

The creation of this collective book represents a significant endeavor, one that could not have been achieved without the invaluable support and contributions of numerous individuals. We extend our deepest gratitude to the Indigenous Peoples—the Kanak, Aboriginal, Sami, Innu, Naskapi, and Inuit—who welcomed us onto their territories. Their willingness to share their insights and experiences regarding mining encounters has been indispensable to our work.

This task would not have been possible without the support of the Social Sciences and Humanities Research Council of Canada (grant number 895-2016-1015) and of our numerous partners that have contributed in kind and in cash: In Canada, Makivvik Corporation, Kativik Regional Government, Salluit Landholding Corporation, United Steelworkers of Canada, Regroupement des femmes de la Côte-Nord, Nunavut Government, Qikiqtani Inuit Association, and Université Laval; in Scandinavia, the Arctic Centre at Umeå University and the Resource Extraction and Sustainable Arctic Communities Project; in New Caledonia, Institut Agronomique Néo-Calédonien and CNRT Nickel et son environnement; in Australia, James Cook University, Darwin University, and the Queensland Resources Council; and in South America, the Consejo Latinoamericano de Ciencias Sociales working group and Pueblos indígenas y proyectos extractivos.

We are equally grateful to Katherine Hastings for her meticulous review of each chapter. Katherine's exceptional expertise and dedication have been pivotal in enhancing the overall quality and coherence of our manuscript, ensuring that it meets the highest standards of scholarship. Our sincere thanks as well to all the colleagues who reviewed the papers contained in this volume.

Lastly, our heartfelt appreciation goes to Sabrina Bourgeois for coordinating the evaluation process. Sabrina's efficiency and organizational skills have been crucial in ensuring the timely submission of all the chapters and the seamless compilation of this book.

Together, these contributions have been instrumental in bringing this project to fruition, and we are profoundly thankful to everyone involved.

Introduction

Thierry Rodon, Sophie Thériault, Arn Keeling, Séverine Bouard, and Andrew Taylor

Mineral extractive activities are forecast to expand greatly due to the increased demand for minerals spurred by the energy transition. Mining results in a wide range of social, environmental, and economic impacts worldwide, depending on mineral types, extraction processes, environmental conditions, and regulatory contexts. Many extractive companies are global actors, operating in numerous regions and often adapting their activities to different sociopolitical, economic, historical, and legal contexts. However, the impacts of mining have global commonalities. In particular, for Indigenous Peoples and local communities who navigate complex relationships with industry and, in our case, settler states, mining projects present opportunities, such as jobs for local residents and economic growth, as well as negative impacts and challenges, including environmental degradation and social disruption (Ali 2003; O'Faircheallaigh and Ali 2008; Horowitz *et al.* this volume).

> For Indigenous Peoples, who are, in most cases, economically and socially marginalized, these impacts and challenges can be particularly significant (Ali 2003; O'Faircheallaigh and Ali 2008). While research into the impacts of—and Indigenous Peoples' engagements with—mining has grown rapidly alongside the boom in extractive industries, much of this work remains focused on specific local, regional, or national contexts. Although these contexts remain critically important, the globalized nature of extraction in the era of the "planetary mine" (Arboleda 2020), together with international struggles for Indigenous rights (Kirsch 2014), suggest the need for more comparative perspectives and analyses of these dynamics (O'Faircheallaigh 2023). As local communities around the world are facing a new wave of extractive development to support energy transitions, there is also an urgent need to support regional and global engagement (and resistance) to extractivism.
>
> (Jalbert *et al.* 2017)

This book focuses on Indigenous Peoples and their encounters with the mining sector across a range of geographical and social contexts. It presents findings and lessons learned from the research projects conducted as part of the Social Sciences and Humanities Council of Canada-funded Knowledge Network on Mining Encounters and Indigenous Sustainable Livelihoods: Cross-Perspectives from

DOI: 10.4324/9781003406433-1

This chapter has been made available under a CC-BY-NC-ND 4.0 license.

the Circumpolar North and Melanesia/Australia (the MinErAL network). This ambitious network brought together 21 co-researchers from six countries, seven Indigenous organizations and governments in Canada (Nunatsiavut Government, Government of Nunavut, Makivvik Corporation, Kativik Regional Government, Qikiqtani Inuit Association, Cree Health Board, and Nunavik Regional Board of Health and Social Services Board), two NGOs (United Steelworkers of Canada and Regroupement des femmes du Québec), three research entities (CNRT Nickel et son environnement, Institut Agronomique Néo-Calédonien, and Darwin University), and four research networks (TriArc,[1] CLASCO,[2] REXSAC,[3] and AMEDEE[4]).

Initially, the MinErAL network focused on mining encounters in the Canadian North, Fennoscandia, Australia, and New Caledonia, all mineral-rich regions with Indigenous populations that are significantly impacted by extractive activities. These regions were also selected based on the co-researchers' established research networks, as well as for ease of access for research training, which is one of the primary goals of the network. As the project evolved, new research partnerships allowed us to open up new fields to investigate, notably in South America (CLASCO) and in tropical regions (AMEDEE).

This book maps the relationships between Indigenous Peoples and local communities, on one side, and industry and governments, on the other, in different (post)colonial contexts, through the concept of mining encounters. Building on the earlier work of Downing *et al.* (2002), we deliberately avoid using the term "impacts" in the book title, preferring the term "encounters," which captures a broader range of interactions, strategies, and tactics and emphasizes the role of Indigenous agency in these interactions, while acknowledging that mining encounters take place within deeply asymmetrical power relations.

Combining comparative and multidisciplinary analysis, the contributors shed light on how the mining industry adapts its practices to the political, social, and legal contexts in which it operates. Understanding these processes and how communities respond to them is critical to documenting where and how encounters with mining can benefit or negatively impact Indigenous Peoples and local communities. The experiences and findings shared by Indigenous and non-Indigenous contributors have enhanced our understanding of evolving practices and the different strategies, tactics, and discourses that Indigenous communities have developed to engage with mining projects. By mobilizing in-depth fieldwork in five regions (Australia, Canada, Sweden, New Caledonia [France], and Brazil), this body of work highlights voices often marginalized in mining development studies, including those of Indigenous Peoples and women, and documents how mining encounters in different regions and countries may benefit or harm Indigenous Peoples.

The MinErAL network has focused on the framework of Indigenous sustainable livelihoods. Building on a holistic approach, this framework provides a better understanding of interrelated health, social, cultural, economic, and ecological issues related to extractive development, and this has been a very attractive approach with our Indigenous partners and students. For instance, Jesse Marnock, an Aboriginal student at James Cook University, decided to focus his Honors BA on developing an

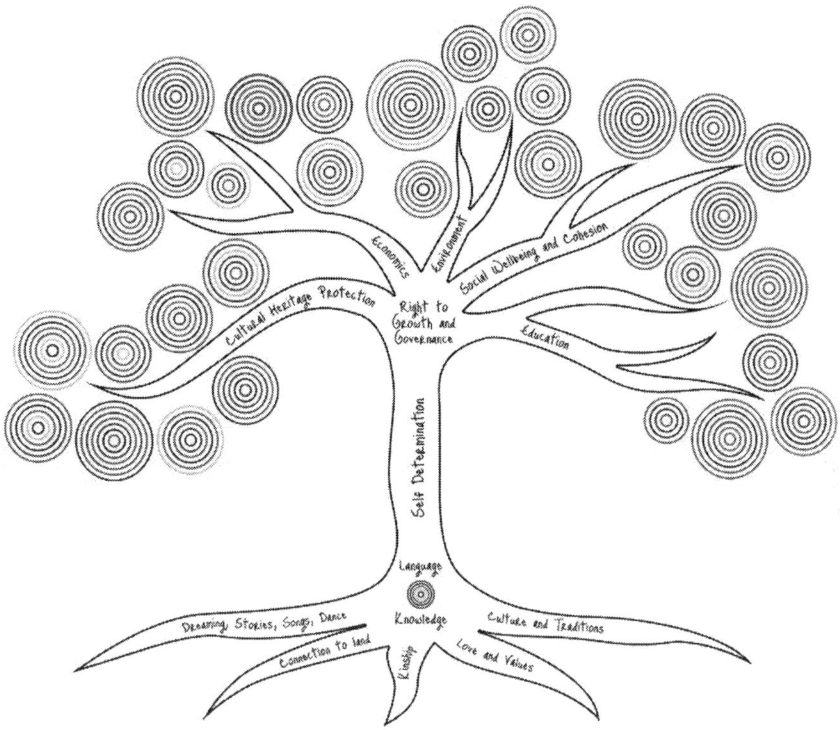

Figure 0.1 Sustainable Indigenous livelihood framework

Source: Jesse Marnock, James Cook University

Indigenous view of sustainable livelihoods (see Figure 0.1 Concept presented at the 2019 MinErAL conference in Cairns, Australia).

In addition to this framework, MinErAL researchers have used a wide range of critical approaches and methodologies for their respective projects. While these approaches are quite diverse, one common thread is a decolonizing or anti-colonial approach to research (Liboiron 2021; Smith 2013). For example, in defining "well-being" and "sustainable Indigenous livelihoods," most of the research involved, or was directly conducted by, Indigenous researchers. In addition, we ensured that network students worked closely with local Indigenous researchers during their fieldwork. However, some research cannot be conducted in partnership, for example, the comparative legal analysis conducted by Thériault and Boirin-Fargues in this volume. Nevertheless, their analysis is anti-colonial in that it addresses concerns and issues raised by Indigenous partners in the course of the project. This "study up" approach focusing on industry and government can be very useful to Indigenous organizations, providing them with tools to negotiate better deals with mining companies and to lobby for improvements in mining-related legal frameworks and policies.

Research accomplishments and challenges

Despite the disruptions caused by the COVID-19 pandemic, "dirt research" that directly engaged local communities and extractive sites has been a key part of our project. Field observations provide opportunities to test and refine theoretical premises or academic claims about the history, environment, and political economy of mining places (Therrien *et al.* 2022). These observations can be made by surveying open pits, traveling underground with miners, flying or driving to remote extractive sites, and exploring boomtowns or abandoned settlements, for example. Using this approach, we follow important predecessors in the study of resource extraction. Specifically, we draw from the "dirt research" tradition, which was pioneered by Canadian political economist Harold Innis. In the 1920s, Innis traveled extensively throughout Canada's resource frontiers, providing valuable insights into the history and geography of European colonization and settlement in northern North America (Watson 2006; Evenden 2013).

Innis's "dirt research" approach has experienced a resurgence, invoked by scholars studying locations as distant as the Pilbara mining region in Australia (Plummer and Tonts 2013), rural Canada (Stanbridge 2014), and northern Sweden (Peters *et al.* 2018). In a sense, this intensive engagement with "place" in the study of extractive industries and communities is nothing new—scholars from many disciplines and approaches have long traveled to extractive sites to make observations and gather data of various kinds, to deepen their understanding of the structures, processes, and practices associated with resource extraction. However, this recent revival of research on "dirt" appears to encompass significant trends in social-scientific fieldwork practices. Current trends in research methodology involve a shift from detached observational or data-gathering exercises to a more engaged approach. This approach, which we have adopted in MinErAL, involves local community members, industry actors, and other figures more directly in the research process, not only as informants but also as collaborators and even co-researchers. We have also supported research training in place-based research methods through two field courses: one in the mining towns of the Québec—Labrador Peninsula, in May 2019, and another in the Norrbotten Province of Northern Sweden, in October 2022.

A second conceptual trend informing this "return" to the field is less concerned with specific sites and landscapes of extraction and commodity production than with placing individual sites within the broader geographies of networks, flows, and infrastructures of extractivism. This trend is perhaps most spectacularly represented in recent books such as *The Global Interior: Mineral Frontiers and American Power* by Megan Black (2018) and *Planetary Mine* by Martin Arboleda (2020). While earlier case- and field-based research sometimes focused too narrowly on local manifestations and encounters with extractive industries, scholars now aim to contextualize local observations within a broader understanding of the legal, political, economic, and knowledge infrastructures at national and global scales that generate these transformative processes (Jacka 2018; Avango 2017). During the two field courses, for instance, students and researchers explored extractive landscapes and developed explicitly comparative perspectives on extractive processes and mining places.

In order to engage researchers and partners spread over three continents, we held an annual meeting in a different region each year. The first MinErAL seminar was held in Umea in 2017 (Sweden), then in Uashat mak Mani-Utenam, an Indigenous community, in 2018 (Québec, Canada), and Cairns in 2019 (Queensland, Australia). The COVID-19 pandemic prevented us from meeting in person in 2020 and 2021, but we still organized an online seminar with REXSAC in April 2021. *Mining the Connections* was our first post-pandemic meeting in 2022 (Québec City). Finally, in 2023, a meeting originally planned for New Caledonia was held in Brisbane (Queensland, Australia) due to the tense situation surrounding mining in New Caledonia.

Managing an interdisciplinary project across three continents presents many challenges. It has not always been easy to maintain the commitment of multiple partners and researchers over 8 years and in the midst of a global pandemic. While change in personnel or orientation among partners is unavoidable, we also had changes in the researchers. For example, the two initial researchers from Australia left the world of academia due to the difficult context of the university system in that country. Although we were able to attract other Australian researchers, this unexpected situation illustrates only too well the kind of challenges that can arise in the course of large-scale projects such as MinErAL.

Mining has complex impacts on societies and environments that can only be understood through an interdisciplinary approach combining environmental, health, and social sciences. However, conducting interdisciplinary research poses significant challenges (Solomon *et al.* 2008; Rossi *et al.* 2021). While the review of academic literature and impact assessment studies helped us outline the multiple impacts of mining from different disciplinary perspectives (see Horowitz *et al.*, this volume), most chapters maintain interdisciplinarity on a more limited scale.

Comparing across different contexts has also been a challenge. Comparison has always been at the heart of social sciences; both Max Weber and Émile Durkheim insisted on the need for comparison. However, comparative research is fraught with methodological problems. According to Mills *et al.* (2006), there are four key issues: (1) case selection, unit, level, and scale of analysis; (2) construct equivalence; (3) variable or case orientation; and finally, (4) issues of causality. In addition to these four issues, it is also difficult to strike a balance between reflecting the intensely local and place-based factors shaping mining communities and the insights gained through comparative analysis of mining communities and landscapes (Peters *et al.* 2018). The level of analysis is also a concern, as mines are glocal entities that are part of global systems of exchange, but with significant impacts on local communities. Finally, in some cases, it has proven to be quite difficult to gather comparable quantitative data, making it impossible to draw equivalences. Despite these challenges, we were able to employ a comparative approach in eight of thirteen chapters of this book.

In the remainder of this introduction, we present the key findings of this research project and explain how each chapter contributed to those findings. We have organized the findings into two broad categories: *Framing mining encounters* and *Framing the relationship*. We conclude by framing a number of issues related to the energy transition and its implications for present and future mining encounters. Although not discussed in this volume, the surge in mining projects associated with the transition to net-zero economies will make our findings even more relevant in the years ahead.

Framing mining encounters

The mining lifecycle is a stylized description of the main phases that are, in general, common to mining projects. Spanning everything from exploration activities through to mine establishment, operations, and eventual closure, each phase can have far-reaching effects across time that may significantly impact Indigenous communities and peoples (Forget and Rossi 2021). The cycle is representative of Western ways of framing how mining projects and individual mines move from exploration to eventual closure. These phases reflect business-driven needs around workforce planning, investment lifecycles, technical and machinery needs, and environmental and other regulatory requirements. Consequently, it is relevant to "reverse-engineer" the mining lifecycle and consider how each phase might impact Indigenous Peoples living in or near the mine.

Mineral exploration processes generally entail methods such as geological surface mapping, drilling, chemical analysis of deposits, and water or soil tests. These activities often cover vast areas in order to discover potential exploitable deposits. Sampling and testing are inherently intrusive, physically disturbing the lands and the vegetation on which exploration work is conducted. Early exploration work may be the first time that local communities and stakeholders become aware of the extent of land disturbance that can be expected during this and future stages of mining. For Indigenous Peoples in particular, the digging of land occurring in or near their traditional territories or sites of cultural significance causes significant disruption and can be potentially traumatic, oftentimes reinforcing ongoing traumas induced by external forces through colonizing processes (Sandlos and Keeling 2012; Fortin, this volume). The arrival of mining in a region may signify, represent, or resurface the trauma associated with colonization as a new wave of colonial-type incursion into the lands of traditional owners or groups (see Fortin in this volume).

During the development and construction stages, significant land disruption and rapid social changes often occur. The physical construction of a mine can bring new infrastructures that are beneficial to remote communities (Zhang *et al.* 2015). These include roads, railways, airstrips, and other facilities that enhance access to services. In Australia, New Caledonia, Sweden, Norway, and Canada, for example, almost all large-scale mines are proximal to or on Indigenous lands with various levels of recognition and protection (for more detail, see Boirin-Fargues and Thériault, and Bourgeois and Zema in this volume). As a result, large-scale mine operations extensively and disproportionately impact Indigenous communities (Ali 2003; Horowitz *et al.*, this volume). Along with physical infrastructure, the construction of mines brings an influx of workers and changes in land use. These have the potential to disrupt Indigenous Peoples' ways of life, including land-based livelihood activities and cultural and spiritual practices (Langton 2010).

From a positive perspective, the development phase may lead to the establishment of new relationships between mining companies and Indigenous communities, with the aim of preventing project delays resulting from conflicts. Negotiating and signing impact and benefit agreements has been shown by O'Faircheallaigh and Rodon in this volume to be the preferred approach in Australia and Canada. These agreements are also beginning to be used in other countries to secure consent for mining projects. However, opposition in some cases is managed through very paternalistic practices (Esteves 2008; Henry *et al.* 2016) and, in rare cases, suppressed by forceful actions, such as with the Goldcorp Mine in Guatemala (Maheandiran *et al.* 2010; Macleod 2016).

The operational phase, which can last anywhere from 5 to 30 years or more, has the most significant direct impacts for Indigenous Peoples and communities. Environmental changes resulting from increased pollution and habitat destruction, for instance, can greatly affect Indigenous communities (Horowitz *et al.*, this volume). However, there may also be economic opportunities such as direct employment or business opportunities (O'Faircheallaigh 2016; Langton 2010; Southcott *et al.* 2022). Although mining itself is often a male-dominated field, in this volume we pay special attention to Indigenous women's employment (see chapters by Cowdery and Taylor, Mills *et al.*, and Vadot *et al.* in this volume). Contracts for Indigenous enterprises, as well as community development programs, are increasingly common, particularly where required by negotiated agreements (Horowitz *et al.*, this volume; Rodon *et al.* 2022; Schott *et al.* 2022). However, as O'Faircheallaigh and Rodon show in this volume, the potential benefits negotiated before the mine opening may not be fully realized once the mine is in operation.

Throughout the mining lifecycle, temporal factors play a crucial role. As the mine progresses, it undergoes changes that can have cumulative effects on Indigenous communities. These effects extend beyond environmental impacts and can influence social and economic aspects, fundamentally altering the livelihood of Indigenous Peoples. For communities, the challenge is to balance the economic benefit of mining with Indigenous livelihoods. Although extractive developments include employment for local Indigenous People in some instances, Cowdery and Taylor in this volume emphasize the very low levels of employment for Indigenous People in communities near large-scale mines. This means local Indigenous Peoples' encounters with the mine construction or operational workforce are almost entirely with nonlocals.

The closure and reclamation phase has been largely overlooked in the literature on mining impacts for communities. There are numerous instances of sudden mine closure resulting in the destruction of the infrastructure and the winding down of life in the mining town (Lim *et al.* 2023; Rodon and Lévesque 2018), with little or no environmental remediation of the mining site. Sudden mine closures are features of a volatile commodity market economy. Keeling *et al.* demonstrate in this volume that socially planned closures facilitate a smoother transition and ensure the implementation of activities for social and environmental transition. However, this practice is limited to a few jurisdictions (Bainton and Holcombe 2018).

Mine closures and their aftermath present major challenges for both the industry and local communities. The cessation of mining activities may result in economic hardship for workers and industry-dependent local communities, destruction or decline of local infrastructure, and even the abandonment of settlements (Carson *et al.* 2020; Tonts *et al.* 2014; Neil *et al.* 1992). But the "afterlives" of extraction may also continue to haunt mining regions, in the form of environmental damage and dislocation, accumulated mine wastes and toxic contaminants, and the long-term effects of disrupted local land and resource use (Keeling and Sandlos 2017). These impacts particularly affect Indigenous Peoples whose territories host extractive industries, but whose communities may have benefited little from the economic activity (Keeling and Sandlos 2015). In cases where negotiated benefits or employment and training opportunities did materialize, their sudden end may present particular challenges of social and economic adjustment (O'Faircheallaigh and Lawrence 2019).

For many Indigenous communities hosting extractive developments, these legacies go to the heart of the negative, colonial relationship around mining in their territories. The "dispossession by accumulation" experienced by Indigenous communities facing displacement by mining's environmental impacts may extend for decades beyond the life of the mine, leaving communities with painful memories of historic episodes of mineral development (Sandlos and Keeling 2016; Perreault 2013). Little wonder, then, that for many Indigenous communities—the inheritors of post-mining landscapes—mine closures intersect with broader questions of Indigenous land rights, self-determination, and social and economic reclamation (Hall and Pryce 2023; Beckett and Keeling 2019; Cohen 2017). As documented in a virtual knowledge exchange forum on mine closure supported in part by the MinErAL network, Indigenous participants from Canada, Australia, and Aotearoa/Zealand emphasized these links between the past experiences and contemporary legacies of mining, and Indigenous visions for repair and recovery (Holcombe *et al.* 2022). As Gija community leader Kia Dowell explained during the forum, in the context of the closing of the Argyle Diamond Mine in Australia,

> In terms of what we dream of . . . we really dream of having our country back. In a way, that is as close as possible to pristine. The reality is that we are very far from that. There's a tailings dam there, we know there is infrastructure that is *in situ* underground which the women have never consented to . . . And the open pit is essentially going to become a huge lake. Which is what we think about: some of the challenges and opportunities

Increasingly, local communities are demanding participation in mine closure planning and objective-setting, aiming to ensure that local knowledge and values are incorporated into reclamation activities. Governments, communities, and industry itself increasingly realize the importance of long-term planning for post-mining recovery and transition-planning that should ideally begin even before the start of operations or, in any case, well before the anticipated final phase (Monosky and Keeling 2021a, 2021b; Kabir 2021; Everingham *et al.* 2020). Such planning activities may extend to questions of ensuring sustainable and community-informed post-mining rehabilitation objectives and land uses (Keenan and Holcombe 2021). While these moves are laudable, greater attention to mine closure planning, reclamation, and socioeconomic transitions is needed to ensure equitable and sustainable post-mining futures.

Finally, resource extraction projects are often located in remote, socially and environmentally fragile territories where state presence and regulatory oversight are often weak. Wastelanding, also known as "sacrificial landscape," describes areas designated or permitted for extractive activities, which allow both the natural environment and human bodies to become polluted at a substantial cost to environmental, social, and cultural values (Voyles 2015). Around the world, Indigenous lands and bodies have frequently been treated as sacrificial landscapes in the extractive context (Gross 2019). This includes the destruction of ecosystems, loss of biodiversity, soil erosion, and water pollution, and for Indigenous communities, displacement, loss of livelihoods, and destruction of cultural sites (Kemp *et al.* 2023; Kirsch 2011). In areas designated as sacrificial, the rights and well-being of local populations can be severely compromised and their bodies allowed to be polluted, as was the case for the Ranger Uranium Mine in Australia's Northern Territory (Graetz 2015), the Ok Tedi Mine in Papua New Guinea (Low and Gleeson 1998), and, in Canada, the Horne mine and smelter in Rouyn-Noranda (Céré 2023) and Giant Mine in Yellowknife (Sandlos and Keeling 2016). These environmental and social costs are a classic case of "negative externalities" unaccounted for by the economic models of mining's costs and benefits. In Indigenous territories, sacrificial landscapes reflect a broader intersection of global demand for minerals and colonialism, which some scholars have dubbed "green colonialism" or "green extractivism" (Dorn *et al.* 2022; Singh 2022; Zografos and Robbins 2020; Scott and Smith 2017).

Framing the relationship

Extractive justice requires the distribution of the benefits and burdens of extractive activities to be considered, as well as the processes and systemic factors that determine these distributions (Atapattu *et al.* 2021; Mohai *et al.* 2009). As we have seen above, Indigenous Peoples are often negatively and disproportionately affected by extractive projects on or near their traditional lands, while benefiting only marginally from these projects (Horowitz *et al.*, this volume; Sehlin MacNeil, this volume). Consequently, the recognition of Indigenous Peoples' territorial rights, right to self-determination, and right to meaningful participation in extractive decision-making processes and benefits is widely seen as a necessary pathway to achieving greater justice for Indigenous Peoples in the context of extractive development (Schilling-Vacaflor and Flemmer 2020; Seck 2016; Anaya 2013).

In recent years, the rights of Indigenous Peoples in the context of extractive industries have been framed more and more in terms of the principle of free, prior, and informed consent (FPIC). This principle, enshrined in the *United Nations Declaration on the Rights of Indigenous Peoples* (UNDRIP) and the *American Declaration on the Rights of Indigenous Peoples* (ADRIP), is increasingly recognized as the global standard for Indigenous Peoples' participation in extractive projects. However, its meaning and implementation are the subject of considerable debate and conflict (Papillon and Rodon 2023; Doyle 2014). As some of the contributions to this book show, states and mining companies generally adopt a procedural or "instrumentalist" conception of FPIC, which in many jurisdictions takes the form of varying degrees of consultation rights and duties, usually at later stages of the mining process (see in this volume Boirin-Fargues and Thériault; Bourgeois and Catarina; Horowitz *et al.*). Formal consultation processes, while aiming to secure the consent of Indigenous Peoples, usually do not allow affected communities to say "no" to specific projects (Boirin-Fargues and Thériault, this volume; Schilling-Vacaflor and Flemmer 2020; Papillon and Rodon 2017).

Faced with increasing pressure from the growing recognition of Indigenous Peoples' rights and the financial and reputational risks associated with conflicts with Indigenous and local communities, the mining industry has responded by developing its own set of voluntary norms and practices to frame its relationship with affected communities and the environment (Devlin 2023; Klein *et al.* 2023; Ruwhiu and Carter 2016; Owen and Kemp 2013; ICMM 2019). These voluntary norms have largely been captured under the umbrella of corporate social responsibility (CSR) and social license to operate (SLO). CSR and SLO have become buzzwords for mining companies claiming to promote the generation of economic, social, and environmental value through their operations. However, these approaches, which can be instrumentalized as a risk management strategy and a branding and marketing tool, create a one-sided relationship and often lead to paternalism, control, and a form of "socio-washing," where companies engage in CSR initiatives more as a public relations strategy than as a genuine effort to contribute positively to the community (Long 2019; Owen and Kemp 2013; Tysiachniouk and Petrov 2018). In such cases, CSR is seen as a means of appeasing community grievances without genuinely addressing them. This can lead to a situation where the company's CSR efforts are seen as mere handouts or superficial gestures, rather than meaningful engagement with the community's needs and interests. Moreover, frequent violations of CSR norms and principles by mining companies operating within Indigenous Peoples' territories have led researchers to question the effectiveness and adequacy of voluntary CSR initiatives to uphold Indigenous Peoples' rights (MacInnes *et al.* 2017; Owen and Kemp 2017). Indeed, despite the growing presence of CSR policies and practices in the mining sector, conflicts between mining companies and Indigenous and local communities persist, and community skepticism of CSR is widespread (Devlin 2023; Long 2019). The recent shift by investors and companies to environmental, social, and governance (ESG) as a new framework for evaluating corporate behavior and determining the future financial performance of companies in terms of ethical impact and sustainability may not address these issues. In fact, researchers have already raised concerns about the lack of standardized criteria and definitions, and the overreliance on self-reported data raises questions about its reliability and transparency, which could lead companies to overstate their sustainability efforts and mislead stakeholders (Abhayawansa and Tyagi 2021; Velte and Stawinoga 2017).

For their part, Indigenous Peoples see the FPIC principle as an extension of their right to self-determination and the right to exercise their inherent authority on their traditional lands according to their own legal and political orders (Klein *et al.* 2023; Montambeault and Papillon 2023; Schilling-Vacaflor and Flemmer 2020; Cambou 2019; Papillon and Rodon 2020). When confronted with the refusal of states to consider FPIC as involving shared decision-making authority over land and resources, and the inadequacy of voluntary CSR initiatives to uphold their rights and interests, Indigenous Peoples are exercising their agency using a wide repertoire of legal and political strategies, including legal action and various forms of resistance and protests, or by enacting their own norms regarding extractive projects on their traditional lands, such as consultation and FPIC protocols, mining policies, and impact assessment frameworks (Bourgeois and Catarina, this volume; Horowitz *et al.*, this volume. See also Montambeault and Papillon 2023; Vanthuyne and Gauthier 2022; Thériault *et al.* 2022; Papillon and Rodon 2020; Schilling-Vacaflor and Flemmer 2020, 2015; Weitzner 2019; Leifsen *et al.* 2017; Lassila 2018).

Indigenous Peoples' resistance to extractive projects that affect them or their territories, as well as the legal uncertainty regarding consultation and FPIC requirements, can lead to direct negotiations between mining companies and affected Indigenous communities (O'Faircheallaigh and Rodon, this volume). As mentioned above, the negotiation of impact and benefits agreements (IBAs) between mining companies and Indigenous communities has become common practice in some jurisdictions, such as Australia and Canada (O'Faircheallaigh and Rodon, this volume). In some circumstances, these agreements can help mitigate the environmental, cultural, and spiritual impacts of extractive development, as well as maximize the benefits generated by the projects for the affected communities. However, the potential for these agreements to achieve these goals hinges on several conditions. Beyond the asymmetrical power relations between mining companies and Indigenous communities, which have been extensively discussed in the literature (Cameron and Levitan 2014; Caine and Krogman 2010), the successful outcomes of IBAs depend on their effective implementation over time, as O'Faircheallaigh and Rodon show in this volume. Moreover, while direct negotiations between mining proponents and Indigenous Peoples can be a way for Indigenous Peoples to assert their right to self-determination and sovereignty over extractive industries on their traditional lands (Vanthuyne and Gauthier 2022), some researchers have argued that proponent-led IBA negotiations, without adequate institutional mechanisms to foster community deliberations on the legitimacy and acceptability of a project, lead, at best, to a "truncated version of FPIC" (Papillon and Rodon 2017. See also O'Faircheallaigh 2020; Hanna and Vanclay 2013). Where extractive projects are conducted by foreign investors, this concern is heightened by the lack of recognition and protection of Indigenous Peoples' rights in international investment law (ILL), as argued by Boirin-Fargues in this volume.

In many countries, the consultation of Indigenous communities regarding mining projects on or near their traditional territories often occurs, in whole or in part, through impact assessment (IA) or environmental assessment (EA) processes (Boirin-Fargues and Thériault, this volume; Myette, this volume; Klein *et al.* 2023;

Papillon and Rodon 2017). IA/EA processes provide an institutional space where Indigenous communities and mining proponents can engage directly on the environmental, social, health, and cultural impacts of a project. Thus, in theory, participation in IA/EA processes may "allow companies to operationalize FPIC, if it is early, continuous, informed, inclusive, timely, transparent, context-sensitive, and non-coercive" (Klein *et al.* 2023). In practice, however, state-led IA/EA processes are typically ill-designed for Indigenous Peoples to form and express their free, prior, and informed consent to extractive projects (Klein *et al.* 2023; Papillon and Rodon 2017; Hanna and Vanclay 2013; Prno 2013). As Myette argues in this volume, EA, conceived as a colonial "knowledge infrastructure," "is a technocratic process that relies on and advances a very specific and narrow understanding of health" that clashes with Indigenous Peoples' broader health ontologies and can be "inappropriate and even harmful for Indigenous communities." In addition, scholars have criticized IA/EA processes as a means of obtaining FPIC, pointing to their formal and often adversarial nature, their rigid timelines, the dominance of scientific and technical knowledge over Indigenous and local knowledges, and their one-time nature at later stages of project design, among other shortcomings (Boirin-Fargues and Thériault, this volume; Allard and Curran 2023; Scott 2023; Klein *et al.* 2023; Papillon and Rodon 2017). Finally, project-level IA/EA processes are "woefully inadequate for considering the potential cumulative and regional impacts of developments on ecosystems and communities" (Scott 2023; see also Atlin and Gibson 2017; Dokis 2015).

In response to the shortcomings of state-led IA/EA processes (see, for example, Myette in this volume), some Indigenous communities, at least in the Canadian context, have developed and, in some cases, applied their own IA processes based on their knowledge systems and legal and political orders (Montambeault and Papillon 2023; Morales 2019; Papillon and Rodon 2017; Scott 2023). While these initiatives are not sanctioned by the state legal system, they provide an interesting pathway for Indigenous communities to exercise their agency over their traditional territories in the context of extractive development. Further research is needed to assess the extent to which Indigenous-led IA has increased Indigenous Peoples' agency in mining development, as well as the limits of such processes in complex settler-colonial geographies, such as where development occurs amid overlapping land claims and inter/intra-community divisions over the desirability of the mining project (Scott 2023).

The most engaging encounters are those between Indigenous self-determination projects and large-scale extractive projects, which Le Meur and Banks (this volume) explore in the Southwest Pacific. This relationship is often overlooked in the academic literature on mining, but Greenland and New Caledonia, two territories on the road to self-determination, have developed strategies based on extractive industries. In the case of New Caledonia, a French territory with a strong Kanak nationalist movement, the development of the Koniambo Nickel project in the Northern Province was part of a political and economic strategy implemented by Kanak independence leaders who sought an industrial partner in Glencore. This project was intended to ensure the economic viability of an independent Kanaky

(Adrian 2019; Demmer 2018) by increasing the added value produced and the indirect spin-offs (compared with the export of raw ore). At the time of writing, Glencore's announcement to withdraw from the project has sent shockwaves through the whole territory. Glencore has decided to leave because of the costs associated with labor, energy, and environmental regulations, which mean they can no longer compete—particularly with China, which is investing in Indonesia. It is not just jobs that are at risk; it is the entire local economy, social services, and, above all, local political leaders, who find themselves destabilized—even discredited—at every level of decision-making (state, provincial, and local) by the vulnerability of their strategy in the face of global industry trends. In this small French territory, the situation has been literally upended, and the political and economic strategies put in place by Kanak leaders and the local agreements between the local communities and Glencore—via Koniambo SAS—are being called into question. While this is one example of mining encounters being completely reconfigured, the decade has been marked by significant learning, particularly on the part of local indigenous communities. In the case of Greenland, the road to sovereignty also lies in its ability to finance its independence (Grydehøj 2020). As in New Caledonia, some Greenlandic politicians had hoped to use mining projects to gain financial independence (Dingman 2014; Erdal 2013). However, there is no consensus among Greenland's political parties or Greenlanders, and for the time being, the government does not support mining as a means of self-government.

Overview of the chapters

In the first section of this book, *Indigenous Peoples, Law, and Politics*, Boirin-Fargues and Thériault highlight the complexities of integrating Indigenous Peoples' rights within mining legal frameworks in Canada and Fennoscandian countries, revealing that, despite some progress, mining development often takes precedence over Indigenous land uses, with limited pathways for Indigenous participation in decision-making processes. In their chapter, Le Meur and Banks explore the dynamics between large-scale mining projects and state/nation-building in the Southwest Pacific, noting how mining operations extend their influence beyond economic impacts, affecting local and national identities and the sociopolitical landscape. Boirin-Fargues examines the lack of accountability of foreign investors in respecting Indigenous rights through ILL, using the *Bear Creek* case as a reflection point. Finally, Bourgeois and Zema delve into the comparative analysis of Indigenous Peoples' participation in mining in Canada and Brazil, highlighting the challenges and emerging strategies for meaningful engagement and negotiation.

In the second section, *Braiding Indigenous Views in the Mining Cycle*, Horowitz *et al.* analyze the multifaceted impacts of large-scale mining on Indigenous communities, emphasizing the need for a multidisciplinary approach to understanding these complex effects. Myette looks at the integration of health impacts in environmental assessments, concluding that their limited scope fails to capture the broad health ontologies of Indigenous Peoples and suggesting a more inclusive approach to evaluating mining's health impacts. O'Faircheallaigh and Rodon outline the importance of effective implementation of agreements between Indigenous communities and the extractive industry, stressing the gap between negotiation and actual benefits realization through two case studies conducted with the Ely Trust in Australia and the Matimekush Band Council in Canada. Finally, Keeling *et al.* compare the social aspects of mine closure and transition in Canada and Australia, pointing out the regulatory and participatory gaps in addressing the socioeconomic impacts on communities.

The third section, entitled *Navigating Relationships with Indigenous Communities*, begins with a chapter by Fortin on a case study from the Cree community of Nemaska, offering insights into that community's perception of mining projects and their influence on social cohesion and cultural vitality, while Sehlin MacNeil explores the phenomenon of lateral violence within Indigenous communities under external pressures from mining activities, suggesting a link between extractive operations and internal conflict.

The final section—*Indigenous Women and Resource Development*—focuses on a topic that has rarely been addressed in the mining context. Cowdery and Taylor analyze employment trends for Aboriginal women in the Northern Territory's mining industry, revealing underrepresentation and challenges to inclusion and equity. Mills *et al.* highlight employment inequities for Inuit women in the case of the Raglan Mine in Nunavik, contrasting perceived benefits with the reality of job segmentation and barriers to long-term employment. Finally, Vadot *et al.* examine the entry and retention of Kanak women in New Caledonia's nickel industry, exploring gender and ethnic divisions in the labor market, and the strategies women employ to navigate these challenges.

Notes

1 The Arctic governance triangle: Government, Indigenous peoples, and industry in change.
2 Consejo Latino americano de Ciencias Sociales (Grupo de Trabajo: Pueblos indígenas y proyectos extractivos).
3 Resource Extraction and Sustainable Arctic Communities: A Nordic Centre of Excellence.
4 Activité minière, environnement, développement, économie et éthique.

References

Abhayawansa, S. and Tyagi, S., 2021. Sustainable investing: The black box of environmental, social, and governance (ESG) ratings. *The Journal of Wealth Management*, 24 (1), 49–54.

Adrian, J., 2019. Le nickel, vecteur d'indépendance ou de dépendances? *In: PIPSA democracy, sovereignty and self-determination in the Pacific Islands*. Conference proceedings, June 25–27, 2019, Nouméa, New Caledonia.

Ali, Saleem H., 2003. *Mining, the environment, and Indigenous development conflicts*. Tucson: University of Arizona Press.

Allard, C. and Curran, D., 2023. Indigenous influence and engagement in mining permitting in British Columbia, Canada: Lessons for Sweden and Norway? *Environmental Management*, 72 (1), 1–18.

Anaya, J., 2013. *Report of the special rapporteur on the rights of Indigenous peoples*, A/HRC/24/41, 24th Session.

Arboleda, M., 2020. *Planetary mine: Territories of extraction under late capitalism*. London: Verso.

Atapattu, S. A., Gonzalez, C., and Seck, S., eds., 2021. *The Cambridge handbook of environmental justice and sustainable development*. Cambridge: Cambridge University Press.

Atlin, C. and Gibson, R., 2017. Lasting regional gains from non-renewable resource extraction: The role of sustainability-based cumulative effects assessment and regional planning for mining development in Canada. *The Extractive Industries and Society*, 4 (1), 36–52.

Avango, D., 2017. Remains of industry in the polar regions: Histories, processes, heritage. *Entreprises et Histoire*, 87, 133–149.

Bainton, N. and Holcombe, S., 2018. A critical review of the social aspects of mine closure. *Resources Policy*, 59, 468–478. Available from: https://doi.org/10.1016/j.resourpol.2018.08.020 [Accessed 12 February 2024].

Beckett, C. and Keeling, A., 2019. Rethinking remediation: Mine reclamation, environmental justice, and relations of care. *Local Environment: The International Journal of Justice and Sustainability*, 24 (3), 216–230. Available from: https://doi.org/10.1080/13549839.2018.1557127 [Accessed 12 February 2024].

Black, M., 2018. *The global interior: Mineral frontiers and American power*. Cambridge, MA: Harvard University Press.

Caine, K. J. and Krogman, N., 2010. Powerful or just plain power-full? A power analysis of impact and benefit agreements in Canada's north. *Organization & Environment*, 23 (1), 76–98.

Cambou, D., 2019. The UNDRIP and the legal significance of the right of Indigenous Peoples to self-determination: A human rights approach with a multidimensional perspective. *The International Journal of Human Rights*, 23 (1–2), 1–17.

Cameron, E. and Levitan, T., 2014. Impact and benefit agreements and the neoliberalization of resource governance and indigenous-state relations in northern Canada. *Studies in Political Economy*, 93 (1), 25–52.

Carson, D. B., Nilsson, L. M., and Carson, D. A., 2020. The mining resource cycle and settlement demography in Malå, Northern Sweden. *Polar Record*, 56. Available from: https://doi.org/10.1017/S0032247420000200 [Accessed 12 February 2024].

Céré, P., 2023. *Voyage au bout de la mine: le scandale de la fonderie Horne*. Montreal: Éditions Écosociété.

Cohen, T., 2017. Bringing country back? *In:* K. Jalbert, A. Willow, D. Casgrande, and S. Paladino, eds., *Extraction: Impacts, engagements, and alternative futures*. London: Routledge, 137–150.

Demmer, C., 2018. Nationalisme minier, secteur nickel et décolonisation en Nouvelle-Calédonie. *Cahiers Jaurès*, 4, 35–52.

Devlin, J. F., 2023. CSR, SLO and local mining communities. *In:* N. D. Brunet and S. Longboat, eds., *Local communities and the mining industry*. London: Routledge, 17–37.

Dingman, E. M., 2014. Greenlandic independence: The dilemma of natural resource extraction. *Arctic Yearbook*, 228, 244.

Dokis, C. A., 2015. *Where the rivers meet: Pipelines, participatory resource management, and aboriginal-state relations in the northwest territories*. Vancouver: UBC University Press.

Dorn, F. M., Hafner, R., and Plank, C., 2022. Towards a climate change consensus: How mining and agriculture legitimize green extractivism in Argentina. *The Extractive Industries and Society*, 11, 101130. Available from: https://doi.org/10.1016/j.exis.2022.101130.

Downing, T. E., 2002. *Avoiding new poverty: Mining-induced displacement and resettlement* (Vol. 52). London: International Institute for Environment and Development.

Doyle, C. M., 2014. *Indigenous peoples, title to territory, rights and resources: The transformative role of free prior and informed consent*. 1st ed. London: Routledge.

Erdal, L., 2013. *Independence on the horizon: A study of the interplay between sovereignty and natural resources in Greenland*. Fridtjof Nansen Institute Report 6/2013.

Esteves, A. M., 2008. Mining and social development: Refocusing community investment using multi-criteria decision analysis. *Resources Policy*, 33 (1), 39–47. Available from: https://doi.org/10.1016/j.resourpol.2008.01.002 [Accessed 12 February 2024].

Evenden, M., 2013. Innis, Biss, and industrial circuitry in the Canadian North, 1921–1965. *In:* W. J. Buxton, ed., *Harold Innis and the North: Appraisals and contestations*. Montreal: McGill-Queen's University Press, 127–148.

Everingham, J., Mackenzie, S., Svobodova, K., and Witt, K., 2020. *Participatory processes, mine closure and social transitions*. Brisbane: Centre for Social Responsibility in Mining.

Forget, M. and Rossi, M., 2021. Mining region value and vulnerabilities: Evolutions over the mine life cycle. *The Extractive Industries and Society*, 8 (1), 176–187.

Graetz, G., 2015. Ranger Uranium Mine and the Mirarr (Part 1), 1970–2000: The risks of "riding roughshod." *The Extractive Industries and Society*, 2 (1), 132–141. Available from: https://doi.org/10.1016/j.exis.2014.10.004 [Accessed 12 February 2024].

Gross, L., 2019. Wastelanding the bodies, wastelanding the land: Accidents as evidence in the Albertan oil sands. *In:* C. N. Westman, T. L. Joly, and L. Gross, eds., *Extracting home in the oil sands*. London: Routledge.

Grydehøj, A., 2020. Unravelling economic dependence and independence in relation to island sovereignty: The case of Kalaallit Nunaat (Greenland). *Island Studies Journal*, 15 (1).

Hall, R. and Pryce, B., 2023. Colonial continuities in closure: Indigenous mine labour and the Canadian state. *Antipode*. Available from: https://doi.org/10.1111/anti.12968 [Accessed 12 February 2024].

Hanna, P. and Vanclay, F., 2013. Human rights, Indigenous peoples and the concept of free, prior and informed consent. *Impact Assessment and Project Appraisal*, 31 (2), 146–157.

Henry, L. A., Nysten-Haarala, S., Tulaeva, S., and Tysiachniouk, M., 2016. Corporate social responsibility and the oil industry in the Russian Arctic: Global norms and neo-paternalism. *Europe-Asia Studies*, 68 (8), 1340–1368.

Holcombe, S., Elliott, V., Keeling, A., Berryman, M., Hall, R., Ngaamo, R., Beckett, C., Moon, W., Hudson, M., Kusab, N., and Ross River Lands Office, 2022. *Indigenous exchange forum: Transitions in mine closure*. St Lucia: Centre for Social Responsibility in Mining, University of Queensland.

International Council on Mining and Minerals (ICMM), 2019. *Integrated mine closure: Good practice guide*. 2nd ed. London: ICMM. Available from: https://guidance.miningwithprinciples.com/integrated-mine-closure-good-practice-guide/ [Accessed 12 February 2024].

Jacka, J. K., 2018. The anthropology of mining: The social and environmental impacts of resource extraction in the mineral age. *Annual Review Anthropology*, 47, 61–77. Available from: https://doi.org/10.1146/annurev-anthro-102317-050156 [Accessed 12 February 2024].

Jalbert, K., Willow, A., Casagrande, D., and Paladino, S., eds., 2017. *ExtrACTION: Impacts, engagements, and alternative futures*. London: Routledge.

Kabir, Z., 2021. The role of social impact assessment (SIA) in the development of a mine closure plan in regional Australia. *Journal of Environmental Assessment Policy and Management*, 23 (4). Available from: https://doi.org/10.1142/S1464333222500156 [Accessed 12 February 2024].

Keeling, A. and Sandlos, J., eds., 2015. *Mining and communities in Northern Canada: History, politics, and memory*. Calgary: University of Calgary Press.

Keeling, A. and Sandlos, J., 2017. Ghost towns and zombie mines: The historical dimensions of mine abandonment, reclamation, and redevelopment in the Canadian North. *In:* S. Bocking and B. Martin, eds., *Ice blink: Navigating northern environmental history*. Calgary: University of Calgary Press, 377–420.

Keenan, J. and Holcombe, S., 2021. Mining as a temporary land use: A global stocktake of post-mining transitions and repurposing. *The Extractive Industries and Society*, 8 (3). Available from: https://doi.org/10.1016/j.exis.2021.100924 [Accessed 12 February 2024].

Kemp, D., Kochan, K., and Burton, J., 2023. Critical reflections on the Juukan Gorge parliamentary inquiry and prospects for industry change. *Journal of Energy & Natural Resources Law*, 41 (4), 379–402. Available from: https://doi.org/10.1080/02646811.2022.2156202 [Accessed 12 February 2024].

Kirsch, S., 2011. Lost worlds: Environmental disaster, "culture loss," and the law. *Current Anthropology*, 42, 167–198.

Kirsch, S., 2014. *Mining capitalism: The relationship between corporations and their critics*. Berkeley: University of California Press.

Klein, L., Munoz-Torres, M. J., and Fernandez-Izquierdo, M. A., 2023. A comparative account of Indigenous participation in extractive projects: The challenge of achieving free, prior, and informed consent. *The Extractive Industries and Society*, 15. Available from: https://doi.org/10.1016/j.exis.2023.101270 [Accessed 12 February 2024].

Langton, M., 2010. The resource curse: New outback principalities and the paradox of plenty. *Griffith Review*, 28, 46–62.

Lassila, M. M., 2018. Mapping mineral resources in a living land: Sami mining resistance in Ohcejohka, northern Finland. *Geoforum*, 96, 1–9.

Leifsen, E., Gustafsson, M.-T., Guzmán-Gallegos, M. A., and Schilling-Vacaflor, A., 2017. New mechanisms of participation in extractive governance: Between technologies of governance and resistance work. *Third World Quarterly*, 38 (5), 1043–1057.

Liboiron, M., 2021. Decolonizing geoscience requires more than equity and inclusion. *Nature Geoscience*, 14, 876–877. Available from: https://doi.org/10.1038/s41561-021-00861-7 [Accessed 12 February 2024].

Lim, T. W., Keeling, A., and Satterfield, T., 2023. "We thought it would last forever": The social scars and legacy effects of mine closure at Nanisivik, Canada's first High Arctic mine. *Labour/Le Travail*, 91, 115–146.

Long, B. S., 2019. CSR and reconciliation with Indigenous peoples in Canada. *Critical Perspectives on International Business*, 18 (1), 15–30.

Low, N. and Gleeson, B., 1998. Situating justice in the environment: The case of BHP at the Ok Tedi Copper Mine. *Antipode*, 30, 201–226.

MacInnes, A., Colchester, M. and Whitmore, A., 2017. Free, prior and informed consent: How to rectify the devastating consequences of harmful mining for Indigenous peoples. *Perspectives in Ecology and Conservation*, 15 (3), 152–160.

Macleod, M., 2016. Development or devastation?: Epistemologies of Mayan women's resistance to an open-pit goldmine in Guatemala. *AlterNative: An International Journal of Indigenous Peoples*, 12 (1), 86–100. Available from: https://doi.org/10.20507/AlterNative.2016.12.1.7 [Accessed 12 February 2024].

Maheandiran, B., Di Federico, J., Aguilera, R. E., and Imai, S., 2010. *Goldcorp and Hudbay Minerals in Guatemala (2010 update)*. Osgoode CLPE Research Paper No. 09/2010. Available from: http://dx.doi.org/10.2139/ssrn.1595157 [Accessed 12 February 2024].

Mills, M., van de Bunt, G., and Bruijn, J., 2006. Comparative research: Persistent problems and promising solutions. *International Sociology*, 21 (5). Available from: https://doi.org/10.1177/0268580906067833 [Accessed 12 February 2024].

Mohai, P., Pellow, D., and Timmons Robert, J., 2009. Environmental justice. *Annual Review of Environment and Resources*, 34, 405–430.

Monosky, M. and Keeling, A., 2021a. Planning for social and community-engaged closure: A comparison of mine closure plans from Canada's territorial and provincial North. *Journal of Environmental Management*, 277. Available from: https://doi.org/10.1016/j.jenvman.2020.111324 [Accessed 12 February 2024].

Monosky, M. and Keeling, A., 2021b. Social considerations in mine closure: Exploring policy and practice in Nunavik, Quebec. *The Northern Review*, 52, 29–61. Available from: https://doi.org/10.22584/nr52.2021.002 [Accessed 12 February 2024].

Montambeault, F. and Papillon, M., 2023. Repoliticising indigenous participation: FPIC protocols in Canada and Brazil. *The International Journal of Human Rights*, 27 (2), 335–358.

Morales, S., 2019. *Indigenous-led assessment processes as a way forward*. Waterloo, ON: Centre for International Governance Innovation (CIGI). Available from: www.cigionline.org/articles/indigenous-led-assessment-processes-way-forward/ [Accessed 12 February 2024].

Neil, C., Tykkalainen, M., and Bradbury, J., eds., 1992. *Coping with closure: An international comparison of mine town experiences*. London: Routledge.

O'Faircheallaigh, C., 2016. *Negotiations in the indigenous world: Aboriginal peoples and the extractive industry in Australia and Canada*. London: Routledge.

O'Faircheallaigh, C., 2020. Negotiated agreements, indigenous FPIC, and the mine life cycle. *In:* I. T. Odumusu-Ayanu and D. Newman, eds., *Indigenous-industry agreements, natural resources and the law*. London: Routledge, 63–78.

O'Faircheallaigh, C., 2023. *Indigenous peoples and mining: A global perspective*. Oxford: Oxford University Press.

O'Faircheallaigh, C. and Ali, S., eds., 2008. *Earth matters: Indigenous peoples, the extractive industries and corporate social responsibility*. Sheffield: Greenleaf Publishing.

O'Faircheallaigh, C. and Lawrence, R., 2019. Mine closure and the aboriginal estate. *Australian Aboriginal Studies*, 1, 65–82.

Owen, J. R. and Kemp, D., 2013. Social licence and mining: A critical perspective. *Resources Policy*, 38, 29–35. Available from: https://doi.org/10.1016/j.resourpol.2012.06.016 [Accessed 12 February 2024].

Owen, J. R. and Kemp, D., 2017. *Extractive relations: Countervailing power and the global mining industry*. London: Routledge.

Papillon, M. and Rodon, T., 2017. Proponent-indigenous agreements and the implementation of free, prior and informed consent in Canada. *Environmental Impact Assessment Review*, 62, 216–224.

Papillon, M. and Rodon, T., 2020. The transformative potential of Indigenous-driven approaches to implementing free, prior and informed consent: Lessons from two Canadian cases. *International Journal of Minority and Group Rights*, 27 (2), 314–335.

Papillon, M. and Rodon, T., 2023. Le consentement préalable, libre et éclairé (CPLE) en contexte canadien. *Les Cahiers du CIÉRA*, 43–50. Available from: https://doi.org/10.7202/1099219ar [Accessed 12 February 2024].

Perreault, T., 2013. Dispossession by accumulation? Mining, water and the nature of enclosure on the Bolivian Altiplano. *Antipode*, 45 (5), 1050–1069. Available from: https://doi.org/10.1111/anti.12005 [Accessed 12 February 2024].

Peters, P., Carson, D., Porter, R., Vuin, A., Carson, D., and Ensign, P., 2018. My village is dying? Integrating methods from the inside out. *Canadian Review of Sociology*, 55 (3), 451–475. Available from: https://doi.org/10.1111/cars.12212 [Accessed 12 February 2024].

Plummer, P. and Tonts, M., 2013. Geographical political economy, dirt research and the Pilbara. *Australian Geographer*, 44 (3), 223–226. Available from: https://doi.org/10.1080/00049182.2013.817034 [Accessed 12 February 2024].

Prno, J., 2013. An analysis of factors leading to the establishment of a social licence to operate in the mining industry. *Resources Policy*, 38, 577–590.

Rodon, T., Keeling, A., and Boutet, J-S., 2022. Schefferville revisited: The rise and fall (and rise again) of iron mining in Québec-Labrador. *The Extractive Industries and Society*, 12. Available from: https://doi.org/10.1016/j.exis.2021.101008%20 [Accessed 12 February 2024].

Rodon, T., Lemus-Lauzon, I., Séguin, J. M., and Schott, S., 2022. Resource revenue allocation strategies and Indigenous community sustainable development. *In:* C. Southcott, F. Abele, D. C. Natcher, and B. Parlee, eds., *Extractive industry and the sustainability of Canada's Arctic communities*. Montreal: McGill-Queen's University Press. Available from: www.mqup.ca/extractive-industry-and-the-sustainability-of-canada—s-arctic-communities-products-9780228011552.php [Accessed 12 February 2024].

Rodon, T. and Lévesque, F., 2018. From narrative to evidence: Socio-economic impacts of mining in Northern Canada. *In:* C. Southcott, F. Abele, D. C. Natcher, and B. Parlee, eds., *Resources and sustainable development in the Arctic*. London: Routledge.

Rossi, M., Forget, M., Gunzburger, Y., Bergeron, K. M., Samper, A., and Camizuli, E., 2021. Trajectories of mining territories: An integrated and interdisciplinary concept to achieve sustainability. *The Extractive Industries and Society*, 8 (1), 1–7.

Ruwhiu, D. and Carter, L., 2016. Negotiating "meaningful participation" for Indigenous peoples in the context of mining. *Corporate Governance*, 16 (4), 641–654.

Sandlos, J. and Keeling, A., 2012. Claiming the new North: Development and colonialism at the Pine Point mine, Northwest Territories, Canada. *Environment and History*, 18, 5–34.

Sandlos, J. and Keeling, A., 2016. Toxic legacies, slow violence, and environmental injustice at Giant Mine, Northwest Territories. *Northern Review*, 42, 7–21.

Schilling-Vacaflor, A. and Flemmer, R., 2015. Conflict transformation through prior consultation? Lessons from Peru. *Journal of Latin American Studies*, 47 (4), 811–839.

Schilling-Vacaflor, A. and Flemmer, R., 2020. Mobilising free, prior and informed consent (FPIC) from below: A typology of Indigenous Peoples' agency. *International Journal on Minority and Group Rights*, 27 (2), 291–313.

Schott, S., Belayneh, A., Boutet, J. S., Rodon, T., and Seguin, J. M., 2022. Mining economies, mining families: The impacts of extractive industries on economic and human development in the eastern sub-Arctic. *In:* C. Southcott, F. Abele, D. C. Natcher, and B. Parlee, eds., *Extractive industry and the sustainability of Canada's Arctic communities*. Montreal: McGill-Queen's University Press, 53–77. Available from: www.mqup.ca/extractive-industry-and-the-sustainability-of-canada—s-arctic-communities-products-9780228011552.php [Accessed 12 February 2024].

Scott, D. N., 2023. Impact assessment in the Ring of Fire: Contested authorities, competing visions and a clash of legal orders. *In:* D. S. Scott, J. Sankey, and L. Tanguay, eds., *Operationalizing indigenous-led impact assessment. Report prepared for the Impact Assessment Agency of Canada*. Available from: https://operationalizingindigenous-ledimpactassessments.com/reports [Accessed 12 February 2024].

Scott, D. N. and Smith, A. A., 2017. "Sacrifice zones" in the green energy economy: Toward an environmental justice framework. *McGill Law Journal*, 62 (3), LJ 861.

Seck, S. L., 2016. Indigenous rights, environmental rights, or stakeholder engagement? Comparing IFC and OECD approaches to implementation of the business responsibility to respect human rights. *JSDLP-RDPDD*, 12 (1), 57–91.

Singh, S., 2022. Green colonialism. *The Fourth World Journal*, 21 (2), 117–122.

Smith, L. T., 2013. *Decolonizing methodologies: Research and Indigenous peoples*. London: Zed Books Ltd.

Solomon, F., Katz, E., and Lovel, R., 2008. Social dimensions of mining: Research, policy and practice challenges for the minerals industry in Australia. *Resources Policy*, 33 (3), 142–149.

Southcott, C., Abele, F., Natcher, D., and Parlee, B., 2022. Extractive industry and the sustainability of Canada's Arctic communities: An introduction. *In:* C. Southcott, F. Abele, D. Natcher, and B. Parlee, eds., *Extractive industry and the sustainability of Canada's Arctic communities*. Montreal: McGill-Queen's University Press, 3–23.

Stanbridge, K., 2014. How to commit Canadian sociology, or "What would Innis do?" *Canadian Review of Sociology*, 51, 389–394. Available from: https://doi.org/10.1111/cars.12053 [Accessed 12 February 2024].

Thériault, S., Bourgeois, S., and Boirin-Fargues, Z., 2022. Indigenous peoples' agency within and beyond rights in the mining context: The case of the Schefferville region. *The Extractive Industries and Society*, 12. Available from: www.sciencedirect.com/science/article/pii/S2214790X21001507 [Accessed 12 February 2024].

Therrien, A., Lépy, É., Boutet, J. S., Bouchard, K. and Keeling, A., 2022. Place-based education and extractive industries: Lessons from post-graduate courses in Canada and Fennoscandia. *The Extractive Industries and Society*, 12.

Tonts, M., Plummer, P., and Argent, N., 2014. Path dependence, resilience and the evolution of new rural economies: Perspectives from rural Western Australia. *Journal of Rural Studies*, 36, 362–375. Available from: https://doi.org/10.1016/j.jrurstud.2014.04.001 [Accessed 12 February 2024].

Tysiachniouk, M. S. and Petrov, A. N., 2018. Benefit sharing in the Arctic energy sector: Perspectives on corporate policies and practices in Northern Russia and Alaska. *Energy Research & Social Science*, 39, 29–34. Available from: https://doi.org/10.1016/j.erss.2017.10.014 [Accessed 12 February 2024].

Vanthuyne, K. and Gauthier, M., 2022. Mining the land while sustaining *liyiyiuituwin*: Exercising Indigenous sovereignty through collaboration in Eeyou Istchee. *Canadian Journal of Political Science*, 55, 279–299.

Velte, P. and Stawinoga, M., 2017. Integrated reporting: The current state of empirical research, limitations and future research implications. *Journal of Management Control*, 28, 275–320.

Voyles, T. B., 2015. *Wastelanding: Legacies of uranium mining in Navajo Country*. Minneapolis: University of Minnesota Press.

Watson, A. J., 2006. *Marginal man: The dark vision of Harold Innis*. Toronto: University of Toronto Press.

Weitzner, V., 2019. Between panic and hope: Indigenous peoples, gold, violence(s) and FPIC in Colombia, through the lens of time. *Journal of Legal Pluralism and Unofficial Law*, 51 (1), 3–28.

Zhang, A., Moffat, K., Lacey, J., Wang, J., González, R., Uribe, K., Cui, L. and Dai, Y., 2015. Understanding the social licence to operate of mining at the national scale: A comparative study of Australia, China and Chile. *Journal of Cleaner Production*, 108, 1063–1072.

Zografos, C. and Robbins, P., 2020. Green sacrifice zones, or why a green new deal cannot ignore the cost shifts of just transitions. *One Earth*, 3 (5), 543–546.

Part I

Indigenous peoples, law, and politics

1 The space left for Indigenous peoples' voices in Canadian and Fennoscandian mining legal frameworks

A comparative analysis

Zoé Boirin-Fargues and Sophie Thériault

Introduction

In recent decades, the mining sector has experienced rapid growth in many parts of the world, driven by rising metal and mineral prices, global demand for critical minerals, and public policies that encourage mining investment and development (Government of Canada 2022; Haikola and Anshelm 2016). While mining development generates economic benefits, the industry has widespread and long-lasting socio-environmental impacts that can disproportionately affect Indigenous peoples' traditional land uses, livelihoods, health, and well-being (Horowitz *et al.* 2018; Ivsett Johnsen 2016; Nachet *et al.* 2022). The intensification of mining activities has led to an increase in conflicts between Indigenous communities and mining proponents, which are occurring alongside a growing recognition of Indigenous peoples' rights at the national and international levels (Horowitz *et al.* 2018; Nachet *et al.* 2022).

In this chapter, we analyze how and to what extent Indigenous peoples' rights are recognized and protected when they are confronted with mining legal frameworks, with a particular focus on Canada and the Fennoscandian countries (Norway, Sweden, and Finland). Governments in these jurisdictions have historically encouraged mining investment and development, notably through their mining laws and policies (Haikola and Anshelm 2016; Nygaard 2016; Petersen St-Laurent and Le Billon 2015). More recently, these countries have also taken steps, albeit of varying scope and intensity, to recognize and protect the rights of Indigenous peoples, including in the context of mining. These regions are therefore fertile grounds for observing the dynamics at play when emerging Indigenous peoples' rights come up against well-established mining legal frameworks.

Using a comparative legal approach based on the analysis of primary and secondary legal materials,[1] we first provide a general overview of the foundations, sources, and modes of the legal recognition and protection of Indigenous peoples' territorial rights in Canada and the Fennoscandian countries. This overview is necessary to understand how the patchwork of Indigenous peoples' rights found in these jurisdictions shapes the avenues available to Indigenous peoples to influence mining decision-making processes. Building on this foundation, we then describe the legal frameworks for mining in the jurisdictions examined, as well as the specific forms that Indigenous peoples' rights take when they intersect with these

DOI: 10.4324/9781003406433-3
This chapter has been made available under a CC-BY-NC-ND 4.0 license.

frameworks. In the third part of the chapter, we argue that, despite recent changes to the mining legal frameworks in Canada and in the Fennoscandian countries to account for Indigenous peoples' rights, in all of these jurisdictions, mining development is still largely prioritized over the land uses of Indigenous peoples, whose pathways to participation in formal mining decision-making processes remain narrow and limited. Thus, we conclude that the increased recognition of Indigenous peoples' rights has not fundamentally altered the preexisting structures and rationales of the mining legal frameworks in the jurisdictions studied.

Overview: Trajectories of the recognition and protection of Indigenous peoples' rights in Canada and Fennoscandia

Indigenous peoples' rights in Canada and Sami rights in the Fennoscandian countries differ significantly in their foundations, sources, and forms, as a result of different historical and colonial trajectories that have been shaped over time by the relationships between these states and Indigenous peoples (Allard and Funderud Skogvang 2016; Grammond 2013).[2]

In Canada, Section 35(1) of the *Constitution Act 1982* "recognizes and affirms" the "existing Aboriginal and treaty rights" of Indigenous peoples. The Canadian state, following British colonial policy, entered into treaty negotiations with many Indigenous groups to settle their land claims, a process that continues today under the federal government's comprehensive land claims policy (Alcantara 2013). The treaties negotiated in the nineteenth and early twentieth centuries granted their Indigenous signatories limited rights and benefits, including the right to hunt and fish on ceded lands, in exchange of the extinguishment of their land rights (Grammond 2013, pp. 291–293). For their part, contemporary land claims agreements, which cover large tracts of land mostly in the northernmost regions of the country, grant their Indigenous signatories a broader range of rights and benefits, including land ownership; harvesting rights; a degree of political and administrative autonomy; and co-management arrangements over land, water, and natural resources (Rodon 2018, Thériault *et al.* 2022).

Indigenous peoples whose land rights have not been extinguished by treaty or other valid means may claim Aboriginal title—"the right to exclusive use or occupation of the land . . . for a variety of purposes"—in regard to lands they occupied prior to the assertion of Crown sovereignty (*Tsilhqot'in Nation v. British Columbia* 2014, para. 24–50, 67, 73; *Delgamuukw v. British Columbia* 1997, para. 117, 143). The process of formally establishing Aboriginal title is long and onerous. To date, only one Indigenous group, the Tsilhqot'in Nation, has successfully established Aboriginal title over a portion of its traditional territory. Indigenous peoples can also claim Aboriginal rights related to the exercise of traditional activities, such as hunting and fishing, which were practiced prior to European contact and are "integral to the distinctive culture" of the group claiming the right (*R. v. Van der Peet* 1996, para. 46). According to the Supreme Court, Aboriginal title and rights are *sui generis* ("of their own kind") (*Tsilhqot'in Nation v. British Columbia* 2014, para. 14, 72; *R. v. Van der Peet* 1996, para. 29–41). As such, these rights possess unique characteristics that cannot "be described with reference to traditional property law concepts" (*Tsilhqot'in Nation v. British Columbia* 2014, para. 72; *Delgamuukw v. British Columbia* 1997, para. 190; Sanderson and Singh 2021; Borrows 2015).

Sami rights in the Fennoscandian states rest on legal foundations different from the rights of Indigenous peoples in Canada. For one, these states, which were deemed owners of the land in the absence of any registered owners (Amatulli 2015, p. 17; Ravna and Bankes 2017, p. 84), have not entered into treaty negotiations with the Sami. In fact, the recognition of the Sami as an Indigenous people in the Fennoscandian countries is quite recent.[3] It was not until the 1980s that discussions around colonialism emerged in Sweden (Fur 2013, p. 22), and the Sami were first recognized as an Indigenous people in the Constitution of Norway in 1988, followed by Finland in 1995 and Sweden in 2011.

Moreover, the recognition of Sami rights to land, or to practice their traditional activities, is based on Scandinavian general property law principles (Raitio *et al.* 2020, p. 6) rather than on *sui generis* rights as they have developed in Canadian common law. In short, according to the principles of immemorial usage in Norway and immemorial prescription in Sweden and Finland, one can acquire property rights or land use rights through "protracted use of the land" (Allard 2016, p. 55), meaning a use that is "intensive, continuous, and exclusive to succeed into a right" (Allard 2011, p. 167). Although the Sami are recognized as an Indigenous people in the constitutions of the three Fennoscandian countries,[4] Sami land rights *per se* are not constitutionally protected. Nevertheless, these rights may benefit from the constitutional protection of general property rights (see, e.g., in Sweden, the *Skattefjäll* (*Taxed Mountain*) case 1981).

While Fennoscandian constitutions do not explicitly protect Sami land rights, the international law as well as domestic human rights law (Allard 2024, p. 185) have played an important role in the recognition of Sami rights (Cambou and Ravna 2024, p. 3, Cambou 2024, p. 54; Heinämäki 2024, p. 89), mainly through court decisions (Torp 2024, p. 73; Allard 2024, p. 197). Of the countries studied, Norway is the only one to have ratified the 1989 *ILO Convention 169* concerning Indigenous and Tribal Peoples in independent countries (ILO 169) (Ravna 2015, p. 66). Norway's recognition of Sami rights has also been influenced by Article 27 of the *International Covenant on Civil and Political Rights* (ICCPR), which protects the cultural rights of persons belonging to minorities. This provision has served to protect Sami reindeer husbandry from measures impacting the practice of their traditional livelihood (*Fosen* case 2021; Cambou 2024, p. 52; Ravna 2023, p. 157).

Among the measures taken by Norway to comply with its international obligations is the adoption of the *Finnmark Act*, which applies to the traditional Sami lands in the County of Finnmark, located in the northernmost part of Norway (*Finnmark Act* 2005, Section 3, Ravna and Bankes 2017, p. 75). The Act provides for the transfer of ownership of land and resources from the state to the Finnmark Estate, which administers the land and natural resources (*Finnmark Act*: Section 6, Nygaard 2016, p. 19; Ravna 2015, p. 68). The Act also establishes the Finnmark Commission, which is tasked with investigating land rights based on prescription or immemorial usage according to the criteria established by the Norwegian Supreme Court (*Finnmark Act* 2005, Section 5; Ravna 2011, pp. 429–432; *Karasjok* case 2024; *Selbu* case 2001; *Svartskog* case 2001). This process has been denounced by several authors as insufficiently implementing *ILO Convention 169* (Koivurova *et al.* 2015, p. 15) and for the limited scope of Sami rights' recognition outside of Finnmark (Akhtar 2022, p. 123; Angell *et al.* 2020, p. 53; Nygaard 2016, p. 20).

The story of Sami rights' recognition is quite different in Sweden. Historically, and to this day, the recognition of Sami rights has mainly concerned reindeer husbandry (Árnadóttir 2017, p. 146; Nilsson 2020, pp. 298–300). This is reflected in the Constitution, which states that the Samis' right "to practise reindeer husbandry is regulated in law" (The Instrument of Government, Chapter 2: Article 17). The *Reindeer Husbandry Act* of 2007 provides for Sami rights to pursue reindeer herding, although the definition of reindeer husbandry as a "collective right" remains unclear (Allard 2013, p. 208). The right to reindeer husbandry consists of the right to use real property, the right to use land and water, and other "husbandry-connected activities and measures" (Allard 2013, p. 208) but does not amount to an exclusive ownership right in the land used (Raitio *et al.* 2020, p. 2).

In addition to this legislation, the Swedish Supreme Court has ruled that the Sami's land rights are based either on the principle of immemorial prescription (Allard and Brännström 2021, p. 58; Bengtsson 2016, p. 67; *Skattefjäll* (*Taxed Mountain*) case 1981) or on the basis of customary law (*Nordmaling* case 2011; Sasvari and Beach 2011, p. 132). Furthermore, in the 2020 *Girjas* case, the Supreme Court ruled that the international rights of Indigenous peoples, including those recognized under *ILO 169*, should be taken into account when interpreting national provisions, including the principle of immemorial prescription, despite the fact that Sweden has not ratified this convention (Allard and Brännström 2021, p. 64; Torp 2024, p. 76). However, because of the elements described above, Sweden has been repeatedly criticized by the United Nations for failing to protect Sami land rights (Lawrence and Åhrén 2016, p. 189), and it is not clear how and if the *Girjas* case will lead to an improvement of all Sami people's rights in Sweden (Torp 2024, p. 80).

In Finland, the protection of Sami rights focuses mainly on the right to "maintain and develop their own culture," which is protected by Section 17.3 of the Constitution. In the two areas designated as Sami in Finland—the "Sami Homeland" and the Skolt area within it[5]—the Sami enjoy mainly linguistic and cultural autonomy (*Act on the Sámi Parliament* 1995: Section 1; Allard 2016, p. 51; Joona and Joona 2011, p. 377). Ninety percent of the Sami Homeland is owned by the state and administered by the Finnish Forest and Park Enterprise (Metsähallitus) (*Act on Metsähallitus* 2004; Heinämäki 2024, p. 93). Metsähallitus is responsible for the management, use, and protection of natural resources in the Sami Homeland, "ensuring the conditions of the Sami people to practice their culture," and in reindeer herding areas, according to the *Reindeer Husbandry Act* (*Act on Metsähallitus* 2004: Section 4[2]). Unlike in Norway, with its Finnmark Estate, Finland has no formal process for recognizing Sami rights to their lands and resources (Heinämäki 2024, p. 87). While the Finnish Supreme Court has issued two rulings in favor of

Sami fishing rights protected under the Sami people's constitutional right to their culture, demonstrating a certain "progress in the judicial recognition of the Sámi People's rights as an Indigenous People" (Scheinin 2024, p. 47),[6] no court decision has yet recognized Sami land rights based on immemorial prescription or customary law. Lastly, unlike in Sweden and Norway, reindeer husbandry in Finland can be practiced by the non-Sami people as well (*Reindeer Husbandry Act* 1990). Given that most Swedish and Norwegian case law concerning the Sami relates to their exclusive rights to reindeer husbandry, this particularity of Finnish law shapes the dynamics at play with regard to the legal protection of Sami rights in Finland in a different way.

In short, the recognition and protection of Indigenous peoples' territorial rights in Canada and the Fennoscandian countries differ greatly, reflecting each country's particular colonial history. These differences, as we will now see, are reflected in the level of protection for Indigenous peoples' rights when transposed to mining legal frameworks.

The translation of Indigenous peoples' rights in Canadian and Fennoscandian mining legal frameworks

Free entry mining systems in Canada and the duty to consult and accommodate Indigenous peoples

Indigenous peoples in Canada have historically been marginalized in mining decision-making processes. Indeed, the legal frameworks governing mining and the rights of Indigenous peoples have largely developed on different "scales of legal regulation" (de Sousa Santos 2020, p. 506), with minimal interaction between them. Recently, however, these legal spheres have collided in the context of increasing conflicts between Indigenous communities and mining proponents, particularly in "free entry mining" jurisdictions.

The principle of "free entry mining" is the foundation of most provincial and territorial mining regimes in Canada.[7] Although the origins of "free mining" can be traced back to medieval Europe (Scott 2008, pp. 209–217), contemporary free entry mining regimes in Canada, as in other former British colonies, evolved from the "mining camp codes" that emerged from the California Gold Rush of 1848–1849 (Barton 2019, p. 188; Lacasse 1974; 38–42; Scott 2008, pp. 218–241). The practices and customs embedded in the miners' codes then migrated from gold rush to gold rush during the nineteenth century in the context of colonial expansion and influenced the initial development of mining laws in many regions of the world, including in Canada (Barton 2019, pp. 185–247; Lacasse 1974, pp. 40–42).

Contemporary, free entry mining regimes in Canada, while differing in technicalities and modes, share a basic underlying structure. One of the core features of these regimes, in addition to the principle of the Crown ownership of mineral rights, is the right of the miner to stake a "claim"— or, where possible, to acquire it online—to secure prior and exclusive access to a specific tract of land for the purpose of exploring for publicly owned minerals. It is important to note that mining claims are acquired unilaterally by the miner, without any prior exercise of government discretion (Barton 2019, p. 529; Lacasse 1974, p. 43). The miner is entitled to acquire a claim once all related legal requirements have been met. As Barton (2019, p. 989) writes: "If land has not been withdrawn from mineral entry, free entry allows the miner to choose when and where to acquire mineral rights from the Crown, and the government has no say in the matter."

The unilateral claim acquisition system is even more consequential when one considers that the miner who discovers a "workable deposit" on their claim is, in principle, entitled to a mining lease, provided that all other legal requirements have been met. The mining lease gives its holder the exclusive right, usually for at least two decades (with the possibility of renewal), to access the land in question for the purpose of extracting minerals (Barton 2019, p. 428; Lacasse 1974, pp. 115–117). The extent of the discretion granted to state authorities to refuse to grant a lease following a discovery remains uncertain in most free entry jurisdictions.

In short, free entry mining regimes aim to promote mining investment and exploration by facilitating, ordering, and securing the acquisition of mining rights and titles. As a result, these regimes have historically paid scant attention to non-mining rights and interests in land, including those of Indigenous peoples. However, some provinces and territories have recently amended—or are in the process of amending—their mining legal frameworks to take into account Indigenous peoples' rights in the mining sector, in particular the constitutional duty to consult and accommodate derived from the interpretation of Section 35 of the *Constitution Act* 1982.

This duty arises where the Crown has knowledge of the potential existence of an Aboriginal right or title, or of a treaty right, and is contemplating conduct—such as granting mineral rights or authorizing mining activities—that may adversely affect it (*Haida Nation v. British Columbia* 2004, para. 35). The intensity of the obligation varies according to the circumstances, ranging from notification and information requirements to formal participation of affected Indigenous groups in the decision-making process. Meaningful consultation may also require the government to change its action or decision in light of the information received (*Haida Nation v. British Columbia* 2004, para. 43–47). However, the duty to consult and accommodate does not amount to an Indigenous right to "veto" the development (*Haida Nation v. British Columbia* 2004, para. 48; *Ktunaxa Nation v. British Columbia* 2017, para. 83).

The constitutional duty to consult has influenced, to varying degrees, the recent evolution of mining decision-making processes in several Canadian jurisdictions. For example, mining laws and regulations in Ontario, Québec, and the Yukon have been amended to implement consultation requirements at various stages of mining development, from early exploration to mine closure (see *Mining Act* [Ontario] 1990; *Mining Act* [Québec] 2013; *Quartz Mining Act* [Yukon] 2003). While these amendments provide opportunities for Indigenous peoples to have some say in mining activities occurring on their traditional lands, including under certain conditions at the early exploration stage, their adequacy and sufficiency for the state to meet its constitutional obligations to Indigenous peoples are the subject of ongoing debate and litigation (Drake 2015; Ezeudu 2020; Thériault 2016). In particular, the lack of consultation requirements prior to the registration of mining claims on lands claimed under Aboriginal title remains highly controversial.[8]

Indigenous participation in mining decision-making also takes place in the context of impact assessment (IA) processes that apply to certain mining projects under provincial, territorial, and federal legislation or under the terms of land claims agreements. Indeed, IA processes are often used to implement, in whole or in part, the duty to consult and accommodate Indigenous peoples (Lambrecht 2013; Papillon and Rodon 2017). However, the thresholds for requiring an IA are set by the relevant legal regimes or, where they exist, by land claims agreements and vary widely from one jurisdiction to another. Moreover, IA processes establish different levels and intensities of Indigenous participation. For example, while the IA processes set out in the *James Bay and Northern Québec Agreement* guarantee Inuit and Cree signatories' representation on the joint environmental assessment boards responsible for evaluating and reviewing mining projects in their traditional territories, Québec's IA process for the southern part of the province does not specifically address Indigenous peoples' rights and interests (Thériault *et al.* 2022). Meanwhile, other IA regimes, including the recently reformed federal and British Columbia processes, include specific provisions aimed at integrating Indigenous knowledge and recognizing the rights of Indigenous peoples in the impact assessment process and decision-making (Allard and Curran 2023).

Lastly, Indigenous groups in Canada are increasingly setting their own standards to define the terms of their engagement with the mining industry. For example, some Indigenous groups have adopted their own mining policies to apply within their traditional territories (e.g., Cree Nation Government 2010) or their own consultation and consent protocols (Leclair *et al.* 2019; Montambeault and Papillon 2023). These Indigenous-led standard-setting initiatives, which will not be explored further within the limited scope of this chapter, could help to transform existing norms and practices in the coming years.

The variable geometry of Sami rights in Fennoscandian mining legal frameworks

As in Canada, Fennoscandian mining legal frameworks have historically been designed to encourage mining development (Romsaas 2000, p. 26; Liedholm Johnson and Ericsson 2015, p. 24). Finland and Sweden shared the same framework until 1809, which included strong elements of a claims system, particularly with respect to discovery rights (Liedholm Johnson and Ericsson 2015, p. 25). While the Finnish and Swedish legal frameworks for mining have developed in different ways since then, Liedholm Johnson and Ericsson (2015, p. 26) note that in both countries, "mineral exploration is regarded as beneficial to the community" and that "[a] guiding principle is and has been that the utilization of the country's mineral deposits is best promoted by providing economic incentives to the discoverer of a deposit." The same "right of the discoverer" is provided by Norwegian law, historically influenced by Saxon and German law, and then by French law (Nagel

1994, pp. 147–148; Romsaas 2000, p. 26). The three mining legal frameworks have evolved throughout history, particularly in terms of the balance struck between landowners, discoverers, and the state's interests (Liedholm Johnson and Ericsson 2015, p. 27; Nagel 1994, p. 148).

Today, as in most Canadian jurisdictions, Finland operates a form of mining claims' regime through the reservation system, that is, a proponent can reserve an area by filing a notification with the mining authority, which, if approved, gives priority in obtaining an exploration permit (*Mining Act*: Section 32; Metsä-Simola *et al.* 2022, p. 8), thus facilitating exploration (Kotilainen *et al.* 2022, p. 4). The mining legal frameworks in Norway and Sweden are not based on the free entry principle but rather on the concession system, although the Swedish mining regime exhibits "strong elements of the claim system" (Liedholm Johnson 2010, p. 25).[9]

Mining legal frameworks in the Fennoscandian countries are generally structured around prospecting, exploration, and extraction. While, with limited exceptions, anyone may carry out prospecting activities in the three countries (although in Sweden, it is possible with the consent of the landowner), an exploration permit is required to obtain an exclusive right to explore (*Swedish Minerals Act*: Ch. 3 Sec.2, Ch. 1 Sec. 4, *Norwegian Minerals Act*: Sec. 8, 13, *Finnish Mining Act*: Sec. 7, 9). Mining regimes then provide a smooth transition toward extraction for miners who find a commercially viable deposit (*Swedish Minerals Act*: Ch. 4 Sec. 2, *Norwegian Minerals Act*: Sec. 29, *Finnish Mining Act*: Sec. 32). Various conditions may be required to be met in order to obtain a mining permit or to proceed with extraction work, including the completion of an environmental impact assessment (Hojem 2015, pp. 26, 52, 58). Moreover, planning and building laws play an important role in the context of mining development, especially in Finland, and even more so in Norway (Koivurova *et al.* 2015, p. 15; Pölönen *et al.* 2020, p. 106).

The Fennoscandian mining legal frameworks provide the Sami with the right to participate in mining decision-making processes, albeit to varying degrees, reflecting the different recognition of their rights within the broader national legal framework, as described above. For instance, Sami rights in the Finnmark Estate in Norway, and the Sami Homeland and Skolt area in Finland, are better protected by the mining legal framework than in other parts of the countries, and reindeer husbandry rights holders are specifically mentioned in the *Swedish Minerals Act* (Árnadóttir 2017; Hojem 2015; Koivurova and Petrétei 2014).

The *Norwegian Minerals Act* and the *Finnish Mining Act* contain a general provision on the protection of the Sami. The *Norwegian Minerals Act*, Section 2, states that "the foundation of Sami culture, commercial activities, and social life shall be 'safeguarded.'" The *Finnish Mining Act* provides in Section 1 that mining activities "shall be adapted in the Sámi Homeland . . . so as to secure the rights of the Sámi as an indigenous people to maintain and develop their language, culture, and traditional livelihoods." It also contains a specific reference to the protection of the livelihood and the culture of Skolt people. The extent to which these general provisions influence, in practice, the way in which mining titles are granted and the conditions under which they are granted would require empirical research beyond the scope of this chapter. On the basis of a documentary analysis alone, other provisions appear to be more relevant, as they establish specific rights and obligations. For example, both mining legislations state that the mining authority may refuse to grant a permit if it affects the rights, status, or livelihood of the Sami (*Norwegian Minerals Act*: Section 17, 30, *Finnish Mining Act*: Section 50). The significance of this discretionary power in Finland is uncertain, as the Act clearly states that the permit may be granted nonetheless if it is possible to remove the "impediment" for the Sami "through permit regulations" (Section 50). There is no explicit reference in the *Swedish Minerals Act* to Sami rights or livelihood as grounds for refusing a permit.

Finland's mining legal framework is considered by some authors to provide a strong protection for the right of the Sami to maintain and develop their language and culture (Koivurova *et al.* 2015, p. 19). For example, it provides that the mining authority, in cooperation with the Sami Parliament and the Skolt village meeting, shall assess the impacts of proposed mining activities on the Samis' rights to maintain and develop their own language and culture before granting a permit, taking into account cumulative impacts (*Finnish Mining Act*: Sections 38; Heinämäki 2024, p. 91). However, the lack of definition of such assessment procedures has been criticized as limiting the impact of this provision "prohibi[ting] the weakening of Sami culture" (Heinämäki 2024, p. 92). The permit authority shall furthermore require a statement from the Skolt Village Meeting[10] on the impact of mining activities on the sources of livelihood and living conditions of the Skolt people (Section 38). Outside the Sami Homeland, a report is also required in cases where there is a significant impact on the rights of the Sami people (Section 34). Finally, the Sami Parliament and the Skolt Sami village meeting have the right to initiate proceedings "against activities having a detrimental impact in the Sami Homeland or Skolt area" (Section 159) and the right to appeal against the granting, expiry, and amendment of exploration and extraction permits (Section 165).

Moreover, the Finnish government is obliged to negotiate with the Sami Parliament if a measure or a decision has negative impacts on the Sami in the Sami Homeland, which would include any measure or decision related to mining (*Act on the Sámi Parliament* 1995, Section 4). However, since only a few exploration projects have taken place in this area (Pölönen *et al.* 2020, p. 104), the strength of this statute has yet to be tested.

In Norway, the *Minerals Act* provides for various rights and obligations toward the Sami in the Finnmark Estate, the region where Sami rights enjoy special recognition and protection in the state legal system. For example, the Sami Parliament can oppose the granting of exploration and extraction permits and appeal to the King if the competent authority decides to grant the permit regardless. In any case, the final decision rests with the King (*Minerals Act*: Section 17, 30), as illustrated in the *Nussir* case (Ivsett Johnsen 2016). Furthermore, in accordance with the *Finnmark Act*, the Sami Parliament has adopted a Mineral Guide that mining proponents must follow (Section 4, Nygaard 2016, p. 20). The guide indicates how the mining proponent and the Sami Parliament must "relate to each other" when mining activities take place in the Finnmark Estate (Nygaard 2016, p. 20).

Both inside and outside the Finnmark Estate, the "Consultation Agreement" (Procedures for Consultations between State Authorities and the Sami Parliament), adopted in 2005 for Norway to implement its obligations under *ILO Convention 169* (Ravna 2020, p. 244), provides that the Norwegian government authorities are obliged to consult the Sami Parliament when a measure affects "all material and intangible forms of Sámi culture, including . . . mineral exploration and extraction activities" (Section 2, Ravna 2020, pp. 244–245). For example, the Norwegian government is consulting the Sami Parliament as it revises its mining legislation (Nygaard 2016, p. 20).[11] Furthermore, since 1 July 2021, the new Chapter 4 of the *Sami Act* provides that government authorities, acting as public authorities and landowners, must consult the Sami Parliament and Sami representatives when a measure may "affect Sami interests directly" in "traditional Sami areas" within and outside of the Finnmark Estate. The Sami Parliament or representatives of local Sami organizations may be designated as the competent authorities to assess whether such impacts on Sami interests may occur. However, the obligation to consult must still be set out in the *Minerals Act* itself.

A distinctive feature of the Norwegian mining legal framework is that municipalities can have a strong influence on mining development, thanks to their power under the *Planning and Building Act* to "protect the natural basis for the Sami culture, economy, and social life" (Koivurova *et al.* 2015, p. 33; Nygaard 2016, p. 20; *Planning and Building Act*: Section 3–1, 3–2). However, the importance of this authority for the Sami depends on whether their interests are aligned with those of the municipality, as illustrated by the *Biedjovágii* case in the Municipality of Kautokeino and the *Nussir* case in the Municipality of Kvalsund, where mining projects were approved or rejected by the municipal authorities according to the preferences of the municipality's majority, rather than according to the interests of the Sami (Espiritu 2015; Ivsett Johnsen 2016).

The Swedish mining legal framework can be considered the weakest, as the obligations toward the Sami are limited in scope and focus exclusively on the holders of reindeer herding rights, reflecting the protection of Sami rights at the national level (Amatulli 2015, p. 60; UNGA 2016, para. 40–83). For example, the role of the Sami Parliament is limited to being informed at the various stages of the mining cycle (Minerals Ordinance: Sections 6, 8, 9b, 22) and to giving an opinion only on applications for exploration permits and when the affected area is used for reindeer husbandry (Minerals Ordinance: Section 3). The holders of reindeer husbandry rights may object to the plan of operation prepared by the mining proponent during the exploration phase (*Minerals Act*: Chapter 3 Section 5a), and the mining proponent must reach an agreement with them on the designation of the land that will eventually be used for extraction (*Minerals Act*: Chapter 9 Section 2). However, the Chief Mining Inspector can ultimately impose his or her decision on the reindeer herders.

Mining development can be stopped by the government if it occurs in reindeer husbandry areas, as reindeer husbandry is considered a "national interest" under the *Swedish Environmental Code* (*Swedish Environmental Code*: Chapter 3, Section 5). However, the same Code also considers mining to be in the national interest. The government's balancing of these interests tends to favor mining, which is considered compatible with reindeer husbandry from the outset, provided that mitigation measures are in place (Allard and Curran 2023, p. 12; Koivurova *et al.* 2015, p. 23). As observed by Lawrence and Kløcker Larsen (2017, p. 1171), this "assumption" of possible coexistence between reindeer husbandry and mining "builds on a die-hard myth of the continuous adaptability of Sami communities and reindeer herding in the face of industrial expansion." The case of Rönnbäcken, where mining was considered more beneficial to society than reindeer husbandry (Ahrén 2016, p. 184; Raitio *et al.* 2020, p. 10), and the concession granted to the UK mining company Beowulf, despite strong Sami opposition (London Mining Network 2022), both illustrate the prioritization of mining interests.

It should be noted that Sweden adopted a consultation statute in January 2022. It remains to be seen to what extent this statute will influence mining reform and increase Sami participation in mining decision-making, although the prospects do not look favorable (O'Faircheallaigh 2023, p. 265).

Conclusion: The prioritization of mining interests and the limited pathways for Indigenous peoples' agency in mining decision-making processes

In both Canada and Fennoscandia, mining activities and the legal tools designed to facilitate them are closely intertwined with historical and contemporary settler colonial processes of Indigenous land dispossession (Hoogeveen 2014; Lawrence and Åhrén 2016; Lindmark 2013; Ojala and Nordin 2019; Tollefson 2021). Although the colonial dynamics in Fennoscandia differ in many ways from those in Canada (and also vary widely within these jurisdictions), one of the driving forces behind the territorial expansion in these regions has been the desire to take control of the resource-rich "frontiers," particularly through mining (Avango 2020; Barton 2019; Fur 2013; Lindmark 2013; Ojala and Nordin 2019). Accordingly, the mining legal frameworks developed in these historical and contemporary colonial contexts, which are primarily designed to promote mining investment and development, largely erase competing rights and interests to land and resources, including those of Indigenous peoples who have occupied and used the land for thousands of years.

In the past few decades, some Canadian jurisdictions and Fennoscandian countries have increasingly recognized the rights and interests of Indigenous peoples within their mining legal frameworks, albeit to varying degrees, as described in this chapter. For example, in Norway's Finnmark Estate and in Finland's Sami Homeland, Sami rights enjoy greater protection in mining contexts than in Sweden, where the recognition of Sami rights is generally limited to reindeer herding (Raitio *et al.* 2020, p. 6). In Canada, an Indigenous group that is party to a land claims agreement may use different institutional pathways to influence mining development than, for example, an Indigenous group claiming Aboriginal title and rights that are not yet formally recognized under state law (Thériault *et al.* 2022).

Despite these variations, however, our study suggests that the increased recognition of Indigenous peoples' rights in Canada and the Fennoscandian countries has not fundamentally altered the rationales and underlying values of mining legal frameworks, which emphasize the importance of mining as a means to support economic growth and prosperity. When considered in the context of mining, Indigenous peoples' rights and interests almost always take the form of varying degrees of participatory rights that are merely juxtaposed with the preexisting structures of mining tenure systems. While providing an opportunity for Indigenous peoples to have a say in mining decision-making processes that may affect their land rights and interests, mining legal frameworks in the jurisdictions studied still largely prioritize mining development over Indigenous peoples' land rights and interests (Raitio *et al.* 2020; Thériault 2016; Scheinin 2024, p. 37; Heinämäki 2024, p. 85).

The effect of these mining legal frameworks is compounded by the limited scope of Indigenous peoples' participation rights in formal mining decision-making processes in the jurisdictions studied. First, government obligations to engage with Indigenous peoples typically occur at an advanced stage of the mining cycle. In addition, governments are generally not required to obtain the consent of Indigenous peoples before authorizing mining activities that may adversely affect their rights and interests (Allard and Curran 2023; Imai 2017, pp. 385–386; Kotilainen *et al.* 2022, p. 4; Lassila 2018, p. 5; Nygaard 2016, p. 23; Raitio *et al.* 2020, p. 12; Scott 2020, p. 278).[12] As a result, while Indigenous peoples can influence *how* mining activities are carried out, they have limited space in the official legal system to influence *whether* these activities should be carried out at all, according to their own laws, knowledge, and priorities (Allard and Curran 2023, p. 6; Raitio *et al.* 2020, p. 12).

More broadly, this situation echoes the limited agency of Indigenous peoples at the strategic level of land and natural resource use planning in the four countries (Allard and Curran 2023; Raitio *et al.* 2020; Metsä-Simola *et al.* 2022, p. 29). Such limited agency does not create an enabling environment for the implementation of Indigenous peoples' rights to self-determination and to their land and natural resources, as enshrined in the *United Nations Declaration on the Rights of Indigenous Peoples* (UNDRIP).

Beyond state-based, mining decision-making processes, Indigenous peoples exercise agency in mining development through a broad repertoire of political and legal strategies grounded in their legal orders and inherent jurisdictions (Thériault *et al.* 2022). For example, in addition to resistance strategies such as blockades and protests (Eisenberg 2020) or counter-mapping early in the mining process (Lassila 2018), some Indigenous peoples in Canada and Fennoscandia have enacted their own norms regarding mining development in their territories (or otherwise impacting their livelihoods) such as mining policies that subordinate mining to the norm of free, prior, and informed consent (Ojala and Nordin 2019; Papillon and Rodon 2020). Indigenous peoples in Canada have also developed consultation and consent protocols for mining activities (Leclair *et al.* 2019; Montambeault and Papillon 2023), while some Indigenous groups have implemented their own environmental and social impact assessment processes for development projects based on their laws and knowledge (Friedland *et al.* 2018; Papillon and Rodon 2020). In Fennoscandia, the Swedish Sami Association has developed guidance for conducting social impact assessments on projects impacting Sami reindeer herding (Larsen *et al.* 2018, p. 377), and the Sami Parliament of Norway has adopted their own guidelines governing mining on Sami territory (Nygaard 2016, p. 20; Holroyd 2023, p. 109). These Indigenous-led norms and processes in the study regions, which could contribute to a radical transformation of the dominant mining paradigm, offer interesting avenues for future research.

Acknowledgments

This research was supported by the Knowledge Network on Mining Encounters and Indigenous Sustainable Livelihoods (MinErAL) through funding from the Social Sciences and Humanities Research Council of Canada. We would like to thank Torvald Falch for his suggestions regarding the analysis of Norwegian law, as well as the evaluators for their insightful comments. We would also like to thank the participants and speakers who shared their knowledge with us during the REXSAC-MinErAL PhD course *Extracted places in transition*, which took place in the Swedish Arctic from 3 to 8 October 2022. Any remaining mistakes are ours alone.

Notes

1 It is important to note that primary legal documents in Norway, Sweden, and Finland are not systematically translated into English, especially court decisions. Since the authors of this chapter are not proficient in the Fennoscandian languages, the legal analysis of this region relies heavily on secondary sources produced by Fennoscandian scholars. We would like to thank Torvald Falch for his suggestions regarding the analysis of Norwegian law. We would also like to thank the participants and speakers who shared their knowledge with us during the REXSAC-MinErAL PhD course *Extracted places in transition*, which took place in the Swedish Arctic, 3–8 October 2022. Any remaining errors are ours alone.

2 In Canada, Section 35(2) of the *Constitution Act* 1982 recognizes three Indigenous groups: the "Indians" (First Nations), the Inuit, and the Métis. These broad categories encompass some 60 nations representing 12 linguistic groups. In Fennoscandia, the definition of who is Sami is primarily based on a linguistic approach, although it differs in the three countries. The Sami identity is particularly controversial in Finland: See Joona 2016; United Nations 2022.

3 The Lapp Codicil part of the 1751 *Stromsad Treaty* contains rights for the Sami, mainly to allow the Sami to continue practicing reindeer husbandry across borders (Amatulli 2015, p. 19). There is a debate regarding the recognition of land rights through this Codicil (Nikolova 2007, pp. 53–54). Some consider that the matter at stake in this treaty was strictly about international public law (border crossing) and therefore did not recognize or deprive the Sami of land rights (Eide 2001, pp. 138–139).

4 The *Constitution of the Kingdom of Norway*: Article 108; The *Constitution of Finland*: Section 17.3; Sweden, *Instrument of Government*: Chapter 1, Art. 2 and Chapter 2, Art. 17.

5 The *Skolt Act* recognizes the Skolt Sami as a distinct Sami group based on their specific language and culture (*Skolt Development Act* 253/1995).

6 Cases in Utsjoki, KKO:2022: 25 (R2019/424), and in Vetsjoki, KKO:2022: 26 (R2019/425).

7 According to Barton (2019, p. 539), attributes of free entry mining regimes are found in most provinces and territories, with the exception of Alberta, Nova Scotia, Prince

Edward Island, and Saskatchewan. Canada is a federation consisting of the federal government and ten provinces, in addition to three territories located in the northernmost region of the country. Under sections 92(5), 92(13), and 92A of the *Constitution Act* 1867, the provinces have legislative jurisdiction over most matters relating to natural resources, including mines and minerals. The three territories—the Yukon, the Northwest Territories, and Nunavut—have historically been administered by the federal government. However, the ongoing process of "devolution" aims to increase their legislative authority over their natural resources, including minerals. Through this process, the Yukon has gained ownership of the resources, and legislative authority over their management (Newman 2018, pp. 12–17; *Yukon Act*: Section 19).

8 In 2012, the Yukon Court of Appeal affirmed that the constitutional duty to consult and accommodate Indigenous Peoples is triggered by the registration of mining claims on lands potentially held under Aboriginal title (*Ross River Dena Council v. Government of Yukon* 2012). A negotiation process involving Indigenous Peoples and the industry is currently underway to reform the mining tenure system in the territory in accordance with this decision. More recently, in 2023, the Supreme Court of British Columbia also declared that the province's online mining claims system, which allows automatic registration of mining claims in Indigenous territories without prior consultation, breaches the constitutional obligations of the Crown (*Gitxaala v. British Columbia [Chief Gold Commissioner]* 2023). Finally, at the time of writing, the Mitchikanibikok Inik (Algonquins of Barriere Lake) are challenging the constitutionality of the Québec *Mining Act* under sections 35 and 52 of the *Constitution Act* 1982, arguing that the free entry mining regime, as currently structured by the Act, does not allow for the implementation of the province's constitutional duty to consult and accommodate when registering or transferring mining claims.

9 The *Norwegian Minerals Act* distinguishes and regulates the minerals owned by the state and the minerals owned by the landowner. Exploration and exploitation permits are required for the minerals owned by the state. An agreement with the landowner is required to explore and extract minerals from private lands, although the absence of such an agreement can be overcome by compulsory acquisition (*Minerals Act*: Section 11, 28).

10 The Skolt Sami are a culturally and linguistically distinct Sami group whose specific rights have been protected by Finland since 1995. The Skolt Village Meeting is the representative authority of the Skolt Sami.

11 However, it is worth noting that the government passed the 2009 *Minerals Act* without the consent of the Sami Parliament (Angell *et al.* 2020, p. 53).

12 In *Tsilhqot'in Nation* (2014), the Supreme Court of Canada held that the government must obtain Indigenous Peoples' consent prior to authorizing development projects on lands held under Aboriginal title (para. 76). However, according to the same decision, if consent cannot be obtained, the government may still approve the project, provided that the procedural duty to consult has been complied with, and that the infringement can be justified under Section 35 of the *Constitution Act* 1982.

References

Act of 27 June 2008, No. 71, Relating to planning and processing of building applications (Norway).

Act on Metsähallitus (Finland, Forest and Park Enterprise Act) (1378/2004).

Act on the Sámi Parliament (Finland) (974/1995).

Ahrén, M., 2016. To what extent can Indigenous territories be expropriated? *In:* C. Allard and S. Funderud Skogvang, eds., *Indigenous rights in Scandinavia: Autonomous Sami law*. London: Routledge, 173–187.

Akhtar, Z., 2022. Sami peoples land claims in Norway, Finnmark Act and proving legal title. *The Indigenous Peoples' Journal of Law, Culture, & Resistance*, 7, 115–138.
Alcantara, C., 2013. *Negotiating the deal: Comprehensive land claims agreements in Canada*. Toronto: University of Toronto Press.
Allard, C., 2011. The Nordic countries' law on Sámi territorial rights. *Arctic Review on Law and Politics*, 2 (2), 159–183.
Allard, C., 2013. Who holds the reindeer-herding right in Sweden? A key issue in legislation. *In:* N. Bankes and T. Koivurova, eds., *The proposed Nordic Saami convention: National and international dimensions of Indigenous property rights*. Oxford: Hart Publishing, 207–227.
Allard, C., 2016. Some characteristic features of Scandinavian laws and their influence on Sami matters. *In:* C. Allard and S. Funderud Skogvang, eds., *Indigenous rights in Scandinavia: Autonomous Sami law*. London: Routledge, 49–64.
Allard, C., 2024. Sámi rights in the sustainable transition—Concluding remarks. *In:* D. Cambou and Ø. Ravna, eds., *The significance of Sámi rights: Law, justice and sustainability for the Indigenous Sámi in the nordic countries*. London: Routledge, 184–202.
Allard, C. and Brännström, M., 2021. Girjas reindeer herding community v. Sweden: Analysing the merits of the Girjas case. *Arctic Review on Law and Politics*, 12, 56–79.
Allard, C. and Curran, D., 2023. Indigenous influence and engagement in mining permitting in British Columbia, Canada: Lessons for Sweden and Norway? *Environmental Management* [Online], 72, 1–18. Available from: https://link.springer.com/article/10.1007/s00267-021-01536-0 [Accessed 27 June 2023].
Allard, C. and Funderud Skogvang, S., eds., 2016. *Indigenous rights in Scandinavia: Autonomous Sami law*. London: Routledge.
Amatulli, G., 2015. *The legal position of the Sami in the exploitation of mineral resources in Finland, Norway and Sweden*. Thesis (Master's). Institute for Human Rights, Åbo Akademi University.
Angell, E., Nygaard, V., and Pelle, P., 2020. Industrial development in the North—Sámi interests squeezed between globalization and tradition. *Acta Borealia: A Nordic Journal of Circumpolar Societies*, 37 (1), 43–62.
Árnadóttir, G. R., 2017. Who is Sami? A case study on the implementation of Indigenous rights in Sweden. *In:* H. Devere, K. T. Maihāroa, and J. P. Synott, eds., *Peacebuilding and the rights of indigenous peoples: Experiences and strategies for the 21st century*. Cham: Springer eBooks.
Avango, D., 2020. Imprints on the resource landscape: The long history of mining in the Arctic. *Journal of Northern Studies*, 14 (2), 67–82.
Barton, B., 2019. *Canadian law of mining*. 2nd ed. Toronto: LexisNexis.
Bengtsson, B., 2016. Reforming Swedish Sami legislation: A survey of the arguments. *In:* C. Allard and S. Funderud Skogvang, eds., *Indigenous rights in Scandinavia: Autonomous Sami law*. London: Routledge, 65–78.
Borrows, J., 2015. Aboriginal title and private property. *SCLR (2D)*, 71 (1), 91–134.
Cambou, D., 2024. The significance of the *Fosen* decision for protecting the cultural rights of the Sámi Indigenous people in the green transition. *In:* D. Cambou and Ø. Ravna, eds., *The significance of Sámi rights: Law, justice and sustainability for the Indigenous Sámi in the nordic countries*. London: Routledge, 52–71.
Cambou, D. and Ravna, Ø., 2024. The significance of Sámi rights in the Nordic countries—An introduction. *In:* D. Cambou and Ø. Ravna, eds., *The significance of Sámi rights: Law, justice and sustainability for the Indigenous Sámi in the nordic countries*. London: Routledge, 1–4.
Constitution Act, 1982, Being Schedule B to the *Canada Act* 1982 (U.K.), 1982, c. 11.
Cree Nation Government, 2010. Cree Nation Mining Policy. Policy 2010–07 [Online]. Available from: www.cngov.ca/wp-content/uploads/2018/03/cree_nation_mining_policy-1.pdf [Accessed 27 June 2023].

de Sousa Santos, B., 1987 (2020). *Toward a new legal common sense: Law, globalization, and emancipation*. Cambridge: Cambridge University Press.
Delgamuukw v. British Columbia, [1997] 3 S.C.R. 1010.
Drake, K., 2015. The trials and tribulations of Ontario's Mining Act: The duty to consult and Anishinaabek law. *JSDLP*, 11 (2), 184–218.
Eide, A., 2001. Legal and normative bases for Saami claims to land in the Nordic. *International Journal on Minority and Group Rights*, 8, 127–149.
Eisenberg, A., 2020. Consent, resistance and the duty to consult. *International Journal on Minority and Group Rights*, 27 (2), 270–290.
Espiritu, A. A., 2015. Kautokeino and Kvalsund compared: Rejection and acceptance of mining in communities in northern Norway. *The Northern Review*, 39, 53–65.
Ezeudu, M.-J., 2020. The unconstitutionality of Canada's free entry mining systems and the Ontario exception. *Asper Review*, 20, 156–183.
Finnmark Act, Lov (Act) 17. juni 2005 nr. 85.
Fosen case (Supreme Court of Norway), 2021. HR-2021–1975-S.
Friedland, H., Leonard, B., Asch, J., and Mortimer, K., 2018. Porcupine and other stories: Legal relations in Secwépemcúlecw. *Revue Générale de droit*, 48 (1), 153–201.
Fur, G., 2013. Colonialism and Swedish history: Unthinkable connections? *In:* M. Naum and J. M. Nordin, eds., *Scandinavian colonialism and the rise of modernity: Small time agents in a global arena*. New York: Springer, 17–36.
Girjas v. the State (Swedish Supreme Court), 2020. (No. T 853–18).
Gitxaala v. British Columbia (Chief Gold Commissioner), 2023 BCSC 1680.
Government of Canada, 2022. Minerals and the Economy [Online]. Available from: www.nrcan.gc.ca/our-natural-resources/minerals-mining/minerals-metals-facts/minerals-and-the-economy/20529 [Accessed 27 June 2023].
Grammond, S., 2013. *Terms of coexistence: Indigenous peoples and Canadian law*. Toronto: Carswell.
Haida Nation v. British Columbia (Minister of Forests), [2004] 3 S.C.R. 511.
Haikola, S. and Anshelm, J., 2016. Mineral policy at a crossroads? Critical reflections on the challenges with expanding Sweden's mining sector. *The Extractive Industries and Society*, 3, 508–516.
Heinämäki, L., 2024. The prohibition to weaken the Sámi culture in international law and Finnish environmental legislation. *In:* D. Cambou and Ø. Ravna, eds., *The significance of Sámi rights: Law, justice and sustainability for the Indigenous Sámi in the nordic countries*. London: Routledge, 84–100.
Hojem, P., 2015. *Mining in the nordic countries: A comparative review of legislation and taxation*. Copenhagen: Rosendahls-Schulz Grafisk.
Holroyd, C., 2023. Corporate social responsibility, Indigenous peoples and mining in Scandinavia. *In:* N. Brunet and S. Longboat, eds., *Local communities and the mining industry*. London: Routledge, 103–122.
Hoogeveen, D., 2014. Sub-surface property, free-entry mineral staking and settler colonialism in Canada. *Antipodes*, 47 (1), 121–138.
Hopsu, I., 2022. Puhe—Renewal Process of the Mining Law in Finland. *Inka Hopsu*. 18 May 2022. Available from: www.inkahopsu.fi/renewal-process-of-the-mining-law-in-finland/ [Accessed 27 June 2023].
Horowitz, L. S., Keeling, A., Lévesque, F., Rodon, T., Schott, S., and Thériault, S., 2018. Indigenous peoples' relationships to large-scale mining in post/colonial contexts: Toward multidisciplinary comparative perspectives. *The Extractive Industries and Society*, 5, 404–414.
Imai, S., 2017. Consult, consent, and veto: International norms and Canadian treaties. *In:* J. Borrows and M. Coyle, eds., *The right relationship: Reimagining the implementation of historical treaties*. Toronto: University of Toronto Press, 370–408.
International Labour Organization, Convention (No. 169) concerning Indigenous and tribal people in independent countries, 27 June 1989, 1650 U.N.T.S. 383 (entered into force 5 September 1991).

Ivsett Johnsen, K., 2016. Land-use conflicts between reindeer husbandry and mineral extraction in Finnmark, Norway: Contested rationalities and the politics of belonging. *Polar Geography*, 39 (1), 58–79.

Jon Inge Sirum v. Essland Reindeer Pasturing District, No 4B/2001 (21 June 2001) ("*Selbu*" case, Norway).

Joona, T., 2016. The definition of a Sami person in Finland and its application. *In:* C. Allard and S. Funderud Skogvang, eds., *Indigenous rights in Scandinavia: Autonomous Sami law*. London: Routledge, 155–172.

Joona, T. and Joona, J., 2011. The historical basis of Saami land rights in Finland and the application of the ILO Convention No. 169. *The Yearbook of Polar Law Online*, 3 (1), 351–388.

Karasjok case (Supreme Court of Norway), 2024. HR-2024-982-S.

Koivurova, T., Masloboev, V., Hossain, K., Nygaard, V., Petrétei, A., and Vinogradova, S., 2015. Legal protection of Sami traditional livelihoods from the adverse impacts of mining: A comparison of the level of protection enjoyed by Sami in their four home states. *Arctic Review on Law and Politics*, 6 (1), 11–51.

Koivurova, T. and Petrétei, A., 2014. Enacting a new mining act in Finland—How were Sami rights and interests taken into account? *Nordic Environmental Law Journal*, 1, 119–133.

Kotilainen, J., Peltonen, L., and Reinikainen, K., 2022. Community benefit agreements in the nordic mining context: Local opportunities for collaboration in Sodankylä, Finland. *Resources Policy*, 79, 102973–102983.

Ktunaxa Nation v. British Columbia (*Forests, Lands, and Natural Resources Operations*), [2017] 2 S.C.R. 386.

Lacasse, J-P., 1974. *Le claim en droit québécois*. Ottawa: Presse de l'Université d'Ottawa.

Lambrecht, K. N., 2013. *Aboriginal consultation, environmental assessment, and regulatory review in Canada*. Regina: University of Regina Press.

Larsen, R. K., Österlin, C., and Guia, L., 2018. Do voluntary corporate actions improve cumulative effects assessment? Mining companies' performance on Sami lands. *The Extractive Industries and Society*, 5, 375–383.

Lassila, M. M., 2018. Mapping mineral resources in a living land: Sami mining resistance in Ohcejohka, northern Finland. *Geoforum*, 96, 1–9.

Lawrence, R. and Åhrén, M., 2016. Mining as colonization: The need for restorative justice and restitution of traditional Sami lands. *In:* L. Head, ed., *Nature, temporality and environmental management: Scandinavian and Australian perspectives on peoples and landscapes*. Milton: Taylor and Francis, 189–210.

Lawrence, R. and Kløcker Larsen, R., 2017. The politics of planning: Assessing the impacts of mining on Sami lands. *Third World Quarterly*, 38 (5), 1164–1180.

Leclair, J., Papillon, M., and Forget, H., 2019. Les protocoles de consultation autochtone au Canada: Un modèle de convergence des systèmes juridiques autochtones et étatiques? *Recherches amérindiennes au Québec*, 49 (2), 25–36.

Liedholm Johnson, E., 2010. *Mineral rights legal systems governing exploration and exploitation*. PhD Thesis. Stockholm: Royal Institute of Technology.

Liedholm Johnson, E. and Ericsson, M., 2015. State ownership and control of minerals and mines in Sweden and Finland. *Mineral Economics*, 28, 23–36.

Lindmark, D., 2013. Colonial encounter in early modern Sápmi. *In:* M. Naum and J. M. Nordin, eds., *Scandinavian colonialism and the rise of modernity: Small time agents in a global arena*. New York: Springer, 131–146.

London Mining Network, 2022. *Indigenous Sami rights under attack from Swedish state and UK mining company*. Available from: https://londonminingnetwork.org/2022/06/indigenous-sami-rights-under-attack-from-swedish-state-and-uk-mining-company/ [Accessed 19 January 2023].

Metsä-Simola, K., Majamaa, H., Sittnikow, M., and Männistö, M., 2022. *Report on the functioning of the reservation mechanism under the Mining Act*. Helsinki: Ministry of Economic Affairs and Employment in Finland.

Minerals Act (Norway) (Act of 19 June 2009, No. 101, Relating to the Acquisition and Extraction of Mineral Resources).
Minerals Act (Sweden) (1991: 45).
Minerals Ordinance (Sweden) (1992: 285).
Mining Act (Finland) (621/2011).
Mining Act (Ontario), RSO 1990, c. M-14.
Mining Act (Québec), LRC. c. M-13.1.
Montambeault, F. and Papillon, M., 2023. Repoliticising indigenous participation: FPIC protocols in Canada and Brazil. *The International Journal of Human Rights*, 27 (2), 335–358.
Nachet, L., Beckett, C., and Sehlin MacNeil, K., 2022. Framing extractive violence as environmental (in)justice: A cross-perspective from Indigenous lands in Canada and Sweden. *The Extractive Industries and Society*, 12, 100949 [Online]. Available from: www.sciencedirect.com/science/article/pii/S2214790X21001118 [Accessed 27 June 2023].
Nagel, A.-H., 1994. Norwegian mining in the early modern period. *GeoJournal*, 32 (2), 137–149.
Newman, D., 2018. *Mining law of Canada*. Toronto: LexisNexis.
Nikolova, B., 2007. *Sámi reindeer herders—Land and identity—Non-recognition of Indigenous land rights—Reasons, effects and potential development*. Thesis (Master's). Lund University International Master Program in Environmental Studies and Sustainability Science.
Nilsson, R., 2020. The consequences of Swedish national law on Sámi self-constitution—The shift from a relational understanding of who is Sámi toward a rights-based understanding. *Ethnopolitics*, 19 (3), 292–310.
Nordmaling case (Swedish Supreme Court) (NJA 2011 s.109).
Nygaard, V., 2016. Do Indigenous interests have a say in planning new mining projects? Experiences from Finnmark, Norway. *The Extractive Industries and Society*, 3, 17–24.
O'Faircheallaigh, C., 2023. *Indigenous peoples and mining: A global perspective*. Oxford: Oxford University Press.
Ojala, C.-G. and Nordin, J. M., 2019. Mapping land and people in the North: Early modern colonial expansion, exploitation, and knowledge. *Scandinavian Studies*, 91 (1–2), 98–133.
Papillon, M. and Rodon, T., 2017. Proponent-Indigenous agreements and the implementation of the right to free, prior, and informed consent: Lessons from two Canadian cases. *Environmental Impact Assessment Review*, 62, 216–224.
Papillon, M. and Rodon, T., 2020. The transformative potential of Indigenous-driven approaches to implementing free, prior, and informed consent: Lessons from two Canadian cases. *International Journal on Minority and Group Rights*, 27 (2), 314–335.
Petersen St-Laurent, G. and Le Billon, P., 2015. Staking claims and shaking hands: Impact and benefit agreements as a technology of government in the mining sector. *The Extractive Industries and Society*, 2, 590–602.
Pölönen, I., Allard, C., and Raitio, K., 2020. Finnish and Swedish law on mining in light of collaborative governance. *Nordic Environmental Law Journal*, 2, 99–134.
Quartz Mining Act, SY 2003, ch. 14.
R. v. Van der Peet, [1996] 2 S.C.R. 507.
Raitio, K., Allard, C., and Lawrence, R., 2020. Mineral extraction in Swedish Sápmi: The regulatory gap between Sami rights and Sweden's mining permitting practices. *Land Use Policy*, 99, 105001 [Online]. Available from: https://doi.org/10.1016/j.landusepol.2020.105001 [Accessed 2 July 2024].
Ravna, Ø., 2011. The process of identifying land rights in parts of Northern Norway: Does the Finnmark Act prescribe an adequate procedure within the national law? *The Yearbook of Polar Law*, 3 (1), 423–453.
Ravna, Ø., 2015. Sami rights to natural resources and lands in Norway. *In:* N. Loukacheva, ed., *Polar law and resources*. Copenhagen: Nordic Council of Ministers, 63–77.

Ravna, Ø., 2020. The duty to consult the Sámi in Norwegian law. *Arctic Review on Law and Politics*, 11, 233–255.
Ravna, Ø., 2023. The Fosen case and the protection of Sámi culture in Norway pursuant to article 27 ICCPR. *International Journal on Minority and Group Rights*, 30 (1), 156–175.
Ravna, Ø. and Bankes, N., 2017. Recognition of Indigenous land rights in Norway and Canada. *International Journal on Minority and Group Rights*, 24 (1), 70–117.
Reindeer Husbandry Act (Finland) (848/1990).
Reindeer Husbandry Act (Sweden) (1971: 437).
Rodon, T., 2018. Institutional development and resource development: The case of Canada's Indigenous peoples. *Canadian Journal of Development Studies/Revue canadienne d'études du développement*, 39 (1), 119–136.
Romsaas, J. P., 2000. Reform of the Norwegian mineral legislation and the interests of the Sami people. *Mineral Resources Engineering*, 9 (1), 25–38.
Ross River Dena Council v. Government of Yukon, 2012 YKCA 14.
Sanderson, D. and Singh, A. C., 2021. Why is Aboriginal title property if it looks like sovereignty? *Canadian Journal of Law & Jurisprudence*, 34 (2), 417–460.
Sasvari, A. and Beach, H., 2011. Short report: The 2011 Swedish Supreme Court ruling: A turning point for Saami rights. *Nomadic Peoples*, 15 (2), 130–135.
Scheinin, M., 2024. Indigenous peoples' right to fish: Recent recognition of Sámi rights in Finland through civil disobedience and criminal trial. *In:* D. Cambou and Ø. Ravna, eds., *The significance of Sámi rights: Law, justice and sustainability for the Indigenous Sámi in the nordic countries*. London: Routledge, 37–51.
Scott, A., 2008. *The evolution of resource property rights*. Oxford: Oxford University Press.
Scott, D. N., 2020. Extraction contracting: The struggle for control of Indigenous lands. *South Atlantic Quarterly*, 119 (2), 269–299.
Skattefjäll (*Taxed Mountain*) case (Swedish Supreme Court). (NJA 1981: 1).
Skolt Development Act (No. 253 of 1995).
Svartskog case (Norway Supreme Court), Rt. 2001 s. 1229.
The Constitution of Finland, 11 June 1999 (731/1999).
The Constitution of the Kingdom of Norway, (LOV-1814-05-17).
The Instrument of Government (Sweden), (1974: 152).
The Swedish Environmental Code, (Ds 2000: 61).
Thériault, S., 2016. Aboriginal peoples' consultations in the mining sector: A critical appraisal of recent reforms in Québec and Ontario. *In:* M. Papillon and A. Juneau, eds., *Canada: The state of the federation 2013: Aboriginal multilevel governance*. Montréal: McGill-Queen's University Press, 143–162.
Thériault, S., Bourgeois, S., and Boirin-Fargues, Z., 2022. Indigenous peoples' agency within and beyond rights in the mining context: The case of the Schefferville region. *The Extractive Industries and Society*, 12, 100979 [Online]. Available from: www.sciencedirect.com/science/article/pii/S2214790X21001507 [Accessed 27 June 2023].
Tollefson, H., 2021. Staking a claim: Mineral mining, prospecting logics, and settler infrastructures. *Canadian Journal of Communication*, 46 (2), 177.
Torp, E., 2024. The interplay of politics and jurisprudence in the *Girjas* case. *In:* D. Cambou and Ø. Ravna, eds., *The significance of Sámi rights: Law, justice and sustainability for the Indigenous Sámi in the nordic countries*. London: Routledge, 72–83.
Tsilhqot'in Nation v. British Columbia, [2014] 2 S.C.R. 257.
United Nations General Assembly, Human Rights Council, 2016. Special Rapporteur on the rights of Indigenous Peoples: Report on the human rights situation of the Sámi people in the Sápmi region of Norway, Sweden and Finland, 33rd Session, Agenda Item 3, HRC/33/42/Add. 3.
Yukon Act, S.C. 2002, C-7.

2 Closure and connection

A Southwest Pacific reappraisal of the mining enclave

Pierre-Yves Le Meur and Glenn Banks

Introduction

This chapter provides an analysis of the interactions and entanglements between large-scale mining projects and state-making/nation-building processes in four different political settings in the Southwest Pacific. Governance and institutions are increasingly seen as a key factor in countries translating resource revenues into broad-based human development, but typically the arguments take little account of the shifting configurations of power, influence, and effect between mining and the political processes tied up in state-making. We argue that in mineral-dependent contexts, not only are state-making and nation-building processes shaped by large-scale projects but also the nature of this influence varies depending on the specific set of complex, mutually shaping relationships that evolve between corporations, governmental bodies and the political sphere, affected communities, and other key actors, particularly NGOs, across multiple scales and levels (Filer and Le Meur 2017; Allen 2018).

We posit that large-scale mining typically generates multilayered enclaves that extend the influence of such operations well beyond the economic form in which this term is usually conceived.[1] The dialectic of closure and connection was always at work in the economic modeling of the enclave, but it was about local disconnection and global connection. What we propose here is to revisit and enrich this dialectic by importing neglected layers, the ideological layer for instance, in the qualitative model we have built.[2] Hence, an understanding of the ways in which mining and nation-building/state-making are mutually constituted requires taking into account the multi-scalar dimensions of the mining enclave across geographical, political, and administrative levels by which the mining enclave in its material, institutional, and ideological dimensions is connected (partially, or even at times, only tenuously) to the surrounding society or polity.

The four cases we provide here from the Southwest Pacific cover two large islands: New Guinea and New Caledonia. In the case of the former, the island is bisected by the border between Papua New Guinea (an independent nation) and the Indonesia region of Papua. The latter currently (2023) comprises six provinces across a population of 4.5 million, a region characterized by a low degree of autonomy—actually best described as a colonial situation with a strong settler

DOI: 10.4324/9781003406433-4

This chapter has been made available under a CC-BY-NC-ND 4.0 license.

colonization dimension—despite recent moves to devolve some revenue and decision-making from the center. The island of New Caledonia is a French territory engaged in a process of decolonization, but within this context, interprovincial issues of autonomy are also highly contested, with the North Province (and Loyalty Islands Province) ruled by pro-independence parties and the South Province by contra-independence parties. Two referendums for self-determination were held, in 2018 and 2020. The "no" vote won over the "yes" by a margin of 56.7%–44.3% (with a high turnout of 81%) in 2018 and by 53.3% over 46.7% (with an even higher turnout of 86%) in 2020. A third and final referendum took place on 12 December 2021. Pro-independence parties called for voters to abstain, and the "no" side won handily (96.5%), however, with only 43.9% turnout (meaning fewer voices for the "no" side than in 2020). This third and final round opened a transition phase of 18 months toward a new institutional status, a transition that was planned beforehand, regardless of the consultation outcome.

All four of the states/territories have been highly dependent on the large-scale mining economy for an extended period of time. Mineral and petroleum exports from Papua New Guinea, for example, have comprised well over 50% of exports since Independence in 1975 (81% in 2021) and between 10% and 35% of GDP (33% in 2019). Oil and gas are also a more recent factor in Papua New Guinea (PNG) and Indonesian West Papua (IWP) (and in the case of PNG, have produced a seismic shift in the political economy and popular ideology), but in both cases, the enclave effects have amplified rather than fundamentally reshaped the effects of the large-scale mining sector. There is currently no oil and gas production in New Caledonia; hence, the reference to oil and gas in the chapter is largely in relation to some of the analytical literature on governance and oil/gas from elsewhere (e.g., Watts 2004; Ferguson 2005; Mitchell 2011). We should also note that the chapter is not primarily focused on the mining economy per se. The "resource curse" debates (Auty 1993; Rosser 2006; Ross 2015; Gilberthorpe and Rajak 2016) are only relevant (and discussed further below) insofar as they shape state/nation-making processes (as the mining activity introduces a bias known as the "resource curse" or "Dutch disease" in the functioning of national economies). One cannot help notice the theological dimension of the notion and its weak explanatory and analytical value, despite its evocative power. As Rosser (2006, p. 12) rightly points out, "those arguing in favor of the notion of a resource curse have merely inferred causality from the evidence of correlation." Franks (2015, pp. 113–115) clearly shows that the resource curse notion is basically a powerful political tool which, through a chain of academics, NGOs, and international institutions, has been enrolled for promoting transparency in the mining sector, eventually leading to the Extractive Industries Transparency Initiative (EITI) (see also Gilberthorpe and Papyrakis 2015).

From here, the chapter proceeds with a review of the existing literature on mining and enclaves and examines existing arguments relating to the state/nation/resource nexus. This allows us to expand on and illustrate the multilayered and multi-scalar model mentioned above. The basic outline of the four case studies is then presented through an introduction to the historical trajectory (colonial/

postcolonial) and contemporary economic, political, and policy context of these entities. This highlights the very different settings within which mining enclaves persist in the region. Each of the cases is then examined according to the tenets of the three-layered framework/model of the enclave/state-making interaction. In the conclusion, we return to the dialectic of closure and connection within which all mining operations exist. Analyzing the interactions between mining and broader state/nation-making projects through this lens offers insights into the continuing debates around ideology, effective governance, and institutions in resource-rich nations.

The cases used in the chapter are doubly comparative, as we explore the issue in two islands (New Guinea/Caledonia) and within the two islands (IWP/PNG, with two different national/colonial contexts, and New Caledonia South and North provinces within a common territorial and colonial frame). The chapter reappraises the enclave debate from an empirically grounded perspective. Furthermore, the injection of a new, sophisticated notion of the mining enclave into the debates around mining and governance includes the examination of the relationship between mining and both state-making and nation-building, a discussion which has typically proceeded separately (however, see Dinnen 2007). It is noteworthy that our approach to state-making is not "state-centered," as it includes the ideological dimension linked to nation-building and resource nationalism, as well as the intrinsic relations with the concepts of community, indigeneity, and landownership (Li 2007; Filer 1997, 2006).

Our argument, in essence, is that much of the existing literature on mineral enclaves is inadequate, as it proceeds from evidence of one form of enclave—most often an economic form—to a conclusion that mining in its totality exists in an enclave. Instead, we argue that a more insightful approach must be taken to understand the workings of multilayered material, institutional, and ideological structures within which mining is encased, with the economy pervading all three layers: the dynamics and shifting priority afforded at points in time to one or other of the layers; the relative strength or porosity of each in their relations with state and society (again, a dynamic rather than static set of relationships); and the sets of forces (historical, internal, and external) that shape their contingent nature.

Literature review: Enclave theory and the state–society–resource nexus

Enclave theory: Boundary work

The notion of "enclave" conveys an obvious sense of boundedness and enclosure. Strictly speaking, it is a physical space (or metaphorically, a social group) bounded by visible and/or recognized limits within a broader entity. Interestingly, the term is also used by geologists to designate a mass of rock included in another mass, and distinct due to its origin or composition. In the field of economic development studies, the notion of "enclave" is theoretically situated.

> An economic enclave can be defined as a physically, administratively, or legally bounded territory whose geography or morphology is intimately related to the following economic characteristics: dependence on one or a few large firms; high specialization in one activity; and weak integration into the local economy, which is used primarily to access some local factors of production.
>
> (Phelps *et al.* 2015, p. 120)

The enclosure of the enclave (often a mining enclave, but it can also be industrial plantation or timber extraction) expresses the economic domination of Western capitalism and its capacity to grab and enclose areas for the sake of extracting profit. In other terms, the closure of the enclave is a matter of fact for the surrounding populations and national territory (but not for the national elites or "comprador bourgeoisie"), whereas, at a higher level, the capitalist enclave is directly connected to the global economy and external capitalist interests. This perspective on the enclave, a matter of "linkages and leakages" (Weisskoff and Wolff 1977), has been intrinsically linked to neo-Marxist and dependency theoretical frameworks from the 1960s onward.

The dependency perspective on the mining enclave (Cardoso and Faletto 1979; see also Long and Roberts 1984 for a nuanced and empirically grounded approach) was replaced in the 1990s by a "neo-dependency theory" nourished more by Foucault than by Marx, as exemplified by Ferguson (2005). This renewed view stresses the governmental effects and the territorial inequalities and discrepancies generated by the implementation and functioning of extractive enclaves.

Parallel to this stream of thought, liberal thinkers and neoclassical economists have questioned the effect, if not the reality, of the enclave as an instrument of uneven exploitation, on the basis of a modernization theory relying on the idea of the snowball or trickledown effect. Discussing whether large mining projects are (extractive) enclaves or (development) catalysts, Auty places the burden of economic dysfunction and adverse impact on "misguided postwar policies that led developing countries' governments to overextend their economic interventions" (Auty 2006, p. 135). Economic geographers have displayed a renewed interest in enclave theory, with a looser connection to dependency or modernization perspectives, stressing the functioning, linkages, and conflicts between the enclave and surrounding arenas, and mobilizing notions of networks and clusters (Arias *et al.* 2014; Phelps *et al.* 2015; Allen 2017; see also Sidaway 2007 at a broader theoretical level[3]). The horticultural metaphor of grafting is also borrowed to illustrate the nature of the relation between the enclave and its surrounding environment (Magrin 2013).

Beyond (or alongside) fierce theoretical debates around the role of economic enclaves within the world system, there has been little study of the materiality of the extractive enclave (also in terms of its social and cultural presence) until recently (i.e., in the last decade or so), and often in reference to oil extraction rather than to mining activities (Appel 2012; Barry 2013). Recent approaches in terms of "minescapes" (a term coined by photographer Edward Burtynsky) attempt to address a conceptual shortcoming of enclave theories as being too narrowly economic. They are based on a "growing awareness of the sociocultural significance of the mining terrain (. . .) and increasing recognition of the interplay of materiality and discourse within the extractive sector" (Ey and Sherval 2016, pp. 176–177). Minescape (by analogy with Appadurai's use of the "scape" suffix, 1996) strives to "problematize extractive landscapes" (Ey and Sherval 2016, p. 178) by emphasizing the cultural and spatial contexts of extractive terrains and postulating the nature of mineral resources as social and political constructs, here echoing Zimmermann (1933). "[T]he use of the term 'minescapes' characterizes extractive space as relationally constructed and assembled" (Ey and Sherval 2016, p. 178).

While drawing attention to the cultural and political embeddedness and the material-discursive nature of mining is welcome, despite its metaphorical power, the notion of minescape misses the point of the role and functioning of the enclave, by throwing out the baby (the dialectic of closure and connection) with the bathwater (a rigid and simplistic notion of enclave). The very notion of enclave and the processes of boundary-making (and of the boundary-transgressing they imply) deserve empirical and theoretical scrutiny as a locus of exchanges. The mining enclave builds on the idea of closure, but it cannot ever be complete, as the extracted ore has to be exported, and one must let workforce and technological resources in. These are the normal flows of human and nonhuman entities crossing the enclave boundaries. There are also uncontrolled "overflows." Historians of European industrialization have elaborated on the notion of industrial overflow from an environmental perspective (Le Roux and Letté 2013). As regards the mining enclave, the discharge of toxic substances and tailings has been a routine phenomenon that companies have long deemed as an "externality" (Kirsch 2014, for the case of Ok Tedi in PNG). Flows and overflows generate boundary work at two levels, as regards the mining perimeter itself and the mining-impact perimeter. Defining the former is a matter of land rightsholder identification and mining-lease attribution, whereas the latter implies political and environmental issues about impacts, responsibility, and local citizenship (Banks 2006). The two perimeters are also closely related to the dialectic between the social relations of production and compensation. The rise of the notion of "mine-affected communities," first in relation to resettlement policies and then due to the internalization of environmental damages in mining regulations and corporate calculus, has changed the balance between compensation and production: "In effect, mine-affected communities had now come to be defined as communities united by the social relations of compensation, not by the social relations of employment, even if jobs are still understood to be part and parcel of a compensation package" (Filer and Le Meur 2017, p. 17).

The state–nation–society–mine nexus: Connection work

The connection between the enclave economy and (failed) development has been increasingly encapsulated in the notion of a "resource curse" from the late 1990s on (Rosser 2006, p. 7). The term seeks to capture the multiple effects and, especially, the negative outcomes of mining on national and local political economies, in terms of economic performance, conflict (civil war), or regime type. At first glance, the four territories studied in this chapter are all highly resource dependent and exhibit elements of the "resource curse" condition. The "resource curse" thesis has gained currency among international development practitioners and researchers, as it provides a simple model against which simple development solutions can be promoted. It conflates heterogeneous processes in a single normative and explanatory framework, thus hindering the understanding of specific historical trajectories. As Michael Rosser rightly puts it, researchers and practitioners "should have been asking what political and social factors enable some resource-abundant countries to utilize their natural resources to promote development and prevent other resource-abundant countries from doing the same" (Rosser 2006, p. 8). We endorse this welcome call for empirical scrutiny and strive to historicize mining development and outcomes as resulting from the interplay between actors and institutions involved in the mining arenas at different levels. Political economy, conflict, state-making, and environmental issues are present as constitutive elements in the four situations, although in varying proportions, and our analysis aims to avoid the pitfalls of the catch-all "resource curse" thesis, let alone its almost religious terminology.[4] Nevertheless, whatever its theoretical shortcomings, the resource curse as an influential discursive practice "has now taken an intriguing new life in the form of policy recommendations, awareness-raising campaigns, and technical programs, implemented by global institutions, national governments, and non-governmental organizations" (Weszkalnys 2011, p. 366), nurturing a form of performative action or "virtualism" aimed at reshaping the world (Carrier and Miller 1998).

The point is about connection here. We have shown the contemporary transformations of the mining enclave due to the rise of environmental concern and global regulations. The post-frontier approach stresses the density of the institutional and regulatory web within which the extractive sector must now operate. "Whereas frontier modernism implies a radical transformation of localities to be discovered, extracted, and transformed by new technologies and external actors alone, post-frontier narratives build on and work through the recognition and incorporation of localities, rights, and environments" (Larsen 2015, p. 2). Hence, a little-analyzed paradox: "Resource frontiers are thriving and have intensified in the very same years that post-frontier regulatory regimes have been consolidating globally (. . .). Intensified nature commodification and the mushrooming of resource frontiers are, in this sense, mediated by, rather than existing in spite of, post-frontier institutions" (Larsen 2015, pp. 10, 15–16). Nuttall discusses the notion but rejects it, although the Greenlandic case he studies appears as a good candidate for the label "post-frontier" in terms of strong mining policy, Indigenous autonomy, and environmental regulations (Nuttall 2017, pp. 35–37). The notion could be tested in New Caledonia (seen as a whole as a single mining enclave) and also at a more local level, where post-nickel political dynamics are debated across complex arenas (Levacher and Le Meur 2022). On the contrary, the Freeport case in West Papua clearly resembles the wild (and colonial) frontier model ridden with violence, dispossession, and large-scale clientelism (Leith 2003; Kusumaryati 2021). The evolutionary connotation of the "post" of "post-frontier" is actually misleading, and we observe, rather, a synchronic tension between frontier and post-frontier processes resulting in different configurations according to the national and local contexts. Comparing the practices of the same companies in different national settings is enlightening in this respect. For instance, transnational companies like Glencore and Trafigura display disastrous records in terms of human rights and the environment in the Democratic Republic of the Congo, while behaving (more) correctly in New Caledonia, where they face a (more) binding regulatory framework.

Another perspective on the state–society–extraction nexus is proposed by Mitchell in his book *Carbon Democracy* (2011). He shows, through a historical perspective, the mutual constitution of the oil industry, the state, and society, particularly by contrasting it with the coal era that had produced another state–society–energy political and material topography. In the same vein but at the subnational level, Penelope Anthias (2019) shows, in the Bolivian Chaco, how Indigenous claims, organizations, and territorial practices are reshaped by the infrastructural, political, environmental, and political impact of the oil industry, generating a new pattern of "hydrocarbon citizenship" connected to different political and administrative levels and arenas. Extracting metals and oil are obviously two different activities, but there are commonalities in the rent-sharing issues and the infrastructural violence they both generate (Rodgers and O'Neill 2012; Appel *et al.* 2015; Anand *et al.* 2018).

These theoretical proposals enrich and go beyond the focus on the "resource–state nexus" (Bridge 2014), all the more by conceiving the state not as a substance or an organization but more according to the "qualities" of particular institutions that are "able to define and enforce collectively binding decisions on members of society" (Lund 2006, p. 685; see Fisher and Timmer 2013 for Oceania). This implies a repatriation of the notion of a "fourth estate" (beside state, corporation, and affected/neighboring communities), "comprising a wide variety of NGOs, financial intermediaries, lawyers, business partners, and consultants" (Ballard and Banks 2003, p. 304) in the analytical landscape, subsumed under the loose category of "society" and including actors influencing local mining arenas at a distance (Filer and Le Meur 2017, p. 20). This leads to another important factor to consider: Scale, not as an "ontological given category" (Marston 2000, p. 220) but as the unstable result of struggles around mineral resources control, ownership, and revenues. State, society, and the nation are mutually constitutive entities (Foster 1995, 2002; Fisher and Timmer 2013), and mining is also part of the process in the settings we study. An obvious feature of mining is its material localization. Mining development is structured by the tension between localized extraction, the material and immaterial flows allowing it to work, and the movements of the mining frontier driven by geological, economic, technological, and political factors. This dialectic of fixity and movement also lies at the core of capitalism. Drawing on Marx, David Harvey argues that there is "a fundamental contradiction between fixity and movement within the theory of accumulation in space and time" (Harvey 2006, p. 101). As regards mining,

> This materiality of the spatially fixed nature of mining resources, the local socio-ecological effects from extraction, i.e., air pollution, water contamination, social displacement, etc., and the immense fixed capital and dangerous labor conditions needed to get the stuff out of the ground, immediately politicize the scale at which extraction takes place. This stands in stark contradiction to the mobility of multi-scalar flows (e.g., capital, people, money, wastes, energy) that fixate on their development.
>
> (Huber and Emel 2009, p. 374)

Obviously, the mining enclave is intrinsically about scale; it is never fully enclosed. Material and immaterial flows and interactions link it to various human and nonhuman actors, processes, and institutions, among them state bodies, markets, environments, and the nation. The mining enclave reconceptualized as the unstable result of closures and connections takes a multilayered form that can be broadly characterized as extending across three domains: The material, the institutional, and the ideological.

The *material enclave* expresses the effects of the necessary localization of mining combined with its infrastructural and technological dimension. It also includes mining material impacts on the environment, its uncontrolled and controversial overflows defining a moving perimeter of impact, and responsibility beyond the mining lease area. Beyond these contested but localized perimeters, the materiality of mining (Bakker and Bridge 2006) directly interferes with the multi-scalar politics of mining control and ownership (Huber and Emel 2009, pp. 373–374).

The *institutional enclave* refers, at first glance, to the fact that a mining site has specific rules that must be abided by once one passes the entry gate. Conversely, the institutional dimension of the enclave reflects the specific sets of arrangements and connections allowing the enclave to function institutionally—namely, the "rules of the game" (North 1990)—from a normative, legal, and political point of view. It includes contrasted and sometimes conflictive forms: National mining and environmental laws, global standards and soft laws, political and administrative decentralization, local community-corporation and impact and benefit agreements, patron–client relations, corruption, and violence. Importantly, there is a mutually constitutive element to this (and other) enclave-constructing projects: The power and influence of the multinational mining companies typically exert leverage to shape and constrain the institutional and regulatory reach of the state over and into the enclave.

The *ideological enclave* is about the place of the mine in the local, national, and global imagination. It encompasses associated issues such as the selective presence/absence of the state (Szablowski 2007; Bainton and Skrzypek 2021); neoliberalism (Ganti 2014) and the rise of the corporate social responsibility (CSR) discourse (Dashwood 2012; Dolan and Rajak 2016); the enrollment of the mining sector in the politics of Indigenous recognition or resource nationalism (Wilson

2015), including conflict and controversy unfolding in various arenas (Long 2001); and spaces of engagement (Cox 1998) about the right scale of control, appropriation, and benefit-sharing (Huber and Emel 2009). The ideological dimension of the mining enclave also encompasses the construction of the "extractive subject" (Frederiksen and Himley 2020) and debates about local/national, corporate, and Indigenous forms of sovereignty and citizenship (echoing Anthias' notion of hydrocarbon citizenship; Anthias 2019; Le Meur and Levacher 2022 for the New Caledonia case; Filer 2006 for PNG). Among other discursive practices and developmental buzzwords, the resource curse thesis also contributes to the ideological construction of the mining enclave.

These three dimensions of mining enclaves can be variously connected with external bodies (state, society, corporation, and community) across different spatial levels from the local to the global (Ballard and Banks 2003; Huber and Emel 2009; Filer and Le Meur 2017). Land also plays a key role by structuring and connecting the three layers. It is the three-dimensional material basis for mining extraction (Elden 2013), associated infrastructure networks, and environmental overflows. Second, from an institutional standpoint, land is a sensitive issue ridden with conflicts, negotiations, and agreements between land rights-holders and companies (with the intervention of governmental bodies and other actors) that eventually define the local and supra-local rules of the game, including compensations and mobilities (Bainton and Banks 2018 for PNG; Le Meur *et al.* 2021 for New Caledonia). Third, land nurtures ideological visions, such as PNG being a nation of customary landowners (Filer 2006), would-be independent New Caledonia as grounded in a specific "link to the earth" inscribed in the 1999 *Organic Law* (Le Meur 2022), or the power of state through dispossession and the denial of Indigenous land rights, as in West Papua (Leith 2003; Kusumaryati 2021). The multilayered/multi-scalar mining enclave contributes to the production, reshaping, or activation of different forms of sovereignty (indigenous, customary, national, corporate, and global) and state-making processes. Our focus then is on the interaction and mutual reshaping between enclaves and state/nation/society-making/building processes. In other words, this analytical exercise does not aim to develop a model of the enclave per se but rather of the linkages between enclave and state/nation/society-making processes. We analyze our four states (or rather four governmental contexts) and two islands through this analytical lens.

The multilayered enclave model in the "four states, two islands" context

In this section, we test the multilayered approach to enclaves that we developed above on four cases: New Guinea, East and West, New Caledonia, North and South. The two territories of the New Caledonia case are presented in a single section, as we will see that this territory, over and above political, ethnic, and economic cleavages, may be analyzed as in a process of generating a single mining enclave (although reworked by scalar politics). The presentation is narrative in each case, as the comparison focuses on trajectories of mine–state–society relations, allowing unexpected convergences to emerge, while the historical perspective enables us to highlight the connections between layers and the mutual construction and reshaping of mine, state, and society.

Papua, Indonesia

At the time of its independence in 1949, the Republic of Indonesia inherited a body of colonial legislation from the Netherlands that vested the state and its institutions with strongly centralized powers. Even where new legislation was introduced, such as the *Basic Agrarian Law* of 1960, the ethos of absolute dominance by the state endured, and the priority of national interests was asserted over traditional forms of land ownership. Indonesia's annexation in 1962–1963 and the formal incorporation in 1969 of the western half of the island of New Guinea (now divided into six provinces) met with growing local resistance and correspondingly harsh repression by Indonesia's security forces, establishing a pattern that persists today (Kusumaryati 2021).

The first mining Contract of Work (CoW) under the new Suharto regime in 1967 was issued to PT Freeport Indonesia (PTFI), a subsidiary of the American Freeport-McMoRan Corporation, for its copper and gold mining operations in the central highlands of New Guinea, although the extensive CoW extended down to the south coast. This represented the first major foreign investment deal for the new regime and predated the formal and contested incorporation of West Papua into Indonesia. As a result, right from the outset, this project was closely tied to the "national project."

Drafted for the government by PTFI lawyers, the CoW emphasized the relationship between the company and the central government in Jakarta, to the virtual exclusion of all other stakeholders, including local communities and provincial authorities. Local Amungme and Kamoro people were not consulted in the negotiations for either CoW, although on occasion, PTFI responded to subsequent protests by brokering its own informal agreements with local communities. When those protests threatened mining operations, PTFI turned to the state and its security forces for protection, resulting in periods of high conflict and human rights abuses in 1977–1978, 1984, and 1994–1996 (Ballard 2001).

By the late 1980s, PTFI had grown to be one of the largest copper–gold mines in the world, a position it has cemented in subsequent decades. PTFI has been able to shore up its local relationships and legitimacy through the judicious distribution of funds to all parties, including the local communities, through "development," "business opportunity," or "recognition" agreements; local governments; and the national security forces. Much of the violence within the enclave has been promoted by the security forces, seeking to further entrench their business operations (including forestry and alluvial gold mining) and themselves locally, and to force PTFI to increase its formal and informal channels of financial support for their alliance (estimated in 2011 to be over US$3 million per month; see Kusumaryati 2021, fn90).

PTFI's Papuan operations are a spectacular example of an enclave economy, with the large CoW corridor of land running from the enormous mining operation in the highlands down through the lowland boomtown of Timika to the coastal port of Amamapare. This area has been a dominant but bounded economic zone (by its geographical isolation) within the island (Ballard and Banks 2009) and remains of regional importance today. Even in terms of labor, the enclave is more porous to Indonesians from outside the Papua region than to locals from the surrounding area, employing twice as many from other parts of the country. The housing for the workforce is also tightly controlled and bounded, with the development of a large town (Kuala Kencana) with restricted access in the lowlands some distance from the major population center (Timika). Access to the CoW area is still limited, with no road access to other centers, and tightly controlled air and sea access to the CoW facilities.

From an institutional perspective, PTFI was regulated under the CoW arrangements that define and ring-fence the operation from most institutional requirements and connections. Fiscal terms, labor and localization, import and export fees, and conditions and regulations are specified here, effectively constructing an institutional enclave for the operation in a way that other mining leases in Indonesia do not. Importantly, the original CoW was designed and co-constructed by PTFI and the Indonesian government of the day, and subsequent maintenance of this institutional regime has required significant political work by the corporation. High-level connections and political sway among Indonesian elites have been central to PTFI's history, and their ability to ride out the political storms of the subsequent democratic era is testimony to the ability of large corporations to maneuver in complex political environments (Leith 2003). This demonstrates how political and institutional legitimacy of the enclave is closely tied to forms of patronage politics. Events of the 1990s and 2000s—particularly, high-profile environmental and human rights campaigns against PTFI (Leith 2003; Abrash Walton 2010)—produced little discernible disruption to "the enclave." These drew new international actors (largely, environmental and human rights civil society organizations) and soft forms of regulation (through international codes of corporate behavior; see Freeman and Uriz 2017, e.g.) into the local arena; however, further concessions to local pressures were largely both tokenistic and divisive. The institutional enclave has been gradually eroded over the decades as political and economic elites have asserted claims to a share—or at least greater oversight—of the operation, although the enclave remained effectively bounded and enclosed.

By the 2010s, however, there were increasingly strident and concerted efforts by the government that sought to explicitly break down the mining enclave, aimed at Freeport and the other major multinational mining operations. These included a central government export ban on concentrates (which sought to break down the economic enclave further by forcing greater local processing/smelting), and an explicit change to foreign ownership rules that insist on an eventual national majority shareholding (Winanti and Diprose 2020). Responding to persistent political and popular resource nationalism within Indonesia (driven by ideological claims touched on below), these eventually led to a settlement that saw 51% of PTFI being controlled by Indonesian interests, a smelter (the second built by Freeport) being constructed by the company in East Java, and Freeport operating under a more generalized Special Mining license in place of the uniquely tailored CoW. While on the surface this represents a significant dismantling of the enclave, Winanti and Diprose (2020, p. 1544) argue that in practice, Freeport "used their market, capital, and discursive leverage to constrain the Indonesian government in its institutional functioning and fully implementing its new policy goals."

There is a fundamental tension to the way that the ideological position of Freeport plays out within Indonesia. On the one hand, the mine has been, and remains, a "vital national object." As noted, it was the first demonstrative agreement signed with a significant foreign investor by the new regime in the 1960s, and it remains a flagship enterprise, a major taxpayer, and now a prominent national asset. The Freeport operation and the extensive infrastructure associated with it have also been used by the Indonesian government as an entry point to assert other forms of state institutions, power, and control (primarily military, but also bureaucratic and developmental) over Indonesia's "unruly" eastern frontier. Hence, mining operations at Freeport are linked to both the historic Indonesian "recolonization" of Papua (from 1962 to 1969) and, throughout the 1990s, the state-driven migration (transmigration) from overpopulated Java to the eastern parts of Indonesia, and especially the Papua region (see McKenna 2016, pp. 150–154). The CoW area and surrounding lands were the site of at least nine transmigration settlement areas, allowing these settlers to draw on the infrastructure, facilities, and resources created by and around the company's operations. Most troubling, the heavy police and security force presence in the CoW has actively used the facilities and infrastructure as a base from which to participate in the suppression of local unrest across the surrounding areas well outside the CoW area.

In direct opposition to the nationalistic use of PTFI by the government as a focal point for their developmental and sovereignty claims within Papua, the local Papuan imaginary of Freeport is central to the political and ideological claims of Indigenous Papuans who have, since the 1960s, contested Indonesia's claims of sovereignty over their lands (Rutherford 2012). As a clear example of resource exploitation and environmental devastation of Indigenous lands, and as the center of a militarized zone in the southern part of the island whose people have suffered numerous widespread human rights abuses, Freeport has been a focus of significant international attention in support of Papuan calls for greater independence. Resistance to PTFI has become a central motif of Papua identity, particularly among those whose lives have been affected by its presence over the past six decades.

Managing the boundaries of the enclave across all levels is an ongoing challenge for Freeport. On the ground, PTFI exerts strong controls over access but has little control over some of its partners, such as the security forces and their local proxies: "Freeport also tends to define the scope of its responsibility for human rights issues rather narrowly as 'the area of company operations,' electing to disassociate itself from the operations beyond this area of the security forces quartered at Timika and to claim no role in the frequent violence between different ethnic communities in the Timika area" (Ballard and Banks 2009, p. 169), even though much of this conflict centers on the distribution of benefits from the company's activities and the other economic activities controlled by the security forces.

Papua New Guinea

Mining in PNG is also strongly linked to colonial and postcolonial projects in PNG. Gold exploration was the key driver behind early European interest in the country and saw the subsequent spread of mining activities in the late nineteenth century and the early part of the twentieth century (Nelson 1976). The growth of the Wau-Bulolo gold dredging enclave in the 1930s, complete with a modern colonial mining town, cemented the role of the industry in the colonial economy, a role that continued up until World War II (Healey 1967).

In the post-WWII period, further expansion of colonial influence was tied to ongoing mining exploration of the Highlands region, an area and population that had only been first "discovered" in the immediate prewar period (Gammage 1998). The Australian government's geological mapping surveys in the 1960s identified the potential of other sites for larger-scale mineral exploration, including at Panguna on the island of Bougainville (Denoon 2000). By the late 1960s, Panguna had become a critical component of the decolonization project, driven by international pressure and Australian concern to leave the nascent nation with an economic future.

However, as part of the nation-building project, Bougainville was also highly problematic, as the ethnically and linguistically distinct Bougainvillean population sought to assert claims to secession from the soon-to-be-established PNG state. This, in turn, led to a compromise of sorts that saw significant autonomy granted to newly established provinces, of which Bougainville was one of the most successful—the first concession to subnational demands for a stake in mineral developments.

Independence in 1975 also provoked nationalistic encounters with mining capital. Both the vigorous renegotiation of the original BCL agreement and the rejection of the agreement proposed for the development of the Ok Tedi Mine in the Western Province (Jackson 1984) provided a sense of PNG state power over the multinational miners. Economic policy at the time was premised on using mining as "an engine of economic development" resources (with the recognition that there would be little direct employment or enclave spill-over) which would be used to support broad-based forms of social and economic development across the nation (O'Faircheallaigh 1984). Most contentiously, and based on the inherited Queensland mining legislation, the state reserved for itself ownership of minerals, despite 97% of the country being under customary forms of land tenure. The rationale for the enclave, and the way in which it would contribute to the broader nation-building project, was thus well established from the start of the independent nation.

By the 1990s, however, the economic enclave model was under serious pressure from local communities impacted by the existing and newly developed mines. On Bougainville, locals had violently ejected BCL, sparking a vicious 10-year civil war between the nation state and a strongly pro-independence majority population of the province. Partly in response to the growing locally driven conflict on Bougainville, a new mechanism (the "Development Forum"; Filer 2012) was used from the development of the Porgera Mine in 1988–1989 onward. This provided a seat at the negotiating table for local interests in relation to new mines and conceded significant revenue flows (largely from the state share), creating new "enclaves" of legitimated landowners. This prompted the growth and elaboration of the "customary landowner" as a particular form of ideological identity within PNG, one that is often regarded with equal parts disdain and jealousy (Filer 2006; Allen 2018). However, even these concessions often did little to quell local resentments, due to weak forms of local governance, corruption, rising local inequalities, and elite capture of revenue and benefit streams (Banks 2019).

In recent decades, the legitimacy of different aspects of the enclave has been progressively undermined. In an economic sense, the growth and sophistication of local industry have meant tighter links between the mines and local and national businesses, but there has been a fundamental failure of the central element of mineral policy in that the state's ability to capture, then redistribute, a significant share of the value of the minerals has been significantly weakened. This is due to unstable commodity prices (and hence highly variable corporate profitability and tax payments, with the period 2014–2018 recording the lowest government tax take from the sector since the 1980s), poor economic governance, and a weak capacity to effectively deliver basic services to the 85% of the population that resides outside the main urban areas (UNDP 2014).

Each of the mines in PNG has produced highly localized, wealthy enclave economies. There are significant, tightly bound, and usually opaque revenue flows to the relatively small numbers of "customary landowners" from royalties and compensation, as well as preferences clauses that direct business contracts and employment toward this same (often highly contested) group. Specific government regulations and agreements around localization promote the latter two. Customary ownership of land creates the legal entitlement that underpins this relationship. Migration into the enclave fueled by these revenue flows undermines the stability of any social and economic structures and processes (Bainton and Banks 2018), and steep inequalities and stratification within this "community" rapidly develop (Banks 2022). The tensions and conflict that often ensue require corporate and (often reluctant) state efforts to manage this social dislocation, although usually to little effect. In places such as Porgera (Golub 2014; Banks 2006), the social, economic, and environmental boundaries of the enclave become increasingly porous and frayed, and there is often considerable local spill-over. One of the more successful forms of "breakout" from the enclave is the growth of mining landowner businesses to become significant national and even international firms, such as has occurred at Lihir, Ok Tedi, and Porgera.

Significant state dependence (post-independence) on "traditional" forms of revenue from mining—corporate taxes, royalties, dividend share from equity, payroll taxes from employees, etc.—has shifted increasingly to a model where revenue flows through more opaque, state-owned entities (EITI 2023). This is due to the success of the state in securing significant equity shares in several of the major mines, which is a result of the persistence, over decades, of a politically charged resource nationalism ideology.

In governance terms, the ability of the state to ensure the industry's adherence to environmental, labor, and social regulations has been undermined by a lack of capacity within the regulatory authorities and a lack of willingness by senior political leaders to push for compliance on most occasions. Below the national level, the problems are amplified by the near-complete absence of capacity, which has not stopped lower levels of government in mining areas from being recipients of large revenue flows from these operations. As a result, the mines operate in a way that is effectively largely decoupled from much state regulation, oversight, and governance. In this context, self-regulation—shaped in the past decade by corporate social responsibility expectations and a growing raft of global guidelines regarding human rights, environmental standards, and social behavior (see Banks 2006)—has played an important role in both decoupling the enclave from the state nexus and opening it up to much broader, global influences. The occasional, and typically short-term, interventions of "issue-based" international actors—often coming from the home countries of the multinational corporations themselves, especially Canada and Australia—have highlighted the absence of traditional, effective governance systems and mechanisms as much as they have the problematic, social and environmental behaviors and outcomes of the corporations (HRW 2011).

There has also been a strong and recurrent belligerent nationalism adopted toward the industry by the state (the effective nationalization of the Ok Tedi Project and the 2022 forced "nationalization" of a majority share in the Porgera Gold Mine [Burton and Banks 2020], for example). In tandem with the growing economic power of a small elite of public and private figures (a political enclave or cabal) based on, at best, questionable access to resource flows from the mining and petroleum sector, the moral and political legitimacy of the mining enclave in the national psyche/landscape/project has largely collapsed. The strong populist nationalism that argues that the country's resources have been pillaged by global forces for little local return—which is held in tension and distinct from the strongly nationalistic political rhetoric of which the bulk of the population is extremely suspicious, due to its apparently corrupt source—is in large part a reflection of this loss of legitimacy for the industry.

New Caledonia, North and South

Mining, especially nickel mining, is inseparable from New Caledonia's colonial history and, more specifically, from settler colonialism. Nickel ore was discovered in 1864 in New Caledonia by engineer Jules Garnier, and nickel mining started in 1873 alongside less successful attempts at extracting gold, copper, and coal (cobalt and chromium have also been exploited). New Caledonia rapidly became the world's largest nickel producer, before Canada took the lead in 1902. The territory experienced rapid boom and bust cycles from 1877, and in the 1880s and 1890s, partly due to the very speculative and financial nature of this pioneer mining phase in an era of liberal world capitalism (Bencivengo 2010; Black 2014). Today, New Caledonia is a true "nickel island," holding around 9% of the world's nickel reserves.

Garnier was also one of the three founding fathers of Société le Nickel (SLN). This company, controlled by the Rothschild Bank from 1888 to 1974, came to dominate the mining landscape in New Caledonia, especially from the 1930s onward after the absorption of its principal rival Ballande (including the main Caledonian processing plant of Doniambo, Noumea, opened in 1910). SLN built a specific form of multi-sited enclave, entangled with the colonial apparatus in different ways. On the one hand, it fulfilled government functions in terms of public infrastructure implementation and management against the backdrop of a financially weak (although socially brutal) colonial administration (Le Meur 2013). On the other hand, SLN political connections with the colonial political and administrative spheres allowed the firm to gain better access to cheap labor, to the great displeasure of its rivals, although it sometimes bypassed the colonial administration to directly access foreign labor, as with Japan in 1892 (Bencivengo 2010; Le Meur 2024). Due to the brutal and continuous demographic decline of Kanak populations, mining companies had to resort to importing their workforce, including convicts (New Caledonia opted for the penal settler colony from 1864 until 1898) and indentured workers from New Hebrides ("blackbirding"), French Indochina, Dutch Indonesia, and Japan. Kanak people subject to the disciplinary *indigénat* regime were also employed in mines, but to a lesser extent, mainly for loading ore carriers.

The aftermath of WWII saw profound mutations in the New Caledonian colonial society and mining economy. The labor issue was transformed by the abolition of *indigénat* in 1946, allowing the Kanak to participate in the political life and monetary economy, while the extractive industry experienced a process of rapid mechanization, decreasing the demand for workforce but increasing environmental damage and thus the material range of the mining enclave (or at least of its impacts). In the 1960s, the French government carried out a form of recolonization of its Pacific territories, to maintain its control over strategic mineral resources (especially Caledonian nickel), and proceeded to conduct nuclear tests in French Polynesia. It drastically reduced the post-*indigénat* political autonomy and promoted mass immigration from France and other Pacific French territories to New Caledonia. During the 1967–1972 nickel boom, 35,000–40,000 migrants settled in New Caledonia, then populated by fewer than 150,000 inhabitants, and the Kanak people became a minority group on their own land.

While SLN remained the hegemonic operator, an array of small-scale local mining companies ("*petits mineurs*") played an important role in cushioning the effects of nickel cycles (Freyss 1995; Bouard *et al.* 2016). The ones that survived the end of the "nickel boom" of 1967–1972 have become medium-size mining companies ruled by wealthy Caledonian families and political and economic entrepreneurs (Lafleur, Pentecost, Ballande, Montagnat, etc.).

Paradoxically, mining only recently emerged as an explicit political issue. The violent years euphemistically known as "the events" (1984–1988)—a combination of anti-colonial struggle and civil war—focused on cultural and political recognition, as well as land redistribution in the historical context of a settler colony and its legacy of spatial and racial segregation. Mining was mainly mentioned as an element of Kanak heritage to be retrieved and as an expression of colonial plunder.

For instance, in Thio, a historical SLN stronghold over the century (and headquarter during the interwar period), the destruction of the company offices appears, oddly enough, as a sort of side effect of the conflict. All this is not to say that Kanak political leaders did not reflect on a mining strategy (Mokaddem 2011).

Mining came to the foreground in the 1990s as an economic means for political emancipation and ideologically "de-enclaving" the mining issue—or "re-enclaving" it at a higher, national level. The pro-independence political parties pushed the mining issue to the fore in the 1990s as the 1988 *Matignon Agreement* and self-determination (initially planned for 1998) were renegotiated (Bouard *et al.* 2016, pp. 33–39).

By 1990, Société minière du Sud Pacifique (SMSP), owned by loyalist entrepreneur and politician Jacques Lafleur, was "nationalized," meaning it was transferred with the financial support of the French government to SOFINOR, the financial holding of the North Province ruled by pro-independence parties. The "mining prerequisite," that is, making the mining issue part and parcel of the decolonization process launched by the *Noumea Accord* of 1998, resulted in the *Bercy Agreement* signed on 1 February 1998. An exchange of nickel deposits was organized between SLN and SMSP, and one cannot help noticing the significant involvement of the French government in the negotiations, exerting pressure on ERAMET (of which SLN has been a subsidiary since 1985), perhaps behind the backs of anti-independence leaders. The "mining prerequisite" was actually discussed between the pro-independence Kanak front (FLNKS) and the French government. These trends, including the deposit exchange, may be interpreted as paving the way toward the consolidation of New Caledonia as a single mining territory.

The nickel strategy was partly formalized in legal instruments, namely, the mining plan (*Schéma de mise en valeur des ressources minières*) set out in the 1999 *Organic Law*, art. 39, and issued in 2009, and the 2009 *Mining Code*.[5] In actual fact, the main operational outcome of this specific form of resource nationalism (Emel *et al.* 2011; Wilson 2015; for the New Caledonia case, Bouard *et al.* 2016; Demmer 2018; Burton and Levacher 2021) was the Koniambo Project, a joint venture between SMSP and Falconbridge (absorbed by the Anglo-Swiss firm Xstrata in 2006 and Glencore in 2013) for extracting and processing the Koniambo nickel deposit in the North Province (Grochain 2013).

The integration of Koniambo into the pro-independence political design shifted the mining enclave boundaries both institutionally (through public equity and local participation) and ideologically (by making it part of the nationalist imaginary). The shift also concerned the land issue that was discussed in the negotiations of the two world-class projects of Koniambo and Goro, as well as in smaller-scale local conflicts. The issue was the recognition of Kanak anteriority and legitimacy on the territories targeted, as a basis for negotiating impact and benefit agreements (Grochain 2013; Le Meur *et al.* 2013; Levacher 2016) and not a claim of land distribution under customary status as under the land reform launched in 1978 (Herrenschmidt and Le Meur 2016; Le Meur 2022).

This original resource nationalism (Burton and Levacher 2021) exemplified by the Koniambo Project defines a specific enclave morphology and functioning. The

51/49 pattern has been extended in an innovative way through two joint ventures, with POSCO in South Korea and, more recently (March 2018), with Yangzhou Yichuan Nickel Industry in China.[6] In both cases, nickel from New Caledonia is processed in plants held 51% by a Caledonian SMSP branch. This resource nationalism, locally known as the "nickel doctrine," also fuels the political debate across the pro/contra-independence divide, which has developed in the specific context of negotiated decolonization and shared sovereignty. Since 2007, a public holding company (STCPI, *Société territoriale de participation industrielle*) has held 34% of SLN's equity share and aims to acquire 51% in the near future, a goal shared by one of the main (and long-dominant) anti-independence parties, Calédonie Ensemble.

Parallel to this trend, another project has emerged in the South Province, also based on the integration of nickel extraction and processing (a hydro-metallurgical process allowing value to be added to cobalt, too). The Goro Nickel project, held by INCO and, from 2006, by the Brazilian giant Vale, has taken the form of a more classic "neoliberal" enclave and stands outside the realm of the *Bercy Agreement*, with Vale holding 95% equity and 5% in public shareholding. The project faced strong local opposition led by environmental and Indigenous movements associated with customary authorities, resulting in an IBA signed in 2008 (Le Meur *et al.* 2013; Levacher 2016; Levacher and Le Meur 2022). In 2020, Vale's decision to sell the refinery[7] sparked a phase of tension. This decision led to a failed negotiation with Australian New Century Resources, and then months of controversy and acute conflict surrounding the make-up of the shareholding. Within a few months, the sale of Vale-NC triggered strong reaction among the local population, pro-independence leaders, Indigenous organizations, and customary authorities of the South New Caledonia and beyond. A collective was swiftly created, called *Usine du Sud = Usine Pays* (The South's Plant is the Country's Plant), along with an organization called Instance coutumière et autochtone de négociation (ICAN; the Customary and Indigenous Negotiating Authority) backed by the pro-independence parties. The industrial site has been shut down since 10 December 2020, and tensions have risen, with demonstrations accompanied by violence and acts of destruction of the industrial infrastructure. The conflict resulted in a government crisis and the resignation of the pro-independence members of the New Caledonian government on 2 February 2021. Following this, a new government was elected by the Congress and, for the first time since the government's creation in 1999, the pro-independence parties are in the majority. In April 2021, pro- and anti-independence parties agreed to sell the plant to a consortium, Prony Resources, composed of Trafigura (19%), one of the biggest transnational commodity trading companies; Compagnie financière de Prony (30%), representing the former company management and a New Zealand investment fund, Agio; and NC shareholders (51%), including the three provinces (30%), company employees (12%), and local customary and individual shareholders (9%). This new avatar of the 51%/49% formula is a departure from the model that prevails in Koniambo, as it introduces a certain amount of popular shareholding. Furthermore, it is accompanied by a controversial innovation regarding the mining lease and land tenure arrangement. Prony Resources, which holds a formal mining permit (APM/*autorisation personnelle minière*), ceded the mining lease to Sud Nickel, a public consortium constituted by the South Province and Loyalty Islands Development Companies (respectively, Promosud, 80%, and Sudil, 10%) and a Risk Reduction Fund (FPRESC/*Fonds de prévention des risques*, 10%). Prony Resources will pay rent to Sud Nickel, which will use it to create a fund for future generations and to support economic diversification. While the amount of the land rent has not been made public, it will be strongly influenced by LME nickel prices, and one can assume that Trafigura will play a key role in this matter. In addition, a trade agreement was signed with Tesla to deliver NHC produced by the hydro-metallurgic processing plant to its different battery production units. The institutional connections allowing the mining enclave to function are thus multi-scalar. They involve new actors in the NC mining arenas (including Tesla and TotalEnergies for the supply of solar energy) and reshape the notion of "mining company," as Prony is disconnected from its former APM through the cession and farm-out of the mining lease.

The current mining arena in New Caledonia is therefore organized around three main poles incorporating nickel extraction and processing—a nationalist enclave in the North (KNS); a still ill-defined, hybrid, and originally neoliberal enclave in the South (Goro); and a multi-site neo-paternalist enclave (SLN). Despite their differences, they are all embedded in a common policy and legal framework defined by the creation of the Mining and Energy Department (DIMENC) in 2004, and by the 2009 Mining Plan and Mining Code, which provide stakeholders with a framework for organizing the extraction, processing, and export of nickel.

The analysis of recent conflicts and their outcomes in terms of negotiations between large-scale, medium/small-scale companies, and governmental bodies (Demmer 2017) confirms a relatively shared desire to control exports and value high-, medium-, and low-grade deposits, although the 49%/51% capital ownership pattern favored by the nickel doctrine is far from hegemonic. Given the strong density of human and material networks that constitute the mining sector in New Caledonia, one cannot help viewing it in terms of a sort of unique mining enclave at the New Caledonia level. However, this configuration is still strongly influenced by French government policy, as witnessed in the negotiation of the *Bercy Agreement* in 1998 and, since then, through the subsidies and loans granted to the three main entities by the French government and through the French Development Bank (AFD/*Agence française de développement*), the most recent being a €40 million loan granted to SLN in February 2023 (*Les Nouvelles Calédoniennes*, 14 February 2023). For the New Caledonia mining enclave to work, financial and institutional connections are still needed involving both the French state and transnational actors. Conversely, local connections play a key role as far as extractive activity is concerned. The mining leases are all located on public land (the New Caledonia public estate), and there are no land rents paid to local landholders. Nevertheless, clans and customary authorities claim their historical status as first-comers as a basis for negotiating compensations and IBA-like agreements with mining companies, whereas Indigenous organizations are pushing for a municipal mining tax.[8] Mining enclaves are multi-scalar in New Caledonia, but in a non-fractal way. The nature of New Caledonia's territory as an enclave is different from the various local enclaves generated by single extractive sites, which echoes Tsing's argument on non-scalability (Tsing 2012). Lastly, the New Caledonia context can be characterized as "post-frontier" (Larsen 2015), given the dense web of legal regulations, political connections, staff mobility, exchange of ore deposits, and circulation of ideas within which mining activity is entangled. This ongoing trend tends to nuance the difference between the North and South provinces' mining contexts and policies in place in the 2000s and early 2010s.

Conclusion: Conceptualizing the state–society–mining nexus through the lens of the three-layered enclave model

What we propose in this chapter is to consider, in all its complexity, the dialectic of closure and connection underlying any enclave situation. Rather than a full-fledged, static model, the enclave is an evolving and empirically oriented analytical instrument that can be used to examine, from a historical standpoint, the dialectic of closure and connection that influences the functioning of a mine at different levels: Materially, institutionally, and ideologically. It considers the embeddedness of the political economy of mining in its multi-scalar settings and paves the way for different comparative analyses.[9]

The dual comparison of Papua (West and East) and New Caledonia (North and South) led us far beyond what we anticipated in terms of convergences. The combined trajectories of mining, colonization, state-making, and migration in these territories display similar patterns, especially during the critical junctures of the late 1960s and 1970s, strikingly for West Papua and New Caledonia. While PNG followed a different path, obtaining independence from Australia in 1975, the specific case of Bougainville and the Panguna Mine converges with the two other situations as regards the combination of large-scale open-pit mines, associated migration, and a conflictual relation with the central state (albeit not a colonial metropolis in this case).

The combination of recolonization and mine-related conflict triggered violence in the three territories in the late 1970s and during the 1980s. The late 1990s and 2000s saw another movement of convergence fueled by the CSR turn in the mining sector and the global mining boom that resulted, at the national level, in the drafting of new regulations and policy frameworks (2006 for PNG, 2004–2009 for New Caledonia, and 2009 for West Papua). These, in turn, can be regarded as various degrees and forms of resource nationalism (Wilson 2015; see Demmer 2018 for New Caledonia, Winanti and Diprose 2020 for Indonesia). In other words, the densification of regulatory ties with the national and global levels reshaped the institutional enclave and, indirectly, its material layer through environmental impact assessments, technologies, and rules that were designed to better contain material overflows. Moreover, the (somewhat ill-defined) resource nationalist discourse framing these policies has been contributing in different ways to the ideological anchoring of the mining enclave in the national imagination.[10]

The land issue is probably the one that best illustrates the entanglements and mutual reshaping of the different layers that make up the mining enclave. Land is, at once, material and immaterial (Strathern 2009), mythical and legal (Abramson 2000), and is best conceived as an assemblage (Li 2014) that includes the subsoil coveted by the mining companies. The creation of the material enclave disrupts everything, including ecosystems, property rights, residential patterns, and the link to the earth. The negotiation of compensation generates conflict and the proliferation of institutional arrangements while reconfiguring attachments to land, identity, and the values of place (Filer 2006; Bainton and Banks 2018; Bainton *et al.* 2021 for PNG; Le Meur *et al.* 2021; Levacher and Le Meur 2022 for New Caledonia).

The different case studies presented highlight the importance of historical trajectories and their ongoing differences and deep similarities. We conclude here by turning our attention to understanding these different forms of state–society–resource relationships mediated through the lens of the enclave introduced above. The key differences between the cases—the "landownership" regime, the levels of institutional connection, and the place of the mine in the national imagination—are clearly depicted in the model we propose, specifically the emphasis on the multi-scalar connections across geographical, political, and administrative levels through which the mining enclave is connected to the surrounding society or polity.

Each of the domains is, itself, multi-scalar and multilayered, and there are synergistic feedback loops between them, as evidenced above. Even an ill-defined ideological resource nationalism can drive economic, ideological, and often institutional changes that, in turn, will have implications for the material position of the enclave. Tensions within the material enclave, such as the impact of the necessary localization of mining, include the use of local labor versus the importing of workers, and the development or otherwise of mining townships, as well as localized environmental damage in the absence of investment in a more effective and contained environmental management system.

While the influence of large-scale mining projects on state/nation-making emerged clearly in the comparative analysis, it rarely functions in a straightforward way, and the functioning of the ideological mining sector enclave is highly dependent on colonial and postcolonial histories (including the role of mining in these histories) and the contemporary state-making projects that have flowed from them. As one aspect of this, the mining enclave as "distorting" in state-building processes implies that, in the institutional sense, enclaves co-constitute governance regimes that function poorly to achieve positive outcomes for state or local actors, the political equivalent of the macro-economic "resource curse" thesis. This is apparent in all the cases, with specific and targeted institutional arrangements—typically co-created with the mining operators themselves—that sit outside national policy frameworks, and often undermine the effectiveness of the institutional frameworks for the sector and the achievement of broader national goals connected to the mineral resource. Locally, similar trends are apparent: The arrangements and local connections allowing the enclave to function institutionally and from a normative/legal/political point of view vary from one operation to the next in PNG and New Caledonia, reflecting particular political and economic moments and interests.

The place of the mine in the national project, or national imagination, is strongly apparent across the different cases, although often in conflictual ways, depending on the politics and the scale. Where some states see large-scale mines as "national projects" that help bring control and development to marginal or unruly subjects, there is also tension generated by the ideological marking of mining as a face-to-face relational encounter between corporations and communities that deliberately by-passes the role of the state ("the company is our government now") and/or promotes a selective, strategic "absence" of the state (see Bainton and Skrzypek 2021), or that promotes the construction of the "extractive subject" over a more national identity, and that has facilitated, in Papua, and to a lesser extent, in New Caledonia, the development of networks of political patrimony that allow the enclosed industry to function more smoothly.

Large-scale mining in the region (which can be broadly characterized as "Melanesia") has produced numerous conflictual and contested events, at both the national and local levels. By drawing from the different domains across which mining enclaves have developed, the multidimensional model proposed above, and its comparative application across four of the major mining territories in the region, helps provide insights into why the sector has been bedeviled by these conflicts. In these cases, the combination of specific forms of enclaves has produced certain seemingly intractable solutions—a locally ineffective or even absent state, environmental overflows, irrepressible landowners, and tightly channeled and controlled benefit streams, which in this configuration make the untangling of causes and the charting of possible ways forward incredibly difficult or, at best, opaque. It is this complexity into which we have sought to provide some insight by developing a multilayered model of the enclave, with a clearer distinction between the different domains that at once develop, connect, and enclose mining and mines in an ever-changing dialectic. This approach has implications for the development of solutions, be they operational or policy-oriented, in that it points to the need to surface the nature and history of the domains that contextualize each of the mining encounters and to approach issues connected to the mines holistically, understanding the affective (as ideological positioning), material, and institutional settings and levers that can realistically be drawn on. A more myopic stance—one that sees enclaves and claims from a merely economic perspective, for example—is far less likely to provide a satisfactory understanding of the social, political, and economic dynamics of the state–society nexus in play at the mines of Melanesia, and hence equally unlikely to produce a meaningful and lasting path forward for the parties involved.

Acknowledgments

Early versions of this chapter benefited from rich discussions with our colleagues and friends Chris Ballard (ANU, Canberra) and Jean-Michel Sourisseau (CIRAD, Montpellier), and the following conference paper marked the starting point of our long-running conversation: Le Meur *et al.* (2013), Two islands, four states: Comparing resource governance regimes in the Southwest Pacific, *in*:

J. Wiertz, ed., Proceedings of the 2nd International Conference on Social Responsibility in Mining (SRMining 2013, Santiago, Chile), Santiago, GECAMIN, p. 191–199. We also wish to warmly thank the participants at the Mining Encounters Research Seminar held at the University of Queensland, Brisbane, in August 2023, for their insightful comments and, especially, Ciaran O'Faircheallaigh and Thierry Rodon, who pushed us to clarify our arguments, as well as the anonymous readers of this book.

Notes

1 The notion of an enclave economy was the subject of numerous studies and discussions from the 1950s to the 1980s, particularly in the context of controversies between proponents of dependency and modernization theories. They mainly used the examples of the plantation economy, industrial and mining enclaves, or oil extraction, analyzed in terms of linkages/leakages and rentier logics (e.g., Hirschman 1958; Weisskoff and Wolff 1977; Beblani and Luciani 1987).

2 We thank the reviewers for having pushed us to better explain why the enclave notion deserves to be rescued and thus to make our argument about the enclave more precise. Economic enclaves are not "economic islands" as suggested by the reviewers. In the economic and development literature, the enclave was never conceived as a fully isolated and closed entity. From a South Pacific perspective, if enclaves are islands, they partake in "a sea of connections" (Fache *et al.* 2022).

3 The concept of enclave helps highlight similarities with other production or activity sites functioning in a relative or strong disconnection with their immediate surroundings (e.g., Redfield 2000, on the French Guiana Spatial Center in Kourou, or Kroeker-Mauss 2014, on conservationist enclaves).

4 Stuart Kirsch tried—not entirely convincingly—to rescue the notion by applying it to the local level in the guise of a "microeconomics of resources," showing, for the Ok Tedi case in PNG, how it "poses a threat to both the subsistence practices and cultural survival of these groups" (Kirsch 2014, p. 228).

5 Only partly, because the mining plan stresses the need to process the richest nickel ores in New Caledonia without favoring the 51%/49% pattern of capital ownership or specific channels for exporting nickel (Laurent and Merlin 2022, pp. 338–339).

6 The latter agreement has not been successful so far. Laurent and Merlin (2022, p. 335) also mention a failed attempt to replicate the model in Queensland, Australia, with the Yabulu processing plant, the main buyer of New Caledonia lateritic nickel.

7 For reasons of weak profitability, and as one of the consequences of the financial and reputational losses following the two tailings dam failures in Brazil (Bento Rodrigues in 2015 and Brumadinho in 2019), which killed 19 and 270 people, respectively.

8 The level of taxation on the mining sector remains weak, and the large-scale processing projects benefit from a tax exemption. There are nevertheless current debates about the creation of a municipal mining tax, and two new mining taxes were adopted by New Caledonia's Congress on 16 October 2023, one on mining extraction and another on mining exports. Municipalities were supposed to receive 60% of the first, but the ratio will be much lower after political negotiations between pro- and contra-independence parties in Congress (www.lnc.nc/article/nouvelle-caledonie/politique/mines/deux-nouvelles-taxes-sur-l-activite-miniere-adoptees-par-le-congres).

9 For instance, despite strong historical differences, Bougainville and New Caledonia display convergences in the mine–land–migration nexus, and similarities in terms of geographic and demographic size; see Regan (2017, 2019) and Maclellan and Regan (2018) for a comparison of the self-determination processes in Bougainville and New Caledonia.

10 We could extend the comparison to neighboring countries such as Timor Leste that exhibit convergent patterns in terms of colonization/decolonization and extraction (Bovensiepen and Nygaard-Christensen 2018), or even beyond (see Nuttall 2017 for Greenland).

References

Abramson, A., 2000. Mythical lands, legal boundaries: Wondering about landscape and other tracts. *In:* A. Abramson and D. Theodossopoulos, eds., *Land, law and environment. Mythical land, legal boundaries*. London: Pluto Press, 1–30.

Abrash Walton, A., 2010. Conservation through different lenses: Reflection, responsibility, and the politics of participation in conservation advocacy. *Environmental Management*, 45, 19–25.

Allen, M. G., 2017. Islands, extraction and violence: Mining and the politics of scale in Island Melanesia. *Political Geography*, 57, 81–90.

Allen, M. G., 2018. *Resource extraction and contentious states: Mining and the politics of scale in the Pacific Islands*. London: Palgrave Macmillan.

Anand, N., Gupta, A., and Appel, H., eds., 2018. *The promise of infrastructure*. Durham, NC: Duke University Press.

Anthias, P., 2019. *Limits to decolonization: Indigeneity, territory, and hydrocarbon politics in the Bolivian Chaco*. Ithaca and London: Cornell University Press.

Appel, H., 2012. Offshore work: Oil, modularity, and the how of capitalism in Equatorial Guinea. *American Ethnologist*, 39, 692–709.

Appel, H., Mason, A., and Watts, M., eds., 2015. *Subterranean estates: Lifeworlds of oil and gas*. Ithaca, NY: Cornell University Press.

Arias, M., Atienza, M., and Cademartori, J., 2014. Large mining enterprises and regional development in Chile: Between the enclave and cluster. *Journal of Economic Geography*, 14, 73–95.

Auty, R. M., 1993. *Sustaining development in the mineral economies: The resource curse thesis*. London: Routledge.

Auty, R. M., 2006. Mining enclave to economic catalyst: Large mineral projects in developing countries. *Brown Journal of World Affairs*, XIII (1), 135–145.

Bainton, N. and Banks, G., 2018. Land and access: A framework for analysing mining, migration and development in Melanesia. *Sustainable Development*, 26 (5), 450–460.

Bainton, N., Burton, J., and Owen, J. R., 2021. Land relations, resource extraction and displacement effects in island Papua New Guinea. *The Journal of Peasant Studies*. Available from: https://doi.org/10.1080/03066150.2021.1928086.

Bainton, N. and Skrzypek, E., eds., 2021. *The absent presence of the state in large-scale resource extraction projects*. Canberra: ANU Press.

Bakker, K. and Bridge, G., 2006. Material worlds? Resource geographies and the "matter of nature." *Progress in Human Geography*, 30 (1), 5–27.

Ballard, C., 2001. The signature of terror: Violence, memory and landscape at Freeport. *In:* B. David and M. Wilson, eds., *Inscribed landscapes: Marking and making place*. Honolulu: University of Hawaii Press, 13–26.

Ballard, C. and Banks, G., 2003. Resource wars: The anthropology of mining. *Annual Review of Anthropology*, 32, 287–313.

Ballard, C. and Banks, G., 2009. Between a rock and a hard place: Corporate strategy at the Freeport mine in Papua, 2001–2006. *In:* B. Resosudarmo and F. Jotzo, eds., *Working with nature against poverty: Development, resources and the environment in Eastern Indonesia*. Singapore: ISEAS, 147–177.

Banks, G., 2006. Mining, social change and corporate social responsibility: Drawing lines in the Papua New Guinea mud. *In:* S. Firth, ed., *Globalisation, governance and the Pacific Islands*. Canberra: ANU E Press, 259–274.

Banks, G., 2019. Extractive industries. *In:* P. Hirsch and W. Rollason, eds., *The Melanesian world.* London: Routledge, 501–516.

Banks, G., 2022. Afterword: Reflecting on resource driven inequalities. *In:* B. Beer and T. Schwörer, eds., *Large-scale capital in small-scale life-worlds: The establishment of novel inequalities in Papua New Guinea.* Canberra: ANU Press, 183–199.

Barry, A., 2013. *Material politics: Disputes along the pipeline.* Chichester, UK: Wiley-Blackwell.

Beblani, H. and Luciani, G., eds., 1987, *The Rentier state.* London: Croom Helm.

Bencivengo, Y., 2010. *La Société Le Nickel. Une entreprise au cœur de la naissance de l'industrie du nickel (1880–1914).* PhD in History. Paris: Université Paris 1 Panthéon-Sorbonne (3 Volumes).

Black, P., 2014. *"Green gold" The contribution of New Caledonia's nickel industry to the age of steel. 1870–1920.* PhD Thesis. Canberra: The Australian National University.

Bouard, S., Sourisseau, J-M., Géronimi, V., Blaise, S., and Ro'i, L., eds., 2016. *La Nouvelle-Calédonie face à son destin: quel bilan à la veille de la consultation sur la pleine souveraineté?* Paris-Pouembout: IAC-Khartala-GEMDEV.

Bovensiepen, J. and Nygaard-Christensen, M., 2018. Petroleum planning as state building in Timor-Leste. *The Asia Pacific Journal of Anthropology*, 19 (5), 412–431.

Bridge, G., 2014. Resource geographies II: The resource–state nexus. *Progress in Human Geography*, 38 (1), 118–130.

Burton, J. and Banks, G., 2020. The Porgera mine in PNG: Some background. *Development Policy Blog.* Available from: https://devpolicy.org/the-porgera-mine-in-png-some-background-20200507–2/.

Burton, J. and Levacher, C., 2021. The state that cannot absent itself: New Caledonia as opposed to Papua New Guinea and Australia. *In:* N. Bainton and E. Skrzypek, eds., *The absent presence of the state in large-scale resource extraction projects.* Canberra: ANU Press, 313–346.

Cardoso, F. H. and Faletto, E., 1979. *Dependency and development in Latin America.* Berkeley, CA: University of California Press.

Carrier, J. G. and Miller, D., eds., 1998. *Virtualism: A new political economy.* Oxford and New York: Berg.

Cox, K. R., 1998. Spaces of dependence, spaces of engagement and the politics of scale, or: Looking for local politics. *Political Geography*, 17, 1–23.

Dashwood, H., 2012. *The rise of global corporate social responsibility: Mining and the spread of global norms.* Cambridge: Cambridge University Press.

Demmer, C., 2017. L'export du nickel au cœur du débat politique néo-calédonien. *Mouvements*, 91, 130–140. Available from: https://doi.org/10.3917/mouv.091.0130.

Demmer, C., 2018. Nationalisme minier, secteur nickel et décolonisation en Nouvelle-Calédonie. *Cahiers Jaurès*, 230 (4), 35–52.

Denoon, D., 2000. *Getting under the skin: The Bougainville copper agreement and the creation of the Panguna mine.* Melbourne: Melbourne University Press.

Dinnen, S., 2007. *The twin processes of nation building and state building.* Canberra: Australian National University, SSGM Briefing Note 1/2007.

Dolan, C. and Rajak, D., eds., 2016. *The anthropology of corporate social responsibility.* Oxford: Berghahn.

Elden, S., 2013. Secure the volume: Vertical geopolitics and the depth of power. *Political Geography*, 34, 35–51. Available from: http://dx.doi.org/10.1016/j.polgeo.2012.12.009.

Emel, J., Huber, M. T., and Makene, M. H., 2011. Extracting sovereignty: Capital, territory, and gold mining in Tanzania. *Political Geography*, 30 (2), 70–79.

Extractive Industries Transparency Initiative (EITI), 2023. 2021 Papua New Guinea EITI Report. Available from: https://eiti.org/documents/papua-new-guinea-2021-eiti-report.

Ey, M. and Sherval, M., 2016. Exploring the minescape: Engaging with the complexity of the extractive sector: Exploring the minescape. *Area*, 48 (2), 176–182.

Fache, E., Kon Kam King, J., Riera, L. and Breckwoldt, A., 2022. A sea of connections: Reflections on connectivity from/in Oceania. *Ambio*, 51, 2333–2341.

Ferguson, J., 2005. Seeing like an oil company: Space, security, and global capital in neoliberal Africa. *American Anthropologist*, 107 (3), 377–382.

Filer, C., 1997. Compensation, rent and power in Papua New Guinea. *In:* S. Toft, ed., *Compensation for resource development in Papua New Guinea*. Boroko and Canberra: Law Reform Commission (Monograph 6) and Australian National University, National Centre for Development Studies. (Pac. Policy Pap. 24), 156–189.

Filer, C., 2006. Custom, law and ideology in Papua New Guinea. *The Asia Pacific Journal of Anthropology*, 7 (1), 65–84.

Filer, C., 2012. The development forum in Papua New Guinea: Evaluating outcomes for local communities. *In:* M. Langton and J. Longbottom, eds., *Community futures, legal architecture: Foundations for Indigenous People in the global mining boom*. London: Routledge, 147–161.

Filer, C. and Le Meur, P.-Y., eds., 2017. *Large-scale mines and local-level politics: Between New Caledonia and Papua New Guinea*. Canberra: ANU Press.

Fisher, D. and Timmer, J., 2013. Preface: Becoming like the state. *Oceania*, 83 (3), 153–157.

Foster, R. J., ed., 1995. *Nation making: Emergent identities in postcolonial Melanesia*. Ann Arbor: The University of Michigan Press.

Foster, R. J., 2002. *Materializing the nation: Commodities, consumption, and media in Papua New Guinea*. Bloomington: Indiana University Press.

Franks, D. M., 2015. *Mountain movers: Mining sustainability and the agents of change*. Abingdon: Routledge.

Frederiksen, T. and Himley, M., 2020. Tactics of dispossession: Access, power, and subjectivity at the extractive frontier. *Transactions of the Institute of British Geographers*, 45, 50–64.

Freeman, B. and Uriz, G. H., 2017. Managing risk and building trust: The challenge of implementing the voluntary principles on security and human rights. *In:* R. Sullivan and M. Robinson, eds., *Business and human rights*. London: Routledge, 243–259.

Freyss, J., 1995. *Economie assistée et changement social en Nouvelle-Calédonie*. Paris: Tiers Monde, IEDES, PUF.

Gammage, B., 1998. *The sky travellers: Journeys in New Guinea 1938–1939*. Melbourne: Melbourne University Publishing.

Ganti, T., 2014. Neoliberalism. *Annual Review of Anthropology*, 43, 89–104.

Gilberthorpe, E. and Papyrakis, E., 2015. The extractive industries and development: The resource curse at the micro, meso and macro levels. *The Extractive Industries and Society*, 2, 381–390.

Gilberthorpe, E. and Rajak, D., 2016. The anthropology of extraction: Critical perspectives on the resource curse. *The Journal of Development Studies*, 53 (2), 186–204.

Golub, A., 2014. *Leviathans at the gold mine: Creating Indigenous and corporate actors in Papua New Guinea*. Durham and London: Duke University Press.

Grochain, S., 2013. *Les dynamiques sociétales du projet Koniambo*. Païta, New Caledonia: Institut Agronomique Néo-Calédonien Editions.

Harvey, D., 2006. *Spaces of global capitalism: Towards a theory of uneven geographical development*. London and New York: Verso.

Healey, A. M., 1967. *Bulolo: A history of the development of the Bulolo region, New Guinea*. Canberra: The Australian National University.

Herrenschmidt, J.-B. and Le Meur, P.-Y., 2016. *Politique foncière et dynamiques coutumières en Nouvelle-Calédonie et dans le Pacifique*. Noumea-Kone: IRD Editions-Province Nord.

Hirschman, A. O., 1958. *Strategy of economic development*. New Haven: Yale University Press.

Huber, M. T. and Emel, J., 2009. Fixed minerals, scalar politics: The weight of scale in conflicts over the "1872 Mining Law" in the United States. *Environment and Planning A*, 41, 371–388.

Human Rights Watch, 2011. *Gold's costly dividend: Human rights impacts of Papua New Guinea's Porgera Gold Mine*. Human Rights Watch.

Jackson, R. T., 1984. *Ok Tedi: The pot of gold*. Port Moresby: University of Papua New Guinea Press.

Kirsch, S., 2014. *Mining capitalism: The relationship between corporations and their critics*. Berkeley: University of California Press.

Kroeker-Mauss, D., 2014. The protected area as enclave: Towards new geographies of tourism and conservation. *Geography Compass*, 8 (11), 796–807.

Kusumaryati, V., 2021. Freeport and the states: Politics of corporations and contemporary colonialism in West Papua. *Comparative Studies in Society and History*, 63 (4), 881–910.

Larsen, P. B., 2015. *Post-frontier resource governance: Indigenous rights, extraction and conservation in the Peruvian Amazon*. Houndmills: Palgrave Macmillan.

Laurent, B. and Merlin, J., 2022. Les ressources de l'indépendance? Propriété des entreprises minières et fabrique de la valeur en Nouvelle-Calédonie-Kanaky. *In:* S. Chailleux, S. Le Berre, and Y. Gunzburger, eds., *Ressources minérales et transitions: Trajectoires politiques du sous-sol français au 21ème siècle*. Brussels: Peter Lang, 323–351.

Le Meur, P.-Y., 2013. Locality, mobility and governmentality in colonial/postcolonial New Caledonia: The case of the Kouaré Tribe (*xuâ Xârâgwii*), Thio (*Cöö*). *Oceania*, 83 (2), 130–146. Available from: https://doi.org/10.1002/ocea.5009.

Le Meur, P.-Y., 2022. Proof and test: The construction of customary land in New Caledonia. *In:* E. Frezet, M. Goetzmann, and L. Mason, eds., *Spaces of custom and law*. Abington: Routledge, 160–178.

Le Meur, P.-Y., 2024. Mining fronts, labor mobilities, and the construction of locality in Thio, New Caledonia. *In:* G. Castillo, D. Brereton, and M. Himley, eds., *Mining, mobility, and social change in the Global South: Regional perspectives*. Abingdon: Routledge, 165–185.

Le Meur, P.-Y., Ballard, C., Banks, G., and Sourisseau, J. M., 2013. Two islands, four states: Comparing resource governance regimes in the Southwest Pacific. *In*: J. Wiertz, ed., *Proceedings of the 2nd international conference on social responsibility in mining (SRMining 2013, Santiago, Chile)*. Santiago, GECAMIN, 191–199.

Le Meur, P.-Y., Horowitz, L. S., and Mennesson, T., 2013. 'Horizontal' and 'vertical' diffusion: The cumulative influence of Impact and Benefit Agreements (IBAs) on mining policy-production in New Caledonia. *Resources Policy*, 38 (4), 648–656.

Le Meur, P.-Y. and Levacher, C., 2022. Mining and competing sovereignties in New Caledonia. *Oceania*, 92 (1), 74–92.

Le Meur, P.-Y., Levacher, C., Bouard, S., Herrenschmidt, J.-B., and Sabinot, C., 2021. Mining and the value of place in New Caledonia: Negotiation, evaluation, recognition. *Extractive Industries and Society*, 8, 44–54.

Le Roux, T. and Letté, M., eds., 2013. *Débordements industriels: environnement, territoire et conflit (XVIIIe-XXIe siècle)*. Rennes: Presses Universitaires Rennes.

Leith, D., 2003. *The politics of power: Freeport in Suharto's Indonesia*. Honolulu: University of Hawai'i Press.

Levacher, C., 2016. *De la terre à la mine? Les chemins de l'autochtonie en Nouvelle-Calédonie*. PhD in Anthropology. Paris: Ecole des Hautes Etudes en Sciences Sociales.

Levacher, C. and Le Meur, P.-Y., 2022. The compensation arenas in South New Caledonia: Minescape, governmentality and politics. *Extractive Industries and Society*, 11. Available from: https://doi.org/10.1016/j.exis.2021.100999.

Li, T. M., 2007. *The will to improve: Governmentality, development, and the practice of politics*. Durham, NC: Duke University Press.

Li, T. M., 2014. What is land? Assembling a resource for global investment. *Transactions of the Institute of British Geographers*, 39, 589–602.

Long, N., 2001. *Development sociology: Actors' perspectives*. London and New York: Routledge.

Long, N. and Roberts, B., 1984. *Miners, peasants and entrepreneurs. Regional development in the central highlands of Peru*. Cambridge: Cambridge University Press.

Lund, C., 2006. Twilight institutions: Public authority and local politics in Africa. *Development and Change*, 37 (4), 685–705.

Maclellan, N. and Regan, A., 2018. *New Caledonia and Bougainville: Towards a new political status?* Discussion Paper 2018/3, Department of Pacific Affairs. Canberra: Australian National University.

Magrin, G., 2013. *Voyage en Afrique rentière. Une lecture géographique des trajectoires du développement*. Paris: Publications de la Sorbonne.

Marston, S. A., 2000. The social construction of scale. *Progress in Human Geography*, 24, 219–242.

McKenna, K., 2016. *Corporate social responsibility and natural resource conflict*. Abingdon: Routledge.

Mitchell, T., 2011. *Carbon democracy: Political power in the age of oil*. London and New York: Verso.

Mokaddem, H., 2011. *Le discours politique kanak: Jean-Marie Tjibaou, Rock Déo Pidjot, Eloi Machoro, Raphael Pidjot*. Kone, New Caledonia: Les Editions de la Province Nord.

Nelson, H., 1976. *Black, white and gold: Gold mining in Papua New Guinea 1878–1930*. Canberra: Australian National University Press.

North, D., 1990. *Institutions, institutional change and economic performance*. Cambridge: Cambridge University Press.

Nuttall, M., 2017. *Climate, society and subsurface politics in Greenland: Under the great ice*. Oxon, New York: Routledge.

O'Faircheallaigh, C., 1984. *Mining and development: Foreign-financed mines in Australia, Ireland, Papua New Guinea and Zambia*. London: Routledge.

Phelps, N., Atienza, M., and Arias, M., 2015. Encore for the enclave: The changing nature of the industry enclave with illustrations from the mining industry in Chile. *Economic Geography*, 91 (2), 119–146.

Redfield, P., 2000. *Space in the tropics: From convicts to rockets in French Guiana*. Berkeley: University of California Press.

Regan, A. J., 2017. Bougainville: Origins of the conflict, and debating the future of large-scale mining. *In:* C. Filer and P.-Y. Le Meur, eds., *Large-scale mines and local-level politics: Between New Caledonia and Papua New Guinea*. Canberra: ANU Press, 353–414.

Regan, A. J., 2019. *The Bougainville Referendum: Law, administration and politics*. Department of Pacific Affairs. Canberra: Australian National University.

Rodgers, D. and O'Neill, B., 2012. Infrastructural violence: Introduction to the special issue. *Ethnography*, 13, 401–412.

Ross, M. L., 2015. What have we learned about the resource curse? *Annual Review of Political Science*, 18, 239–259.

Rosser, A., 2006. *The political economy of the resource curse: A literature survey*. Working Paper 268. Brighton: Institute of Development Studies.

Rutherford, D., 2012. *Laughing at leviathan: Sovereignty and audience in West Papua*. Chicago, IL: The University of Chicago Press.

Sidaway, J., 2007. Enclave space: A new metageography of development? *Area*, 39 (3), 331–339.

Strathern, M., 2009. Land: Intangible or tangible property? *In:* T. Chesters, ed., *Land rights*. Oxford: Oxford University Press, 13–38.

Szablowski, D., 2007. *Transnational law and local struggles: Mining, communities and the World Bank*. Oxford: Hart Publishing.

Tsing, A. L., 2012. On nonscalability: The living world is not amenable to precision-nested scales. *Common Knowledge*, 18 (3), 505–524.

UNDP, 2014. *Papua New Guinea national human development report 2014: From wealth to wellbeing: Translating resource revenue into sustainable human development*. New York and Port Moresby: UNDP.

Watts, M., 2004. Resource curse? Governmentality, oil and power in the Niger Delta, Nigeria. *Geopolitics*, 9 (1), 50–80.

Weisskoff, R. and Wolff, E., 1977. Linkages and leakages: Industrial tracking in an enclave economy. *Economic Development and Cultural Change*, 25, 607–628.

Weszkalnys, G., 2011. Cursed resources, or articulations of economic theory in the Gulf of Guinea. *Economy and Society*, 40 (3), 345–372. Available from: https://doi.org/10.1080/03085147.2011.580177.

Wilson, J. D., 2015. Understanding resource nationalism: Economic dynamics and political institutions. *Contemporary Politics*, 21 (4), 399–416.

Winanti, P. S. and Diprose, R., 2020. Reordering the extractive political settlement: Resource nationalism, domestic ownership and transnational bargains in Indonesia. *The Extractive Industries and Society*, 7 (4), 1534–1546.

Zimmermann, E., 1933. *World resources and industries: A functional appraisal of the availability of agricultural and industrial resources*. New York: Harper & Brothers.

3 Foreign investor accountability for the violation of Indigenous peoples' rights in international investment law and arbitrations

Reflections from the *Bear Creek* case

Zoé Boirin-Fargues

Indigenous peoples' rights have been increasingly recognized at the international level in the recent past, especially since the adoption of the *United Nations Declaration on the Rights of Indigenous Peoples* (UNDRIP) in 2007 (Puig 2021, p. 41; United Nations Human Rights Office of the High Commissioner 2013). Yet, development projects often have a profound negative impact on Indigenous peoples' traditional ways of life, languages, and territories (Puig 2021, p. 4; Foster 2013, p. 381; Colombia Center on Sustainable Development and Office of the High Commissioner for Human Rights 2016; Horowitz *et al.* 2018). These projects are in many cases conducted by foreign investors who enjoy the protection of international investment law (IIL).

IIL encompasses international investment treaties and investment chapters in international free trade agreements—referred to as International Investment Agreements (IIAs)—as well as international investment arbitral decisions. IIAs are signed between states to protect the investments made by their nationals in the territories of the other state party or parties to the IIA (Alvarez 2011, pp. 39–40). Although IIAs are state-to-state agreements, they define the rights of foreign investors, but not of affected communities (Perrone 2021, p. 179). They usually provide for a dispute settlement process—International Investment Dispute Settlement (IISD)—by which a foreign investor can submit a request for arbitration before an arbitral tribunal against the host state. The arbitral tribunal is an *ad hoc* tribunal constituted to deal with a particular dispute. Arbitral tribunals are designed to be a "neutral forum" (Giorgetti 2018, p. 3) by ensuring that arbitration is exempt from any political and diplomatic matters (Innerebner 2019, p. 139). The composition of arbitral tribunals is defined by the specific set of arbitration rules chosen by the parties as provided in the IIA.[1] Each party typically appoints at least one arbitrator. The selected arbitrators or a neutral appointing authority then select the third arbitrator who will be the chairperson of the tribunal (Giorgetti 2018, p. 6). Regarding the requirements for the arbitrators' selection, arbitration rules generally provide for a requirement of impartiality and independence and consider availability and experience as relevant (Prusinowska 2020, pp. 153–154). The tribunal's mandate is to assess whether or not the state violated the IIA and, in case of violation, to define the compensation to be paid by the host state to the foreign investor. The decisions of these *ad hoc*

DOI: 10.4324/9781003406433-5

This chapter has been made available under a CC-BY-NC-ND 4.0 license.

tribunals do not bind other investment arbitration tribunals, which creates uncertainty around the decisions of arbitral tribunals (VanDuzer *et al.* 2013, p. 139).

The implementation of IIAs has raised many issues and concerns, including the limitation of the regulatory power of states (Martini 2017, p. 533; Innerebner 2019, p. 141) as well as the very restricted access to arbitration proceedings for local populations affected by the investment, who are limited to submitting *amicus curiae*[2] to the arbitral tribunal (Simons and VanDuzer 2020, p. 281). Furthermore, IIL has been criticized for providing strong legal protection for foreign investors, while failing to consider the local context and thus contributing to reinforcing inequalities, especially when development projects are undertaken on Indigenous peoples' territories (Goff 2021, para. 43; Perrone 2022b, p. 839; Puig 2021, p. 36). Many publications have focused on these issues, and countless advocates have pointed out that IIL should take into account human rights and, in recent years, Indigenous peoples' rights (Puig 2021; Borrows and Schwartz 2020; Binder 2015; Colombia Center on Sustainable Development and Office of the High Commissioner for Human Rights 2016; Giupponi 2018; Krepchev 2015; Debevoise Ostby 2003; Vadi 2011; Boirin-Fargues 2024). Specifically regarding Indigenous peoples' rights, the new *United States–Mexico–Canada Agreement* (USMCA) signed in 2008[3] provides a general exception that allows states to take measures that would otherwise be considered as violating the agreement, to protect Indigenous peoples' rights.[4] It is therefore acknowledged as the most advanced treaty (Bellegarde 2018), and its existence is a sign of a certain trend toward the inclusion of Indigenous peoples' rights within IIL. However, this development benefits only Indigenous peoples in North America.[5] For the vast majority of IIAs, Indigenous peoples are rendered invisible by IIL, which assumes that states represent Indigenous communities' interests before those of foreign investors (Perrone 2021, p. 173).

In this context, we undertook a research project[6] aimed at evaluating how and under what circumstances IIL has referred to Indigenous peoples. To do this, we looked at IIAs and existing (and publicly available) international investment arbitration decisions.[7] From the data collected, different research strands emerged. In particular, the reference to corporate social responsibility (CSR) in certain treaties and arbitration cases drew our attention, as CSR has a significant presence in the relationships between foreign investors and Indigenous peoples, in particular in the context of mining projects (Cragg and Greenbaum 2002, p. 320; Kemp and Owen 2013, p. 523; Luning 2012, p. 205). Indeed, there are many guidelines and toolboxes designed to aid mining companies in obtaining a "social license to operate" in the context of projects conducted on the territories of Indigenous peoples (Dashwook *et al.* 2014).

While there is no universally accepted definition of CSR (Dahlsrud 2008, p. 1; Wettstein 2012, p. 745; Goyette-Côté 2016, p. 26), it can be viewed as a "societal "mode of governance" that acts in place of state-based forms of corporate control" (Kaplan and Kinderman 2019, p. 132) and whose implementation depends on the activism of nonbusiness actors (Kinderman 2018, p. 366; Ramasastry 2015, p. 237). It can also be considered as a set of voluntary practices, norms, and principles for businesses that are not legally binding or sanctioned by law, to guide them in the conduct of their activities. Along with the Business and Human Rights (BHR) movement (Ramasastry 2015, p. 237), CSR has helped spur the recognition that investors or corporations should respect human rights. In contrast to BHR, which promotes binding obligations toward business, CSR as a nonbinding set of practices is widely criticized as not ensuring the accountability of businesses for the violation of human rights (Tamvada 2020, p. 5).

This chapter examines how IIL has adapted to the increasing recognition of corporations' responsibility to respect human rights—today, embedded in the *UN Guiding Principles on Business and Human Rights* (*Ruggie Principles*)[8]—and, in particular, Indigenous peoples' rights. I purposely conduct a legal analysis of the content of IIAs and arbitral proceedings to address the following questions: How does IIL regulate the behavior of foreign investors to ensure they do not directly or indirectly violate Indigenous peoples' rights? What decisions have been made by arbitral tribunals when the question of foreign investor responsibility for such violations was raised before them? How does CSR play a role in this matter? What can we learn from the arbitral decision in the *Bear Creek Mining Corporation v. Republic of Peru (Bear Creek)* case[9] under the *Canada–Peru Free Trade Agreement*?

While the influence of the arbitral tribunal's composition on its decision is an interesting aspect to consider, I choose to maintain the focus of this chapter on analyzing the content of IIAs and arbitration documents. This facilitates the identification of legal improvements that can be expected in the matter regardless of the identity of the arbitrators.

I will first observe whether IIL provides for mechanisms to make foreign investors accountable for the violation of Indigenous peoples' rights (I). Second, I will discuss whether CSR can be a tool to ensure some level of accountability by foreign investors for their actions toward Indigenous peoples, building on the *Bear Creek* case (II). I conclude by providing some insights on how IIL could evolve to keep ensuring the protection of foreign investors while also fostering the respect of Indigenous Peoples' rights.

I. IIL under review: Absence of regulation of the behavior of foreign investors toward Indigenous peoples

The accountability of foreign investors for the violation of Indigenous peoples' rights is not a legal question addressed by IIAs (A), nor is it a question that can be easily addressed by arbitrators under the majority of IIAs (B).

A. Absence of mechanisms to ensure the accountability of foreign investors for the violation of Indigenous peoples' rights

IIAs are signed by states to ensure protection for the investments made by their nationals—or by companies whose headquarters are located within their jurisdiction—in other countries. As international treaties, IIAs are signed by states and impose obligations only on states. They generally do not provide for any obligations to investors (Perrone 2022a, p. 375).[10]

Typically, an IIA will include core provisions such as national treatment,[11] full protection and security,[12] most favored nation treatment,[13] and minimum standard of treatment, which generally includes the Fair and Equitable Treatment (FET) standard[14]—itself including legitimate expectations, nondiscrimination, fair procedure, transparency, and proportionality (Stepanov 2018, p. 52). For the purpose of this article, FET can be summarized as follows: Host states have the obligation to ensure that foreign investors are all treated in the same way by the host state; that, if the host state takes a measure that affects the foreign investment to a certain degree, this measure shall respect certain criteria and that the foreign investor should receive compensation (protection against illegal expropriation); and that the host state cannot treat foreign investments in an arbitrary way.

One of the intentions behind the development of IIAs post second world war was to ensure that the protection of foreign investors would not be affected by political or societal issues. Thus, IIL was built with the goal of protecting foreign investment from political and arbitrary measures taken by the host state (Vandevelde 2005, p. 170; Miles 2013, p. 10; Alshahrani 2020, p. 129). For example, with the protection provided by the IIA, a foreign investor would receive compensation if the host state nationalized the investment, perhaps after a change of government, or made drastic changes to its laws and/or policies that subsequently had a detrimental impact on the investment. It is therefore not surprising that, with very limited exceptions,[15] the vast majority of IIAs do not include provisions that protect Indigenous peoples' rights (Boirin-Fargues 2024, p. 160). Hence, the legal matter triggering an arbitration, which is necessarily defined in reference to the IIA, will not be based on Indigenous people's rights. Take, for example, a conflict between a Canadian mining company and Indigenous peoples in Peru, where the Indigenous peoples affected by the mine oppose its development and/or its continued operation, and where the mining company violates Indigenous peoples' rights (recognized internationally and nationally) such as the right to free, prior, and informed consent (FPIC). In such a situation, Peru could decide to revoke the Canadian company's mining title to prevent the conflict from escalating, without compensation or with only very limited compensation. The Canadian mining company would then submit an arbitration request under the *Bilateral Investment Treaty* (BIT) signed between Canada and Peru (2006), claiming that Peru had failed to respect its obligations under this treaty—in this case, had failed to compensate the company for the expropriation (among other potential alleged breaches). In settling this dispute, the arbitral tribunal would be bound by the terms of the BIT. Hence, the tribunal would focus exclusively on the expropriation criteria as defined by the BIT. The fact that the Canadian mining company violated the rights of Indigenous peoples would not enter into consideration *per se*. Consideration of Indigenous peoples' rights would form part of the arbitral tribunal's decision: (1) In its analysis regarding expropriation, more specifically to assess if the expropriation was undertaken in order to protect a public interest (preventing escalation of the conflict), and (2) if the BIT contained a clause "in accordance with (national) law" and if the investment was acquired in violation of Indigenous peoples' rights, for example, if

Peruvian law provided for a specific regime protecting Indigenous peoples' rights to land on which the investment was made.

Thus, violations of Indigenous peoples' rights by foreign investors must be addressed before domestic law in host states and in investors' home states under domestic law, and not under the IIA. This raises other issues, such as the capacity of the host state to hold foreign investors accountable (VanDuzer *et al.* 2013, p. 255), and the potential limited protection of Indigenous peoples' rights provided by national laws.

In brief, Indigenous peoples and, more specifically, Indigenous peoples' rights, are unlikely to be considered in the arbitral tribunal's analysis. The only legal opportunity for integrating Indigenous peoples' rights in an arbitration would be through customary international law. Indeed, in accordance with the principle of systemic interpretation contained in Article 31(3)(c) of the *Vienna Convention on the Law of Treaties*, a treaty shall be interpreted, taking into account "any relevant rules of international law applicable in the relations between the parties".

Thus, I examined arbitration cases in which international customary law has been invoked by states to trigger foreign investor responsibility for the violation of Indigenous peoples' rights.

B. Limited window to take into account investor violation of Indigenous peoples' rights in international investment arbitration

In several arbitration proceedings, states have presented arguments to probe the behavior of foreign investors toward Indigenous peoples and to request that the investor's protection under the IIA in question be withdrawn where the investor had violated the rights of Indigenous peoples.

An eloquent illustration of such arguments is found in the *South American Silver Limited* case, in which Bolivia invoked the "clean hands" doctrine. This doctrine derives from the general principle of good faith in international law. According to the "clean hands" doctrine, a foreign investor's claim would not be admissible if they acted illegally or unfairly. However, the customary international law basis of this doctrine remains controversial (Dumberry and Dumas-Aubin 2013, p. 1), as illustrated by this case.

In the *South American Silver Limited* case, Indigenous peoples affected by the investments opposed the project, and Bolivia revoked the investor's mining concession. The Bermudian company claimed that Bolivia had violated its obligation under the *Bolivia–United Kingdom BIT*, arguing that Bolivia had unlawfully expropriated its investment. In its counterargument, Bolivia contended that the company had violated Indigenous peoples' rights as recognized in national and international law.[16] Thus, according to Bolivia, the company did not have clean hands and could therefore not benefit from the protection of the BIT. The company, on the other hand, argued that the very existence of this doctrine as a general principle of international law was not established, and thus, the principle could not be applied in the case.[17]

The arbitral tribunal, therefore, faced the question of whether or not this doctrine, as a general principle of international law, would be applicable to the case. The tribunal rejected Bolivia's argument, considering that the latter had not provided sufficient evidence to establish that the "clean hands" doctrine was recognized as a general principle of law[18] or part of international public policy.[19]

The "clean hands" doctrine was also used by Panama in the *Álvarez y Marin Corporación v. Panama* case. In that case, Panama argued that the investment in question was constituted illegally, as the investor had not respected the legal requirements regarding the acquisition of properties on Indigenous peoples' lands. The tribunal did not refer to the principle of clean hands, but ruled in favor of the state, recognizing the illegality of the investment because of the investor's violation of national law.[20] These cases suggest that the "clean hands" doctrine is unlikely to be applied by an arbitral tribunal.

However, the *Álvarez y Marin Corporación v. Panama* case highlights the possibility that, where there is a violation of Indigenous peoples' rights as protected by the host state domestic law, states may be successful in arguing that the investor should be denied access to the protection of an IIA. However, as this case suggests, such a scenario implies several elements: (1) The presence of clearly defined national provisions protecting Indigenous peoples' rights; (2) the constitution of the foreign investment requires the implementation of these provisions; and (3) in the arbitration case, the state must specifically invoke the violations of these national provisions protecting Indigenous peoples' rights.

Thus, it is clear that the potential violation of Indigenous peoples' rights by foreign investors could, in very rare cases, trigger the legality of foreign investments under IIAs.

The case law mentioned above also suggests that national laws aiming to implement UNDRIP in national contexts, such as in Canada and Colombia, will have limited impact on the protection of foreign investments under IIA. Indeed, for them to have an impact, the investment would have to be constituted—and not implemented—in violation of clearly defined national provisions. The latter would likely not be a general principle, such as the FPIC of Indigenous Peoples, for which no precise legal definition exists (Papillon and Rodon 2017, p. 216).

Furthermore, we observe that the scenario of the *Álvarez y Marin Corporación v. Panama* case depends entirely on national laws that, in most jurisdictions, do not offer legal protection to Indigenous peoples' rights. It would therefore only protect, in very limited instances, Indigenous peoples' rights across the globe.

Consequently, it is likely that the violation of Indigenous peoples' rights by foreign investors will not be addressed by arbitral tribunals without an established accountability mechanism under IIA. This mechanism will be particularly welcomed as it will include the question of the violation of Indigenous peoples' rights directly before the arbitral tribunal, overcoming possible obstacles at the national level (e.g., lack of access to remedies for Indigenous communities and lack of recognition of their rights in certain countries).

I will now turn to how IIL has been addressing CSR and explore whether CSR could compensate for the lack of binding accountability mechanisms for foreign investors.

II. CSR as voluntary practice, a limited trend in IIL

The data collected during our research suggests that there is an emerging trend to include provisions on CSR in IIAs (A). Beyond CSR provisions, while referencing CSR standards could contribute to regulating the behavior of companies, the *Bear Creek* case is a good illustration of the limitations of CSR as a tool to ensure foreign investors' accountability (B).

A. CSR in IIAs: A new yet limited trend

We observed an increasing number of IIAs using "societal language," which focuses mainly on environmental and labor issues but has recently been expanding to include references to human rights and anti-corruption issues (Hepburn and Kuuya 2011, p. 600). A significant number of IIAs impose obligations on state parties either not to weaken existing environmental and labor laws ("non-derogation clauses") or to provide high-level protection on these matters.[21]

Other provisions directly reference CSR. We observe that among 3,268 existing IIAs,[22] there 49 treaties that countain a CSR clause,[23] all of which were adopted after 2013, and 28 of them after 2018.[24] This date is not surprising, since IIAs only began to be widely adopted around the 1990s, and CSR started to grow in importance in the 2000s (Kinderman 2018, p. 366).

Building on the distinction between BHR—promoting binding obligations toward businesses—and the CSR movement, which promotes a nonbinding approach to regulate businesses, as established by Anita Ramasastry (2015, pp. 237–238), we found three types of CSR clauses in IIAs: (1) BHR clauses, (2) type 1 CSR clauses, and (3) type 2 CSR clauses.[25]

The first type of clause—BHR clauses—is very rare. For example, Article 18 of the *Morocco–Nigeria Free Trade Agreement* contains such a clause. The *Pan-African Investment Code* (PAIC) provides, in Article 24 entitled "Business Ethics and Human Rights," several principles that "should govern compliance by investors with business ethics and human rights," including the principle under which investors shall "support and respect the protection of internationally recognized human rights" and shall "ensure that they are not complicit in human rights abuses."[26] In Article 22 on "corporate social responsibility," the agreement provides that investors "shall, in pursuit of their economic objectives, ensure that they do not conflict with the social and economic development objectives of host states and shall be sensitive to such objectives."[27]

As emphasized by Laurence Dubin, the content of these clauses is unclear in terms of what is expected of foreign investors and how the clauses could concretely ensure the accountability of these investors. There is the issue related to the content of "human rights" and to what extent they would incorporate Indigenous peoples' rights, especially regarding land tenure (Dubin 2018). Such clauses could, however, influence the interpretation of the FET standard and, more specifically, the legitimate expectations of investors, or even prevent the foreign investor from accessing the protection of the IIA if its investment was made in violation of human rights.

As for the first type of CSR clause (type 1 CSR), corporations are encouraged by states to self-regulate; CSR is then implemented through the corporations' voluntary practices. For example, Article 12 of the BIT signed between Argentina and Qatar in 2016 provides that: "Investors operating in the territory of the host Contracting Party should make efforts to voluntarily incorporate internationally recognized standards of corporate social responsibility into their business policies and practices."[28]

The second type of CSR clause (type 2 CSR) provides that the host and home states shall promote CSR, with CSR being implemented through domestic frameworks.[29] The only CSR clause explicitly referring to Indigenous peoples' rights belongs to this category. Article 14.17 of the USMCA "reaffirm[s] the importance" of "encourage[ing]" business to "voluntary[ily]" incorporate CSR standards into their business practices and procedures. The article provides for a non-exhaustive list of CSR standards, for example, the *OECD Guidelines for Multinational Enterprises*, and assesses that these "standards, guidelines, and principles may address" different areas, including the rights of "indigenous and aboriginal peoples."[30]

We observe that the USMCA does not refer to the *Ruggie Principles*, which are considered a key – yet limited – international reference tool regarding the responsibility of businesses to respect human rights (Simons 2012, p. 9). Such reference to the *Ruggie Principles* would have provided for a stronger standard in terms of promoting foreign investor accountability, as they define the business responsibility to respect human rights (Schwartz 2020, p. 267; Gunn 2020, p. 194). Furthermore, there is no explicit reference to Indigenous peoples' rights, including to UNDRIP.

Both types of CSR clauses only call for voluntary approaches to CSR, as they do not provide any binding obligations toward foreign investors.

Regarding the potential of these clauses to be taken into account in the arbitral proceedings, it seems that only type 1 CSR clauses could have an impact, as the second type addresses states only. One can imagine that when assessing the legitimate expectations of foreign investors, arbitrators could consider the CSR practices that foreign investors were encouraged to adopt and whether the investor complied with such standards or practices. Legitimate expectations are typically part of the FET component of IIAs. Their content is not fixed and is defined according to the conduct of one party that can create "reasonable and justifiable expectation that the conduct, when relied upon, will not be unjustifiably or unconscientiously departed from in circumstances where such a departure will cause material detriment to the investor (or investment)" (Laryea 2020, p. 100).

However, it is unlikely that arbitrators would do so given the weakness of the language of CSR clauses, especially as the majority of such clauses do not refer to Indigenous peoples. Thus, the CSR clauses would hardly compensate for the lack of binding, accountability mechanisms for foreign investors under IIAs in the case of the violation of Indigenous peoples' rights by foreign investors.

Beyond CSR clauses, CSR has been invoked in a case involving Indigenous peoples' consultation in a mining project. The next section considers the *Bear Creek* case and how CSR was used by the parties in their arguments and how it was integrated into the arbitral tribunal's award.

B. The **Bear Creek** *case: A legalistic approach weakening CSR's potential in arbitration*

From the data we collected, one case drew our attention to the importance gained by CSR in the regulatory landscape of mining development.

The *Bear Creek* case involved a Canadian mining company (Bear Creek) that undertook development of a silver mine—known as the Santa Ana project—in the Puno region of Peru in 2007. The company obtained the special authorization requested from foreign investors under the Peruvian Constitution to operate a mine. As part of the evaluation of the Environmental and Social Impact Assessment (ESIA), the government approved the ESIA's community participation plan and executive summary and required the company to establish community participation mechanisms and engage in consultations.[31]

The project met with strong opposition and protests by the local population, including from Indigenous groups,[32] although Bear Creek and Peru, before the arbitral tribunal, do not agree on how strong and unanimous opposition to the project actually was.[33] Following various meetings, including with protestors, Peru issued a supreme decree revoking Bear Creek's special authorization to operate the mine, which led to the termination of the mining project.[34]

Bear Creek submitted a request for arbitration against Peru, claiming that Peru had violated its rights under the *Canada–Peru Free Trade Agreement*, namely, the protection against unlawful expropriation, the obligation to provide FET, and the obligation to grant protection and security and not to "impair investment with unreasonable and discriminatory measures."[35]

What is most interesting is that both parties—Bear Creek and the Peruvian government—referenced CSR, along with the concept of a "social license to operate," in their respective arguments. However, they attributed different weight to the role of these two concepts in determining the foreign investor's responsibility surrounding the relationships with local and Indigenous communities.

Bear Creek's argumentation: A socially responsible corporation fulfilling its obligations under Peruvian law

In its *Memorial on the Merits* and its *Counter-memorial*, Bear Creek emphasized the importance of its CSR approach and how important community relationships were for the company.[36] It explained the company's involvement in the ESIA process in the Santa Ana project. Without referring to any international standards,[37] the company underlined the fact that it had conducted "informal" workshops alongside the consultations that were part of the ESIA process.[38]

More importantly, from Bear Creek's point of view, the company had to respect only Peruvian law, which provided for the obligation to obtain approval from the Peruvian authorities for its participation plan. Bear Creek argued that it had obtained this approval, which "confirmed that Bear Creek had implemented adequate community relationship programs and maintained good relationships with the communities, and that no social conflicts or issues existed in connection with [the project]."[39] As no social conflicts existed, there was no public interest justifying the expropriation of the company's investment.

Bear Creek further argued that Peru's refusal to grant the ESIA was politically motivated, as local politicians were promoting an anti-foreign investment and anti-mining agenda,[40] and that the government had failed to provide support for Bear Creek and had expropriated the investment: (a) Not for a public purpose, (b) without compensation, and (c) arbitrarily and without due process.[41]

As we shall see, the main point of disagreement between Bear Creek and Peru rested on the source and content of Bear Creek's obligations regarding local and Indigenous communities. While Peru considered that Bear Creek's obligations went beyond Peruvian law, Bear Creek focused solely on the obligations provided for by national law, which the company argued it had fulfilled.[42]

In summary, although the company put forward its CSR commitment and emphasized its actions beyond the obligations defined by Peruvian law, its line of arguments can be considered as being embedded within what I have called a "legalistic approach," that is, an approach that is based only on binding law as a source of law.

Peru's line of arguments: Peruvian law and international best practices require foreign investors to obtain consent from the communities themselves

Contrary to Bear Creek, Peru argued[43] that the expropriation pursued a public purpose as it aimed to "help end a wave of violence," since Bear Creek's project was facing strong opposition from local and Indigenous communities.[44]

In contrast to the "legalistic approach" developed by Bear Creek, Peru adopted a different perspective regarding the company's obligations.

First, according to Peru, the responsibility to obtain the support of the communities affected by the project rested exclusively with Bear Creek.[45] Peru argued that Peruvian law[46] requires mining companies to obtain the social license to operate and that this social license to operate could not be granted by the state, but only by the communities themselves.[47]

Thus, the sole fulfillment of the obligations defined by Peruvian law could not be sufficient.

Peru argued that national law could not provide for the adequate obligations for mining companies to obtain such consensus and that "it was never meant to do so."[48] The components of national law, beyond "some minimum technical requirements,"[49] were "broad":

> The law guides companies to communicate effectively with the local communities without specifying or imposing particular requirements that may not be appropriate to reach community consensus for every project. But the goal of reaching such a consensus may involve steps that go beyond technical, bare essentials of the law, and may also require measures that surpass what was originally anticipated when the company began the process.[50]

Peru also made reference to a Peruvian guide that "although [. . .] not legally binding, [. . .] nonetheless represent[ed] a number of best practices that [were] tailored to community relations in the Peruvian context,"[51] which Bear Creek did not respect by not taking into account the indirect impact on communities.[52]

Furthermore, as a "world-class [Canadian] mining company",[53] Bear Creek should have been aware of international best practices in the field, including the standards recommended by the International Council on Mining and Metals (ICMM).[54] These standards provide that a mining company must build consensus among communities affected by its project[55] and require a mining company to go beyond national requirements regarding the consultation of Indigenous peoples:

> International best practices [. . .] also suggest that the domestic laws provide a floor that a company *must* meet, rather than the ideal for which a company should aim. In other words, a company cannot expect that indigenous peoples will support a potentially intrusive mining project simply because federal law sanctions it or because the company has complied with notice and hearing regulations.[56]

Thus, regarding the argument made by Bear Creek that for "any requirement to go beyond the strict letter of the law" is an "absurd take on community relations . . . ," Peru qualified it as a "disdain for internationally accepted industry practice."[57]

From the state's perspective, the obligation to go beyond national law to reach consensus among local and Indigenous peoples was also supported by Canada's *Strategy to Advance Corporate Social Responsibility in Canada's Extractive Sector Abroad*, Canada being Bear Creek's home state. Peru argued that Canadian companies are expected to apply a "more rigorous standard" "[w]here host country requirements differ from the international standards . . ."[58] and that the government of Canada "demands that its companies live up to these types of elevated international standards."[59]

Thus, Bear Creek could not suppose that complying with minimum requirements of Peruvian law would be sufficient, and in the absence of community support,[60] Bear Creek could not obtain approval to construct the mine.[61]

Finally, Peru contended that Bear Creek could not have legitimate expectations of receiving the approval unless it received a social license to operate.[62] It should have expected the suspension of the review of its ESIA:

> Even a neophyte junior company like [Bear Creek] knew, or should have known, that the regulatory process for mining projects is complex and prone to delay. [Bear Creek] also should have known that mining is controversial in Peru, particularly within indigenous populations like the Aymara communities that lived near the proposed Santa Ana project;"[63] "delays of natural resource extraction projects due to community opposition are a common occurrence around the world,"[64] it is an "ordinary business risk" that does not trigger legitimate expectations.

The award: The prevalence of the "legalistic approach"

In the award, the arbitral tribunal, composed of three arbitrators appointed by the parties,[65] did not refer to Canadian or international standards as relevant sources of law. Instead, it adopted a "legalistic approach": It ignored the concept of CSR and, just as Bear Creek had, considered the concept of a "social license to operate" to the extent that this concept was provided by Peruvian law.

The tribunal thus asked the following two questions: (1) "What actions were legally required of [Bear Creek] in seeking to obtain a social license, and did [Bear Creek] take these actions?" and (2) "[w]hat were the state authorities' responsibilities in relation to obtaining a social license?"[66]

The tribunal recognized that actions beyond legal requirements "would have been possible and feasible"[67] and that "[e]ven though the concept of "social license" is not clearly defined in international law, all relevant international instruments are clear that consultations with indigenous communities are to be made with the purpose of obtaining consent from all the relevant communities."[68] Meanwhile, the tribunal found that

> [w]hile [Bear Creek] could have gone further in its outreach activities, the relevant question for the Tribunal is whether Respondent [Peru] can claim that such further outreach was legally required and its absence caused or contributed to the social unrest, *so as to justify [the contested measure].*[69] *[Our emphasis]*

From the evidence, the tribunal concluded that all actions by Bear Creek were coordinated with Peru authorities ("[w]ith their approval, support, and endorsement, and that no objections were raised"). Therefore, Bear Creek

> [c]ould take it for granted to have complied with all legal requirements with regard to its outreach to the local communities. [Peru] after its continuous approval and support of Claimant [Bear Creek]'s conduct cannot in hindsight claim that this conduct was contrary to the ILO Convention 169 or was insufficient, and caused or contributed to the social unrest in the region.[70]

The tribunal thus concluded that Bear Creek had been subject to an unlawful expropriation.

The *Bear Creek* case is emblematic of the limited scope of CSR in terms of triggering foreign investors' legal accountability independently of the behavior of states toward Indigenous peoples' rights. The foreign investor emphasized that it did "more than required" and that governmental approvals were signs that it had respected the national law requirements. In contrast, Peru argued that these requirements are technical processes but that their fulfillment does not mean that the Canadian company obtained the social license to operate, which is an obligation according to Peruvian law and international standards, in addition to the Canadian ones.

The majority of the arbitral tribunal structured its legal analysis narrowly around requirements defined by Peruvian law regarding the social license to operate. In doing so, it left Indigenous peoples' rights vulnerable to the state's behavior. Indeed, it is because Peru gave all the approvals required under Peruvian law that the revocation of Bear Creek's mining title could not be justified (Marcoux and Newcombe 2018, p. 657).

This case illustrates that while CSR might be raised by a state in arbitral proceedings, it is unlikely to be taken into account outside of what is legally required under the host state's national law, thus leaving the potential responsibility of foreign investors for the violation of Indigenous peoples' rights unaddressed. This case also confirms the observation that arbitrators tend to adopt what I call a "legalistic approach"; in accordance with their mandate, they rule within the strict limits of the IIA's terms,[71] which do not provide any rights to Indigenous peoples.

To nuance the previous paragraph, it is worth noting that there was one dissenting opinion within the arbitral tribunal. In the partial dissenting opinion, Professor Sands considered that, beyond state monitoring of the foreign investor, the latter had an obligation to obtain the social license: "[T]he Canada-Peru FTA is not, any more than ICSID, an insurance policy against the failure of an inadequately prepared investor to obtain such a license." Thus, according to Professor Sands, as the mining project contributed to the protests, the amount of damages to be paid by Peru to Bear Creek should have been reduced (Marcoux and Newcombe 2018, p. 656).

The *Bear Creek* case illustrates the limits of IIA for triggering foreign investor responsibility while such responsibility is not provided by national law or when the state itself contributed to violating Indigenous peoples' rights by approving, or failing to denounce, the behavior of foreign investors. To ensure that foreign investors can be systematically held accountable for the violation of Indigenous peoples' rights, independently of the behavior or national laws of states, IIAs should define a binding obligation that would ensure that foreign investors are held accountable when they violate Indigenous Peoples' rights.

Conclusion

It is clear today that the growing tendency in IIL is to go beyond the sole protection of foreign investments regardless of other legal issues, as illustrated by the adoption of USMCA. However, IIL lacks a binding mechanism that would ensure that foreign investors cannot benefit from the strong legal protection of IIAs while violating Indigenous peoples' rights.

We have seen that the development of CSR clauses would not ensure foreign investors' accountability for such violations. Indeed, CSR is approached as a set of voluntary practices with legal effects that do not go beyond what is provided in national laws. Furthermore, beyond CSR clauses, before an arbitral tribunal, the legal effects of CSR may depend on the state's behavior, as illustrated by the *Bear Creek* case.

Some arguments have been made before arbitral tribunals attempting to link the violation of Indigenous peoples' rights to the determination of whether foreign investors should benefit from the protection of IIAs. These arguments have failed, with the tribunal adopting what I call a "legalistic approach." Such an approach can likely be explained either by the limited mandate of arbitrators, which is to apply IIAs (which do not refer to Indigenous Peoples' rights), or by the absence of undisputed, accepted principles of international law, such as the "clean hands" principle.

In this context, and in light of the *Bear Creek* award, ensuring foreign investor accountability via binding obligations toward foreign investors is necessary.

Several authors suggest subordinating the protection of IIAs for foreign investors to the fulfillment of certain obligations (Hepburn and Kuuya 2011, pp. 599–600). More specifically with regard to Indigenous peoples' rights, Risa Schwartz considers that IIAs should provide an obligation to respect compliance with "a minimum level of Indigenous rights' requirements as a condition precedent for claiming rights afforded under the treaty" (Schwartz 2020, p. 268). Anthony Vanduzer, Penelope Simons, and Graham Mayeda proposed a sample provision that provides an obligation for foreign investors to respect internationally recognized human rights, including Indigenous peoples' rights under UNDRIP (VanDuzer *et al*. 2013, pp. 313–316). To overcome the disparity between countries of the protection of Indigenous people's rights, these rights would need to be defined in IIAs with reference to UNDRIP. It would also be important to allow Indigenous peoples broader access to arbitral tribunals. These changes would certainly require amending the arbitrator selection process, to ensure they have the expertise to make decisions in matters that go beyond IIL. None of these improvements can happen without strong advocacy work with governments, to trigger political will in that direction.

Acknowledgments

I would like to thank Professor Sophie Thériault and Professor Penelope Simons for taking the time and effort to review the manuscript. I sincerely appreciate their valuable comments and suggestions, which helped improve the quality of the manuscript. I would also like to thank the MinErAL network for funding the research, as well as Professor Erick Duchesne for his review and for his support in conducting the research that led to this chapter.

Notes

1 Regarding arbitration rules, BITs usually refer to the Convention on the Settlement of Investment Disputes between States and Nationals of other States (ICSID Convention) or the Arbitration Rules of the United Nations Commission on International Trade Law (UNCITRAL Arbitration Rules).
2 Under most arbitral proceedings, arbitral tribunals have discretion over accepting an *amicus curiae* submitted by a nonparty to the arbitration, such as Indigenous peoples.
3 United States–Mexico–Canada Agreement (USMCA), 30 November 2018.
4 *Idem*, Art. 32.5.
5 For an overview of this agreement linking to Indigenous peoples' rights, see Schwartz 2020, p. 266–270.
6 Under the MinErAL research network: www.mineral.ulaval.ca/

7 We used six databases: Investor–State LawGuide (www.investorstatelawguide.com/); Investment Policy Hub (https://investmentpolicy.unctad.org/); Agreement Text Research Law (https://research.un.org/en/docs/law/treaties); Investment Claim (https://oxia.ouplaw.com/); CISG database (https://iicl.law.pace.edu/cisg/cisg); and Edit (https://edit.wti.org/document/investment-treaty/search). The first step consisted of looking for any awards, arbitration procedural documents, or treaties referring to certain keywords or certain Indigenous peoples—identified because they are the only Indigenous people in a region—in English, French, Spanish, Portuguese, and Russian [in English: Aboriginal(s), First Nation(s), First People(s), Indian(s), Indigenous, Cree(s), Inuit(s), Maori(s), Meti(s), Sami(s), Saami(s), Tribal(s), and native (people(s). In French: autochtone(s), indien(s), indig(è)ne(s), M(é)ti(s), Premier(s) peuple(s), Premi(è)res nation(s), tribu(s). In Spanish: indígena(s), indio(s), nativa(s), nativo(s), guaraní(s). In Portuguese: aborigina(s), india(s), inca(s), maia. In Russian: коренные народы, Коренные малочисленные народы Севера, Сибири и Дальнего Востока Российской Федерации. We also looked for the word "Alaska" while analyzing the content of investment treaties signed by the United States.]. We selected these languages because they are the most commonly used languages in the field of international investment and trade law. Then, we conducted a broader literature review.

8 *Report of the Special Representatives of the Secretary*—Guiding Principles on Business and Human Rights: Implementing the United Nations "Protect, Respect and Remedy" Framework, 21 March 2011, A/HRC/17/31.

9 *Bear Creek Mining Corporation v. Republic of Peru*, ICSID Case No. ARB/14/21.

10 With the exception of the *Morocco–Nigeria Investment Treaty*, which provides a direct obligation for the investor to respect human rights.

11 The host state pledges "not to treat foreign operators in a less favorable manner than domestic operators in a similar situation." De Nanteuil 2020, p. 257.

12 It "protects the physical security of the investor or investment" and "the investor or investment against the inaction or passivity of the host state." *Idem*, p. 334–335.

13 The host state has "the obligation to grant foreign investors a treatment that is no less favorable than that granted to the most favored one." *Idem*, p. 270.

14 Although the relationship between MST and FET is disputed. *Idem*, p. 285–286.

15 Among more than 3,268 IIAs, we found (1) the USMCA's general exception allowing the three states parties to take a measure aiming to protect Indigenous peoples' rights; (2) 12 exceptions included by New Zealand that are close to general exceptions for the protection of Maori rights; and (3) 94 specific exceptions—mainly signed by Canada—allowing the state to give preference to Indigenous businesses, signed by Canada. See Boirin-Fargues (2024).

16 *Respondent Counter-Memorial* (31 March 2015), para. 268.

17 *Claimant's Post-Hearing Brief* (31 October 2016), para. 56.

18 *Award*, para. 445–450.

19 *Award*, para. 443 and 451–452. Note that the arbitral tribunal explicitly recognized the bad practices of the company toward Indigenous peoples, without stating that the company violated their rights. As in other cases like *Grand River*, it seems that arbitrators are not comfortable or have neither the mandate nor the legitimacy to rule on issues not directly linked to the implementation of ILL (see *Award*, para. 501).

20 *Álvarez y Marin Corporación v. Panama*, Award, para 118.

21 For example, the USA–Peru FTA and Canada–Peru FTA.

22 UNCTAD Investment Policy Hub, 2023. *International Investment Agreements Navigator* [online]. Available from: https://investmentpolicy.unctad.org/international-investment-agreements [Accessed 29 August 2023].

23 We have only counted provisions that explicitly refer to CSR in signed investment treaties or treaty chapters dedicated to investment. Note that we also found references to CSR in the preamble of 11 treaties and other provisions of 5 Free Trade Agreement or other partnership agreements.

24 Data available upon request.
25 We made this distinction building on Laurence Dubin's work; however, we disagree on the qualification of "CSR clause" for the first type, which we entitled "BHR clause" (Dubin 2018).
26 Article 24 (a) and (b).
27 Article 22 (2). Laurence Dubin also cites the *Brazil–Malawi Investment Cooperation and Facilitation Agreement* (2015) (Article 9), although we find that its language makes its attachment to the BHR movement questionable.
28 The Reciprocal Promotion and Protection of Investments between the Argentine Republic and the State of Qatar (signed on 6 November 2016).
29 Article 9.17 of the *Comprehensive and Progressive Trans-Pacific Partnership*: "The Parties reaffirm the importance of each Party encouraging enterprises operating within its territory or subject to its jurisdiction to voluntarily incorporate into their internal policies those internationally recognized standards, guidelines and principles of corporate social responsibility that have been endorsed or are supported by that Party."
30 "The parties reaffirm the importance of each Party *encouraging* enterprises operating within its territory or subject to its jurisdiction to *voluntarily incorporate* into their internal policies *those internationally recognized* standards, guidelines, and principles of corporate social responsibility that have been endorsed or are supported by that Party, *which may include* the OECD Guidelines for Multinational Enterprises. These standards, guidelines, and principles *may address* areas such as labor, environment, gender equality, human rights, indigenous and aboriginal peoples' rights, and corruption" [Our emphasis], Article 14.17.
31 *Award*, para. 168.
32 *Award*, para. 173–174, 178, 182–183, 188, 189–190, 195, 198.
33 See for example: *Award*, para. 173, 192.
34 *Award*, para. 202.
35 *Memorial on the Merits*, 29 May 2015, para. 113.
36 Memorial on the Merits, 29 May 2015, para. 57; Reply on the Merits and Counter-Memorial on Jurisdiction, 8 January 2016, para 66.
37 Reply on the Merits and Counter-Memorial on Jurisdiction, 8 January 2016, para. 72–105.
38 *Memorial on the Merits*, 29 May 2015, para. 61.
39 *Idem*, para 62.
40 *Idem*, para. 74.
41 *Idem*, para. 126–181.
42 According to the company, by claiming that Bear Creek should have gone beyond what is required by Peruvian law, Peru confirmed that Bear Creek actually complied with its Peruvian legal obligations. *Reply on the Merits and Counter-Memorial on Jurisdiction*, 8 January 2016, para. 68.
43 Peru also claimed that the investment was not lawfully constituted and was made in bad faith; therefore, the tribunal had no jurisdiction (*Counter-Memorial on the Merits and Memorial on Jurisdiction*, 6 October 2015, para 10).
44 *Idem*, para. 11.
45 *Idem*, para. 58.
46 That "informs Bear Creek that it must obtain community support before it can develop its mine" (*Idem*, para. 59 to 65).
47 *Rejoinder on the Merits and Reply on Jurisdiction*, 13 April 2016, para. 121. Thus, any approval by Peruvian authorities "would be without significance, because the only real measure of the sufficiency of community outreach is whether, in the end, the social license is obtained from the communities," which was not the case. *Idem*, 13 April 2016, para. 164.
48 Peru explained that "the law provides for the *processes* by which the company, with the oversight of the [governmental authority], can work to try to achieve approval from the

relevant stakeholders [. . .] *Peru's legal requirements* for the citizen participation component are a *floor—a bare minimum for the company's efforts—that maintains flexibility for the government and for the mining company*." *Idem*, para. 135 [Our emphasis].

49 Peru details all instruments provided by Peruvian law that are tools that do not guarantee the obtainment of a social license: (1) The citizen participation plan is, according to Peru, "simply a plan developed by the company that the [Peruvian authority] approves to set the company on the right track;" (2) the public hearing under the EIA process is "a formal opportunity for the mining company to present its project and the key points of its EIA to the communities, and to answer the concerns of the population with respect to the project [. . .]," but "it does not constitute a community vote whether to approve or disapprove the project." Peru added: "[a]fter the public hearing, the company must continue building a consensus for a much more intrusive stage of the project (namely, construction and exploitation);" "completion of the Public Hearing [. . .] merely confirms that the company complied with the technical requirements for the hearing." *Idem*, para. 168–170, 173.
50 *Idem*, para. 129.
51 *Idem*, para. 132.
52 *Idem*, para. 137.
53 Counter-Memorial on the Merits and Memorial on Jurisdiction, 6 October 2015, para. 1.
54 *Idem*, para. 67–70.
55 *Idem*, para. 58, 66–71. In a footnote, Peru further stated that the company "should have been aware that a failure in this regard could lead to extreme conflict." *Idem*, Footnote 68, p. 27.
56 *Idem*, para. 66.
57 Rejoinder on the Merits and Reply on Jurisdiction, 13 April 2016, para. 123.
58 Counter-Memorial on the Merits and Memorial on Jurisdiction, 6 October 2015, para 66.
59 *Rejoinder on the Merits and Reply on Jurisdiction*, 13 April 2016, para. 140. Peru also again made reference to a report issued by the Canadian Foreign Affairs and Trade Ministries that promotes CSR practices, that is, "voluntary activities undertaken by a company, over and above legal requirements, to operate in an economically, socially, and environmentally sustainable manner" and underlines that Canada called "on mining companies, including junior mining companies like [Bear Creek] to 'do better than the minimum' legal standard in the host country." *Idem*, para. 124.
60 Counter-Memorial on the Merits and Memorial on Jurisdiction, 6 October 2015, para. 24.
61 "In fact, Claimant appears to believe that it cannot bear any blame for [. . .] the 2011 disruptions in Puno because Claimant complied with the minimum requirements of Peruvian law that govern a company's social interactions with the communities where a project is located. This reflects a misunderstanding of Peruvian law as well as international norms of social responsibility. [. . .] as the strength of the Puno protests make manifestly clear—Claimant failed to obtain the all-important 'social license' from the communities surrounding Santa Ana that is necessary to operate any mine and particularly a large, open-pit operation." *Rejoinder on the Merits and Reply on Jurisdiction*, 13 April 2016, para. 120.
62 *Idem*, para. 534.
63 *Idem*, para. 535.
64 *Idem*, para. 536.
65 *Award,* 30 November 2017, para 6–8.
66 *Award*, 30 November 2017, para 402.
67 *Idem*, para. 404.
68 *Idem*, para. 406.
69 *Idem*, para. 408.
70 *Idem*, para. 412.

71 In the article in which we present the data collected in greater detail, we discuss the issue regarding decisions in which arbitrators would go beyond their initial mandate, considering, in particular, their legitimacy as one-time, ad hoc-appointed arbitrators (Boirin-Fargues 2024).

References

Agreement between Canada and the Republic of Peru for the Promotion and Protection of Investments (signed on 14 Nov. 2006, entered into force on 20 June 2007).

Alshahrani, S. M., 2020. What should we know about the origins of international investment law? *International Journal of Legal Information*, 48 (3), 122–131.

Alvarez, J. E., 2011. *The public international law regime governing international investment*. Netherlands: Hague Academy of International Law.

Bellegarde, P., 2018. By including Indigenous Peoples, the USMCA breaks new ground, *Macleans*. 4 October. Available from: www.macleans.ca/opinion/by-including-indigenous-peoples-the-usmca-breaks-new-ground/ [Accessed 9 May 2023].

Binder, C., 2015. Investment, development and indigenous peoples. *In:* S. W. Schill, C. J. Tams, and R. Hofmann, eds., *International investment law and development: Bridging the gap*. Cheltenham: Edward Elgar Publishing, 433–451.

Boirin-Fargues, Z., 2024. Indigenous people's rights in international investment: An ongoing compartmentalization. *Journal of World Investment & Trade*, 25 (2), 153–200.

Borrows, J. and Schwartz, S., eds., 2020. *Indigenous Peoples and international trade building equitable and inclusive international trade and investment agreement*. Cambridge: Cambridge University Press.

Columbia Center on Sustainable Investment and the United Nations Human Rights Office of the High Commissioner, 2016. *International investment and the rights of indigenous peoples*. Workshop Outcome Document [Online]. Available from: https://www.google.com/url?sa=t&source=web&rct=j&opi=89978449&url=https://ccsi.columbia.edu/sites/default/files/content/Workshop-on-International-Investment-and-the-Rights-of-Indigenous-Peoples-Outcome-Document-November-2016.pdf&ved=2ahUKEwiPyOTN39aGAxWEMlkFHWtWBRcQFnoECBoQAQ&usg=AOvVaw1LFDY3yACnyZSxy0E85oNN

Cragg, W. and Greenbaum, A., 2002. Reasoning about responsibilities: Mining company managers on what stakeholders are owed. *Journal of Business Ethics*, 39 (3), 319–335.

Dahlsrud, A., 2008. How corporate social responsibility is defined: An analysis of 37 definitions. *Corporate Social Responsibility and Environmental Management*, 15 (1), 1–13.

Dashwook, H., Prakash, A., and Griffin, J., 2014. Sustainable development and industry self-regulation: Development in the global mining sector. *Business & Society*, 53 (4), 551–582.

De Nanteuil, A., 2020. *International investment law*. Cheltenham: Edward Elgar Publishing.

Debevoise Ostby, A., 2003. Will foreign investors regulate Indigenous Peoples' right to self-determination? *Wisconsin International Law Journal*, 21, 223–250.

Dubin, L., 2018. Les clauses RSE dans les traités d'investissement. International Institute for Sustainable Development, Investment Treaty News [Online]. Available from: https://www.iisd.org/itn/fr/2018/12/21/corporate-social-responsibility-clauses-in-investment-treaties-laurence-dubin/

Dumberry, P. and Dumas-Aubin, G., 2013. The doctrine of "clean hands" and the inadmissibility of claims by investors breaching international human rights law. *In:* U. Kriebaum, ed., *Transnational dispute management special issue (TDM1): Aligning human rights and investment protection*.

Foster, G. K., 2013. Investors, states and stakeholders: Power asymmetries in international investment and the stabilizing potential of investment treaties. *Lewis and Clark Law Review*, 17 (2), 361–421.

Giorgetti, C., 2018. Selecting and removing arbitrators in international investment arbitration. *Brill Research Perspectives in International Investment Law and Arbitration*, 2 (4), 1–98.

Giupponi, B. O., 2018. Free, prior and informed consent (FPIC) of Indigenous Peoples before human rights courts and international investment tribunals: Two sides of the same coin? *International Journal on Minority and Group*, 25, 485–529.

Goff, P. M., 2021. Bringing Indigenous goals and concerns into the progressive trade agenda. *Papers in Political Economy* [Online], 65. Available from: https://doi.org/10.4000/interventionseconomiques.12777 [Accessed 9 May 2023].

Goyette-Côté, G., 2016. Responsabilité sociale des entreprises dans le secteur minier Quelles contributions pour le développement? *In:* B. K. Campbell and M. Laforce, eds., *La responsabilité sociale des entreprises dans le secteur minier: réponse ou obstacle aux enjeux de légitimité et de développement en Afrique?* Québec: Presses de l'Université de Québec, 19–52.

Gunn, B. L., 2020. International investment agreements and Indigenous Peoples' rights. *In:* J. Borrows and S. Schwartz, eds., *Indigenous peoples and international trade: Building equitable and inclusive international trade and investment agreements*. Cambridge: Cambridge University Press, 194–216.

Hepburn, J. and Kuuya, V., 2011. Corporate social responsibility and investment treaties. *In:* M.-C. Cordonier Segger, M. W. Gehring, and A. Newcombe, eds., *Sustainable development in world investment law*. The Hague: Kluwer Law International, 585–609.

Horowitz, L. S., Keeling, A., Lévesque, F., Rodon, T., Schott, S., and Thériault, S., 2018. Indigenous peoples' relationships to large-scale mining in post/colonial contexts: Toward multidisciplinary comparative perspectives. *The Extractive Industries and Society*, 5 (3), 404–414.

Innerebner, L. F., 2019. Politicization of a future international investment tribunal's appointment and how to avoid it. *Trento Student Law Review*, 1 (1), 137–154.

Kaplan, R. and Kinderman, D., 2019. The business-class case for corporate social responsibility: Mobilization, diffusion, and institutionally transformative strategy in Venezuela and Britain. *Theory and Society*, 48, 131–166.

Kemp, D. and Owen, J. R., 2013. Community relations and mining: Core to business but not "core business." *Resources Policy*, 38 (4), 523–531.

Kinderman, D., 2018. *Corporations and global standards of corporate social responsibility: Handbook of the international political economy of the corporation*. Cheltenham: Edward Edgard Publishing.

Krepchev, M., 2015. The problem of accommodating Indigenous land rights in international investment law. *Journal of International Dispute Settlement*, 6, 42–73.

Laryea, E. T., 2020. Legitimate expectations in investment treaty law: Concept and scope of application. *In:* J. Chaisse, L. Choukroune, and S. Jusoph, eds., *Handbook of international investment law and policy*. Singapore: Springer, 97–120.

Luning, S., 2012. Corporate social responsibility (CSR) for exploration: Consultants, companies and communities in processes of engagement. *Resources Policy*, 37 (2), 205–211.

Marcoux, J.-M. and Newcombe, A., 2018. Case comment *Bear Creek Mining Corporation v Republic of Peru*: Two sides of a "social license" to operate. *ICSID Review*, 33 (3), 653–659.

Martini, C., 2017. Balancing investors' rights with environmental protection in international investment arbitration. *The International Lawyer*, 50 (3), 529–584.

Miles, K., 2013. *The origins of international investment law*. Cambridge: Cambridge University Press.

Papillon, M. and Rodon, T., 2017. Proponent-Indigenous agreements and the implementation of the right to free, prior and informed consent in Canada. *Environmental Impact Assessment*, 62, 216–224.

Perrone, N. M., 2021. *Investment treaties and the legal imagination*. Oxford: Oxford University Press.
Perrone, N. M., 2022a. Bridging the gap between foreign investor rights and obligations: Towards reimagining the international law on foreign investment. *Business and Human Rights Journal*, 7, 375–396.
Perrone, N. M., 2022b. Local communities, extractivism and international investment law: The case of five Colombian communities. *Globalizations*, 19 (6), 837–853.
Prusinowska, M., 2020. Analysing appointments in international arbitration. *In:* F. Baetens, ed., *Identity and diversity on the international bench: Who is the judge?* Oxford: Oxford University Press, 142–163.
Puig, S., 2021. *At the margins of globalization: Indigenous Peoples and international economic law*. Cambridge: Cambridge University Press.
Ramasastry, A., 2015. Corporate social responsibility versus business and human rights: Bridging the gap between responsibility and accountability. *Journal of Human Rights*, 14 (2), 237–259.
Schwartz, S., 2020. Developing a trade and indigenous peoples chapter for international trade agreements. *In:* J. Borrows and S. Schwartz, eds., *Indigenous peoples and international trade building equitable and inclusive international trade and investment agreement*. Cambridge: Cambridge University Press, 248–273.
Simons, P., 2012. International law's invisible hand and the future of corporate accountability for violations of human rights. *Journal of Human Rights and the Environment*, 3 (1), 5–43.
Simons, P. and VanDuzer, J. A., 2020. *Using international investment agreements to address access to justice for victims of human rights. In:* O. E. Fitzgerald, ed., *Corporate citizen: New perspectives on the globalized rule of law*. Montreal: McGill-Queen's University Press.
Stepanov, I., 2018. *Eli Lilly and beyond: The role of international intellectual property treaties in establishing legitimate expectations in investor–state dispute settlement*. Baden-Baden: Nomos.
Tamvada, M., 2020. Corporate social responsibility and accountability: A new theoretical foundation for regulating CSR. *International Journal of Corporate Social Responsibility*, 5 (1), 1–14.
United Nations Human Rights Office of the High Commissioner, 2013. Indigenous peoples and the United Nations human rights system. Fact Sheet No. 9 Rev. 2.
Vadi, V. S., 2011. When cultures collide: Foreign direct investment, natural resources, and Indigenous heritage in international investment law. *Columbia Human Rights Law Review*, 42, 797–889.
Vandevelde, K. J., 2005. A brief history of international investment agreements. *U.C. Davis Journal of International Law & Policy*, 12 (1), 157–194.
VanDuzer, J. A., Simons, P., and Mayeda, G., 2013. *Integrating sustainable development into international investment agreements*. London: Commonwealth Secretariat.
Wettstein, F., 2012. CSR and the debate on business and human rights: Bridging the great divide. *Business Ethics*, 22 (4), 739–770.

4 Power relationships, institutions, and mining

Comparing Indigenous Peoples' participation in Canada and Brazil

Sabrina Bourgeois[1] *and Ana Catarina Zema*

Introduction

In recent decades, Indigenous Peoples have succeeded in advancing a series of demands, including the right to self-determination and the right to control development in Indigenous territories (Anaya 2004). The literature suggests that the content of these emerging norms and their implementation in institutional practices remain conflicting (Montambeault and Papillon 2022). For states, the recognition of Indigenous rights has included an emphasis on Indigenous Peoples' right to effective and meaningful participation in the consultation process in national laws (Haugen 2016), but for Indigenous Peoples, this has not necessarily translated into greater control over development (O'Faircheallaigh 2012). Their participation and ability to negotiate—and even to benefit economically and socially from mining development—therefore vary from one country to another, and even within the same country. These peoples, who have historically been excluded from development (Anaya 2013), experience the highest levels of poverty today (World Bank 2021), which can influence their ability to benefit from and mitigate the socioeconomic impacts of mining activities.

This chapter explores the challenges of the effective participation of Indigenous Peoples in Canada and Brazil. We consider that the issues raised are very similar in the Americas (Montambeault and Papillon 2022; Leclair *et al.* 2020; Gilberthorpe and Hilson 2014). Focusing on power relations and institutional practices for mining development, we will answer the following questions: In what ways do Indigenous Peoples in Canada and Brazil participate (or not) in mining development? How do Indigenous Peoples "inhabit" complex "institutional environments" and respond to institutional pressures and rationales?

By introducing the inhabited institutions approach (Hallet and Ventresca 2006) in the mining context, we propose a new angle of analysis. According to this perspective, institutions are "inhabited" by individual and organizational actors who demonstrate agency in order to interpret and create a shared understanding of the legal framework and norms they share (Ulmer 2019). In this sense, micropolitics, traditionally conceptualized as a struggle for resources and conflicts of interests, can be understood as "meaning as a battleground" (Hallett 2010, p. 68). According to this perspective, Indigenous Peoples, mining companies, and governments must therefore adapt to a set of formal (e.g., legal framework and policies) and informal

DOI: 10.4324/9781003406433-6
This chapter has been made available under a CC-BY-NC-ND 4.0 license.

rules (e.g., practices, ways of doing things, policies, and norms put forward by actors) that govern their social interactions. Institutions, social interactions, and organizations are thus connected, and together, they create meaning while reacting and adapting to the actions of actors (Hallett and Hawbaker 2021), whether in situations of conflict or cooperation.

The negotiation spaces associated with mining development are understood here as an "inhabited institution" not only by companies and state agents but also by Indigenous Peoples who, through the processes of consultation and negotiation, take part in social interactions. We show that while institutions give some actors disproportionate access to decision-making, Indigenous Peoples are asserting their agency, creating new ways of interpreting their right to participate, and challenging the way the state and corporations define their rights. This is despite the fact that their actions are constrained by an institutional environment.

Drawing on Rodon (2018), we propose a narrow perspective on institutions, that is, formal and informal governance structures created by obligations and agreements negotiated among the Crown, the mining industry, and Indigenous Peoples. Our analysis is therefore limited to the main sites of interaction between these actors, which we define as "institutional fields" in which they negotiate, collaborate, appropriate, or continually contest the meaning of norms and the ways of doing things. We will turn our attention to consultation policies, as well as impact and benefit agreements (IBAs) in Canada. In Brazil, we will focus on consultation protocols, which stem from formal and informal processes and help regulate mining activities and implement Indigenous rights.

This research is based primarily on work done by the researchers between 2019 and 2023 in Canada and Brazil. The main data methods are literature review; semi-structured interviews with governments, Indigenous communities, and the mining industry; and participant observation. The literature review was conducted to document institutional contexts and case studies. Fieldwork (interviews and documentary observation) allowed for a better understanding of the perspectives and interactions of the actors involved, as well as their strategies. The two case studies used are Tata Steel Minerals Canada (TSMC)'s mining project in Canada (Goodwood Pit) and the Belo Sun Mining Project in Brazil.[2]

We will begin with an overview of the inclusion of Indigenous Peoples in the formal and informal frameworks of mining development in Indigenous territories in Canada and Brazil to understand how this "institutional environment" influences the behavior of actors. Then, we will develop two case studies in order to consider the social interactions between the actors involved and their influence on the outcome of negotiations related to mining development. We will conclude with reflections on the implications of these findings.

Overview of Canada and Brazil's institutional contexts: Indigenous rights and mining development

The claims of the First Peoples in Canada and Brazil are part of the international movement for the recognition of the rights of Indigenous Peoples, while both the *Constitution Act of Canada* (1982) and the *Constitution* of *Brazil* (1988) recognize their distinct rights from the rest of their respective populations. In Canada, Indigenous People represent 5% of the Canadian population, with the majority located in urban areas, while in Brazil, they make up about 0.8% of the country's total population. Over the years, these countries have developed a complex system for recognizing and managing the rights of Indigenous Peoples.

In Canada, which has a British legal tradition, it is mainly case law that has defined and elaborated Indigenous rights (Grammond 2003), particularly after the failures of political negotiations (Rodon 2019).[3] The division of land is negotiated through treaties,[4] that is, agreements between one or more Indigenous groups and the Canadian government. Treaties may cover self-government and participation in decisions on natural resource management. By laying the groundwork for joint participation in resource management on treaty lands, these co-management boards act as forums for discussion and negotiation between governments and Indigenous Peoples (White 2020).

In the absence of a treaty, Indigenous participatory rights are limited to the Crown's duty to consult and, if necessary, to accommodate Indigenous People,[5] a duty that varies in nature and intensity depending on the acknowledgment of a credible land claim and the relative seriousness of the project's impact on that claim. If this duty to consult is triggered, even in the absence of consent as an outcome, the state's decision on whether or not to approve a project varies depending on the relevant legislation or case law (Haugen 2016).

In Brazil, Indigenous rights were defined by the *Constitution Act* of 1988. It introduced important paradigmatic shifts by devoting a chapter to "Indians," thus acknowledging the multiethnic composition of Brazil and officially abandoning the integrationist position of previous legislations. Indigenous Peoples have the right to their lands, and the Brazilian state only "recognizes and guarantees" rights that existed before the creation of the state (Brazil 1988, Art. 231). The demarcation and registry of Indigenous lands serve only to "inform third parties" (Brazil 1988, Art. 231).

Article 231 of the Federal Constitution recognizes the autonomy of Indigenous Peoples and their original rights to their lands and guarantees their right to "exclusive usufruct"[6] of the riches of the soil, rivers, and lakes therein (paragraph 2, Article 231). In the full exercise of this autonomy, they have the right to decide on the development model they wish to follow for the future.

With regard to the exploitation of water resources and mineral wealth, the third paragraph of Article 231 provides that the exploitation of water resources, including energy potential and the exploitation of mineral resources on Indigenous lands, may be carried out only with the authorization of the National Congress, after hearing the affected communities. Their share of the benefits from extraction must also be ensured (Brazil 1988). However, while Indigenous Peoples have the exclusive right to use soil resources, the subsoil, mineral resources, and water resources continue to belong to the Union (Article 176 of the Federal Constitution), and the Congress may authorize, in exceptional cases, the exploitation of these natural resources "in the national interest" (Article 49). Such authorization is only possible if it does not violate Article 231. Thus, Indigenous Peoples' rights to land; exclusive usufruct of lands, rivers, and lakes; and the right to their social organization and culture are "constitutional limits to the exploitation of water or mineral resources" in/on Indigenous lands (Zema 2014, p. 271).

The emergence of the international regime of Indigenous rights is having an increasing impact on the governance of state resources. In 2002, Brazil ratified the International Labor Organization (ILO) Convention 169 on Indigenous and Tribal Peoples of 1989, then signed the *United Nations Declaration on the Rights of Indigenous Peoples* (UNDRIP) in 2007 and the *American Declaration on the Rights of Indigenous Peoples of the Organization of American States* (OES) in 2017. Despite the country's adherence to these international instruments on the human rights of Indigenous Peoples and the recognition of the rights of Indigenous Peoples contained in the Brazilian Constitution, the right of Indigenous Peoples to be consulted and to give their free, prior, and informed consent (FPIC) remains, to date, weakly institutionalized and is often contested "due to the competing interpretations put forward by the different actors (Indigenous, states, and enterprises)" in Brazil (Montambeault *et al.* 2019, p. 38). Canada has supported the UNDRIP "without qualification" since 2016, subsequently undertaking to implement it in 2021.[7] While discussions continue regarding its full implementation, the duty to consult with Indigenous Peoples remains the closest thing to FPIC (Boutilier 2017).

In recent years, innovative practices have emerged such as consultation protocols proposed by Indigenous groups and agreements between Indigenous Peoples and mining companies (Papillon and Rodon 2019; O'Faircheallaigh 2013; Povo Juruna 2017). While they have the potential to mitigate the social, environmental, and cultural impacts of mining projects and ensure that local communities benefit from these projects, their actual contributions to community development depend on many factors and are not guaranteed (Sosa and Keenan 2001).

Lately, Indigenous Peoples in Brazil have faced a highly adverse context of violence, invasion of their territories, and the denial and destruction of their constitutional rights (APIB 2021). Former president Jair Bolsonaro declared that he would end Indigenous Peoples' rights to their lands, and in a move to promote the opening of Indigenous lands to mining, his government introduced Bill 191 in 2020 to regulate the exploitation of mineral resources on Indigenous lands. This bill has been deemed unconstitutional by experts in Indigenous Peoples' law (MPF 2020). However, the political might of mining companies is visible in the Brazilian Congress, and the co-optation of Brazilian institutions by the mining industry makes it difficult for Indigenous Peoples to resist mining on their lands and have their rights enforced.

For its part, the Canadian government is committed to reconciliation with the Indigenous Peoples. In the mining sector, companies are now negotiating IBAs to address uncertainties around the rights of Indigenous Peoples. These private agreements, negotiated between a mining company and an Indigenous community, deal with things like royalties, local employment targets and contracts, and the creation of committees. In exchange for their support for the project, IBAs have the potential to allow Indigenous Peoples to negotiate benefits and mitigation measures and to participate in decision-making processes related to their lands (Fidler and Hitch 2007; Bradshaw and Wright 2013; O'Faircheallaigh 2013). There were over 400 active agreements between mining companies and Indigenous nations in 2020 (Natural Resources Canada 2020). We drew on two case studies to analyze how Indigenous Peoples "inhabit" complex institutional environments and respond to institutional pressures and rationales in Canada and Brazil.

Case study: TSMC's operations in the mining region of Schefferville

Overview of TSMC's operations in Canada

Incorporated in 2010, TSMC is a partnership between Tata Steel Ltd. (82%), an Indian steel producer, and the government of Québec (18%).[8] TSMC's Direct Shipping Ore (DSO) project is located in the Labrador Trough, a historic iron ore production region in the northern part of the provinces of Québec and Newfoundland and Labrador.[9]

The DSO pits straddle the provincial border across the Québec–Labrador Peninsula in the vicinity of the mining town of Schefferville. The company began mining the Goodwood deposit in 2017. On the Labrador side, TSMC owns several pits and facilities that began production in 2013. However, we will focus on the case of Québec (Thériault *et al.* 2021).[10]

On the Québec side of the border, the deposits are located approximately 45 km from three communities: The mining town of Schefferville, the Innu community of Matimekush-Lac John, and the Naskapi Nation of Kawawachikamach. As the project is in Nunavik, above the 55th parallel, the Inuit of Nunavik also have rights to the territory under the *James Bay and Northern Québec Agreement* (JBNQA), a comprehensive land claims agreement. By focusing on the access of these communities to formal decision-making processes and their agency outside of these frameworks, this case study will illustrate the ways in which they inhabit institutions to strengthen their agency.

Indigenous participation in formal decision-making processes

In Canada, all natural resource development projects are subject to environmental and social assessments, which ultimately allow the proponent to obtain a certificate of authorization from the appropriate government authorities. Since it is located north of the 55th parallel and is under Québec provincial jurisdiction, the project was submitted to the authority of the Kativik Environmental Quality Commission (KEQC), a co-management committee between the Inuit of Nunavik and the government of Québec that assesses development projects under provincial jurisdiction, in accordance with the provisions of the JBNQA (Section 23). The committee guarantees the participation of Nunavik Inuit in the assessment of projects located on their territory through its mixed composition of members appointed by the government of Québec and the Kativik Regional Government (KRG), who have the authority to approve projects.

The Naskapi Nation of Kawawachikamach, although a signatory to the *Northeastern Québec Agreement* (NEQA), a modern treaty, was not represented on the committee. Under the NEQA, the Naskapi have the potential to be involved in the assessment of projects only in a specific territory, that is, the region of Moinier. This is a situation criticized by the Naskapi, who believe that the JBNQA and NEQA should be amended so that

> the Nation's unique perspective and particular interests are taken into account throughout the consultations in general, and in environmental assessment processes in particular. As things stand, the JBNQA and NEQA create inequality for the Naskapis. There is no reason why the Naskapis should be left out of the decision-making processes affecting the Naskapi Sector and Naskapi traditional territory.
>
> (Kawawachikamach 2017, p. 12)

As a result, while TSMC's activities are in close proximity to the Naskapi, their participation was limited to a consultation session held by KEQC on 12 April 2012 in Kawawachikamach.

As for the Innu of Matimekush-Lac John, the community closest to the TSMC sites, their rights have been extinguished by the JBNQA. They were considered an "interest group" during the consultation session in Kawawachikamach, and although they were invited to participate in the consultations, they did not respond to the invitation (KEQC 2012, p. 11).

The Goodwood Project was subsequently approved by KEQC in August 2012. While KEQC's assessment highlights the proximity of Kawawachikamach and Matimekush-Lac John, as well as the potential impacts of the project, the communities were excluded from the decision-making process. As a result, their rights were limited to consultation.

Norms appropriation and avenues for strengthened participation

In the Canadian context, environmental and social assessment procedures are not the only forms of participation available to Indigenous Peoples, who can mobilize a broad repertoire of strategies to engage or resist mining (Conde 2017), both within and outside formal participation processes (Thériault *et al.* 2021).

One of the main ways for Indigenous Peoples in Canada to exercise their power in mineral development is through the negotiation of IBAs with the industry. The communities closest to the projects, in Québec and Labrador, have all negotiated an agreement with TSMC.

On the Québec side, the Naskapi Nation were the first Indigenous group to reach an agreement, in June 2010. According to an elected official from the Naskapi Nation of Kawawachikamach, the agreement was, however, not put to a public referendum and was not respected by TSMC. Ongoing issues related to its implementation (e.g., royalties and employment) have also prompted the Nation, some 10 years later, to build its capacities in the area of mining development. By developing its constitution and a mining policy, and by collaborating with neighboring nations, the Naskapi Nation is now exploring a number of ways to assert its sovereignty over its traditional territory, including by helping foreign mining companies to understand its values and Indigenous rights.

At the same time, in conjunction with the Innu Strategic Alliance,[11] from June to September 2010, the Innu of Matimekush-Lac John erected barricades to limit access to the TSMC and Labrador Iron Mine (LIM), both located near Schefferville, to "stop the new mines from opening" (Cassell 2013). By asserting their "existing Innu traditional judicial system" on the land (Schertow 2010), the Innu sought to address several objectives, including opposing the quota of jobs reserved for residents of Newfoundland and Labrador and modifying the site in order to maintain their traditional activities. For Chief Réal McKenzie of Matimekush-Lac John, these issues are intrinsically linked to his people's sovereignty over their territory (Nachet *et al.* 2022). These barricades pushed the province to reduce its quotas and also led the federal government to fund IBA negotiations between the community and TSMC.

Matimekush-Lac John signed an agreement in June 2011, while the Innu community of Uashat Mak Mani-Utenam, whose families also have traditional territories near Schefferville, signed an agreement in February 2012. Their agreements included the creation of a health, safety, and environment committee that allows the Innu to be included in TSMC's discussions on these issues. Finally, the Nunavik Inuit argued that the project "would not be acceptable until an impact and benefit agreement is signed between the project proponent and Inuit" during the public consultations held in Kuujjuaq (KEQC 2012). This is all the more important because, as an employee of Makivik Corporation pointed out, there is not much economic development in the North apart from mining development, which makes it an important lever for supporting Inuit territorial governance. However, TSMC preferred to sign contracts with Inuit companies since they considered that the impacts of the project "did not justify the signing of an agreement with Inuit political authorities" (KEQC 2012, p. 11).

This case study demonstrates that in Canada, Indigenous Peoples can assert their concerns and interests in relation to mining development in several institutional fields. By sitting on the KEQC, the Nunavik Inuit took part in the decision to approve the mining project and, together with Québec, issued the conditions of the certificate of authorization that included their participation in several monitoring committees. The Naskapi and Innu, meanwhile, have asserted their sovereignty over their respective territories by putting up blockades or negotiating IBAs. For the companies, the decision to negotiate an IBA is motivated by the direct and tangible impacts of their projects on the surrounding Indigenous communities rather than by the formal recognition of rights by the state—established by the JBNQA in the Schefferville region and contested by the Naskapi and Innu. The perspective of inhabited institutions allows us to highlight the ways in which Indigenous Peoples have deployed strategies to make their voices heard and to communicate their view of how their territory should be developed.

Case study: Belo Sun Project and the Juruna Protocol

Belo Sun Mining Corp.'s Volta Grande Project has been under development since 2012 in the municipality of Senador José Porfírio in the state of Pará, in the Brazilian Amazon. The proponent plans to mine 73.7 tons of gold at this site during the 12-year mining phase (Belo Sun 2021), giving it the potential to become the largest open-pit gold mining company in Brazil. Located in the Amazon, its facilities would be within one of the world's most important biodiversity sites, one that is already facing the impacts of the operation of the Belo Monte hydroelectric dam complex (AIDA 2021).

The Yudjá (Juruna), Arara, and Xikrin Indigenous Peoples, who have already been severely affected by the impacts of the construction of the Belo Monte dam, fear the new impacts that the Belo Sun Mining Project could create. For many years, they fought the construction of the dam. At the root of many of the problems associated with the construction of Belo Monte is the total absence of a process of free, prior, and informed consultation. While the government has never denied its obligation to consult with affected Indigenous Peoples, it has limited dialog processes with them to simple information-sharing meetings that took place only after the hydropower plant was approved (Garzon 2019).

Since Belo Sun's arrival in the area, Indigenous Peoples and local communities have been worried about the new impacts the mining project could have on an ecosystem that has already been severely disrupted by the Belo Monte dam. Several civil society organizations[12] have joined in denouncing the socioenvironmental non-feasibility of the project. These organizations, local communities, and the Brazilian authorities responsible for the protection of human rights and environmental legislation have produced a series of reports[13] demonstrating that "Belo Sun has not fulfilled a number of legal obligations as part of the environmental licensing process" (AIDA 2021, p. 2). Their analysis of the social and environmental risks is considered insufficient in addition to the "non-respect of the right to consultation and free, prior, and informed consent (FPIC) of Indigenous Peoples and other traditional populations" (AIDA 2021, p. 2). Indeed, shortly after the presentation of the mining project in 2012, the Federal Prosecutor's Office (MPF)[14] of Pará found inconsistencies in Belo Sun's environmental impact report. In its environmental impact assessments, the mining company completely ignored the vast presence of Indigenous Peoples in the area.

A legal battle ensued. Up until 2021, there were seven legal actions brought by the MPF and the Public Prosecutor's Office of the State of Pará against Belo Sun that call into question not only the direct and indirect impacts on Indigenous lands in the region but also the acquisition of federal lands by the mining company in the settlements of the National Institute of Colonization and Agrarian Reform (INCRA), the eviction of residents from their homes, and the degradation of the environment (Angelo 2020). The seven actions "require the suspension or total cancellation of the licensing process" and a significant proportion of court rulings have been unfavorable to Belo Sun's projects (AIDA 2021, p. 4).

The first court decision against the project dates to 2014. It suspended the preliminary license (LP) obtained by Belo Sun until the company submitted an Indigenous Impact Assessment (Estudo do Componente Indígena—ECI in Portuguese). Belo Sun has, historically, attempted to deny the impacts of the project on these communities and ignored the consultation process mandated by the Brazilian Constitution and ILO Convention 169 (AIDA 2021, p. 4). The company has argued that the Indigenous lands were more than 10 km away from the mining project. This distance criterion, established by Ministerial Agreement No. 60 of 24 March 2015,[15] is often used to avoid having to consult with Indigenous Peoples in the case of development projects near their lands. In the opinion of the Amazon Cooperation Network (RCA), this distance criterion applies to environmental impact studies and not to consultation processes, which "should be subject to international standards on Indigenous rights, i.e., to the criteria established by ILO Convention 169" (Montambeault *et al.* 2019, p. 40).

Belo Sun filed an appeal and, in February 2017, was granted an installation license (IL) that was suspended 2 months later. An initial decision in April 2017 suspended the IL "until free and informed consultation with the Indigenous Peoples has taken place" (AIDA 2021, p. 4). The suspension was confirmed by a new decision in December 2017, ordering the company to consult with and respect the protocols of the affected Indigenous Peoples. "This decision remains valid today" (AIDA 2021, p. 6). Belo Sun's defeat in 2017, when the court reaffirmed the obligation to carry out studies on the socioenvironmental impacts on the Indigenous Peoples concerned, remains the most significant legal victory in terms of the recognition of the value of FPIC and autonomous consultation protocols in Brazil (Garzon 2019). This victory stems from the Juruna Protocol.

The Protocol of the Yudjá (Juruna) People, based on ILO Convention 169, was written in order to "inform the government of the appropriate way to engage with [them] about decisions that affect [them], [their] territory, and [their] rights" (Povo Juruna 2017, p. 13). In the protocol, the Yudjá describe themselves as the masters of the Xingu River and declare that they will not accept any project that would take them away from the river or make it impossible for communities to remain on the river. "We know that we have the right to be consulted, to defend our lands and traditions, to fight for dignified living conditions, and to choose our development priorities. Neither the government nor any company can deny these rights" (Povo Juruna 2017, p. 15).

They set out the principles that must guide the entire consultation process:

> **Respect**. Respect our rules, our customs, and our time. (. . .) **Transparency.** So we all know what's going on. (. . .) **Good faith and honesty**. So that we have confidence in the process of dialogue and the drafting of agreements. (. . .) **Without physical or moral pressure**. We will not accept the presence of private security guards or police forces that want to intimidate our people. Nor do we accept attempts to make deals with leaders or individuals in exchange for favors or goods.
>
> (Povo Juruna 2017, pp. 22–24)

These principles are reflected in "very specific rules within the protocol," such as "adherence to the calendar of traditional activities" before meeting dates are set, "the obligation to record and publish all consultation meetings," and the inclusion of an "independent technical opinion" (Garzon 2019, p. 39).

The Juruna Protocol is divided into three parts. The first part describes who the Juruna are, their territory, the current sociopolitical context, and the reasons why they decided to publish their own consultation protocol. The second part describes the rules that determine "what should be submitted for consultation, when, how, with whom, and why" (Garzon 2019, p. 40).

The third and final part reiterates the Indigenous rights set out in the 1988 Federal Constitution, and "the articles of ILO Convention 169 and the *United Nations Declaration on the Rights of Indigenous Peoples* that refer to the right to consultation and consent" (Garzon 2019, p. 40).

A fundamental rule of the Juruna Protocol is the "voluntary nature of their participation in a consultation process," as well as the notion that the Juruna "do not recognize any obligation to take part in a consultation that is solely in the interest of the government" (Garzon 2019, p. 41). Exhausted from having participated in meetings in which "their views are not taken into account," the Yudjá make it clear in their protocol "that they do not want to waste any more time" (Garzon 2019, p. 41).

At the end of the protocol, the Yudjá describe the "consultation plan" as "a document that contains, in detail, the activities, time, and resources needed to discuss the information necessary for the consultation process" and emphasize that for the development of this plan, they can rely on technical and legal advice (Povo Juruna 2017, p. 32).

The protocol of the Yudjá People, understood as an "inhabited institution," shows how, thanks to their agency, they have been able to bring about changes in the institutional rationale and power relations, managing with this protocol to enforce their interpretation of FPIC and thereby guaranteeing their participation in the decision-making process for the implementation of the Volta Grande Project.

Conclusion

In recent years, a growing number of scholars have turned their attention to the results of negotiations on mining development on Indigenous territory in the Americas. In examining case studies from Canada and Brazil, our main concern was to gain a clearer understanding of how Indigenous Peoples inhabit the institutional context to advance their concerns. These interactions with the state and mining companies occur in complex, formal and informal environments that define the rules of the game differently in Canada and Brazil. As such, the participation of Indigenous Peoples can be viewed as a "principle that distributes power in a more balanced way" (Ramírez 2013, pp. 255–256) by organizing or influencing interactions between actors—which implies the need to think of participation as an aspect that plays a fundamental role in power relations.

Case studies indicate that Indigenous participation in mining development still faces a number of challenges, despite the apparent willingness of the industry and the international community to strengthen Indigenous inclusion (Owen and Kemp 2014; Anaya 2004). The state plays an important role in the control of resources since it defines the "terms on which resources will be accessed, produced, transported, and marketed" (Howlett 2010, p. 100) by essentially constructing the formal institutional framework. The imbalance of power relations may appear when there is control of information on the part of companies or on the side of state agents when there is a predisposition to favor mining projects without considering the will of the Indigenous People or by denying them the right to FPIC.

By engaging in a process of norm-setting through practice, Indigenous Peoples challenge state-centered views and reaffirm their status as self-determined peoples with the capacity—and legitimacy—to make decisions about their traditional territories (Papillon and Rodon 2016). The cases of Canada and Brazil demonstrate the efforts of Indigenous Peoples to implement their right to self-determination: For example, the Innu when they decide to put up blockades and then negotiate directly with the company, and the Juruna with their protocol, which is a vivid manifestation of self-determination and unity within the community. In other words, when Indigenous Peoples are excluded from formal decision-making procedures, they can adopt several strategies in order to make their voices heard.

Our analysis has helped understand the circumstances under which Indigenous Peoples consent to or oppose extraction (Conde and Billon 2017; Willow 2020). While Indigenous agency is limited by existing legal, political, and economic structures, these structures also determine which strategies (co-management, lawsuits, protocols, blockades, agreements, etc.) can or cannot be mobilized (Leclair *et al.* 2020; Conde 2017; Howlett 2010; Thériault *et al.* 2021). In Brazil, for example, the difficulties the Yudjá People have faced in the implementation of Indigenous rights, as well as the violence and the repeated decisions by governments to exclude them from decision-making processes, have thrust them into a dynamic of contestation against the project. In Canada, environmental and social assessment processes ensure Indigenous consultation and, in the presence of a co-management committee, increased participation in decision-making. This dynamic reinforces the rationale for negotiation and benefit-sharing, which can also be mobilized outside of formal processes via the negotiation of IBAs, for example.

To conclude, mining development has proven to be a relevant topic of study to contribute to our understanding of inhabited institutions (Hallett 2010; Ulmer 2019). Mining projects and the negotiations associated with them are institutional fields in which the interactions among Indigenous Peoples, companies, and government—"the beating heart of institutions" (Hallett and Ventresca 2006, p. 2215)—seek to impose their perspectives and interests. This approach allowed us to identify negotiating spaces and to highlight the way in which the actors give meaning to the institutional rationales and rules associated with mining development. Future research will allow us to gain a better understanding of the ways in which Indigenous forms of participation are helping broaden and create new spaces for negotiation based on Indigenous positions on mining development.

Acknowledgments

The authors would like to thank the anonymous reviewer for taking the time and effort to review the manuscript, as well as editors Thierry Rodon, Sophie Thériault, Séverine Bouard, Arn Keeling, and Andrew Taylor for their support. This study was made possible thanks to the anonymous respondents and our Indigenous partners, the Innu community of Matimekush-Lac John, the Naskapi Nation of Kawawachikamach, and the Yudjá (Juruna) People. We would like to thank them for their teaching, for welcoming us on their traditional lands, and their participation in the authors' research. We would also like to thank the respondents from Inuit organizations, the mining industry, and the governments who shared their insights with the authors.

Notes

Corresponding author. The chapter was written mostly between 2021 and 2023. Since then, this author has been employed by the government of Canada. To avoid any misrepresentation, we would like to emphasize that the views and opinions expressed in this article are those of the authors and do not reflect the official policy or position of any Canadian federal departments.

1 The empirical research of the Canadian case is based on the researcher's doctoral research, and her analysis is partly based on a previous scientific article (Thériault *et al.* 2021).

2 In recent decades, the *Constitution Act of Canada 1982*, which recognizes "the existing Aboriginal and treaty rights of the Aboriginal peoples of Canada" (s. 35.1), has contributed to a veritable "legal revolution."

3 It is the treaties between the Crown and Indigenous groups that clarify the management and sharing of land in Canada. They are indispensable tools for understanding Canada's relationship with Indigenous peoples. Modern treaties, also known as comprehensive land claim agreements, follow the recognition of Indigenous rights by the Supreme Court of Canada and the subsequent development of the Comprehensive Land Claims Policy (Eyford 2015). They encourage the negotiation of very detailed agreements that always contain clauses for the relinquishment of land rights but may also deal with Indigenous governance institutions, self-government, participation in resource management decisions, etc.

4 Three specific cases have contributed to setting new requirements for Indigenous communities with regard to the consultation and participation of these communities in the management of land and natural resources. *Haida Nation v. British Columbia* (Minister of Forests; SCC 2004, p. 73), *Taku River Tlingit First Nation v. British Columbia* (Project Assessment Director; SCC 2004, p. 74), and *Mikisew Cree First Nation v. Canada* (Minister of Canadian Heritage; SCC 2005: 69) form what has been dubbed "the Supreme Court of Canada's initial trilogy" (Newman 2014).

5 The term "exclusive usufruct" refers to the exclusive right of the Indigenous community to use all the riches of the soil, but the subsoil, mineral, and water resources belong to the Union (Art. 176 of the FC), and Congress may authorize the exploitation of these natural resources in the national interest (Art. 49 of the FC). It should be noted that this authorization is only possible if it does not violate Article 231. Indigenous rights to land, to exclusive usufruct of the riches of soil, rivers, and lakes represent constitutional limits to the exploitation of water or mines (ZEMA 2014, p. 270).

6 In 2016, the government of Canada announced its full support for UNDRIP. As a result, the federal government introduced Bill C-15, the *United Nations Declaration on the*

Rights of Indigenous Peoples Act, in December 2020, which received royal assent and came into force on 21 June 2021.

7 TSMC is an indirect subsidiary of Tata Steel Ltd. TSMC initially entered into a joint venture with New Millennium Capital Corp. for the development of the DSO Project in Canada. In January 2016, Tata Steel and the government of Quebec also signed an agreement in principle at the Davos Economic Forum, which led to a contribution of $175 million from Quebec to support the DSO Project (Lévesque 2016 in Perrault 2016). Québec's financial contribution in 2016, through Investissement Québec's Capital Mines Hydrocarbures fund, also made it possible to acquire an equity stake. In 2020, New Millennium sold its shares to TSMC when it became Abbax.

8 See Boutet (2015) for more information on the mining heritage of the Schefferville region.

9 See Thériault *et al.* (2021) for more details on the overlap of traditional Indigenous territories and the Quebec–Labrador border in the context of mining development.

10 The Innu Strategic Alliance is a coalition of five Innu communities (Matimekush-Lac John, Uashat Mak Mani-Utenam, Ekuanitshit, Pessamit, and Unamen Shipu) formed in 2010 to defend their common rights and interests.

11 Such as the Socio-Environmental Institute (ISA), the Xingu+ Network, the Xingu Forever Alive Movement, International Rivers, Above Ground, MiningWatch Canada, and the Interamerican Association for Environmental Defense (AIDA).

12 One of the most important studies is Steven Emerman's study, Evaluation of the Tailings Dam, Cyanide Use and Water Consumption at the Proposed Volta Grande Gold Project, Pará, Northern Brazil. This study showed that the tailings dam of the Volta Grande Project was not designed according to seismic safety criteria, which is a violation of Brazilian regulations on tailings dams. Belo Sun neither included in its studies an analysis of local seismicity nor simulated the response of the structure to a hypothetical seismic acceleration. Neither did it present a risk analysis of the geological faults mapped in the vicinity of the project site. The study shows that seismic activity at the mine could lead to the failure of the tailings dam. See Emerman (2020) for more details.

13 The MPF is an autonomous agency responsible for the defense of inalienable social and individual rights and for the prosecution of criminal offenses at the federal level, as well as of civil and administrative wrongdoing related to the federal government and corresponding public interest.

14 This is a ministerial agreement that establishes the administrative procedures that govern the actions of the National Foundation of Indigenous Peoples (FUNAI), the Palmares Cultural Foundation (FCP), the National Institute of Historical and Artistic Heritage (IPHAN), and the Ministry of Health with regard to environmental authorization under the Brazilian Institute of the Environment and Renewable Natural Resources (IBAMA).

15 These dynamics can nevertheless limit the agency of Indigenous nations when they oppose extractive projects that are considered to be at odds with their self-determination and/or land uses. See Isaac (2022) regarding the engagement of the Wet'suwet'en Nation.

References

Anaya, J., 2013. *Report of the Special Rapporteur on the rights of Indigenous Peoples, James Anaya—Extractive industries and Indigenous Peoples*. New York: Human Rights Council.

Anaya, J. S., 2004. *Indigenous peoples in international law*. New York: Oxford University Press.

Angelo, M., 2020. Batalha judicial tenta impedir mineradora canadense Belo Sun de explorar ouro em terras indígenas no Brasil. *Mongabay*.

Articulação dos Povos Indígenas do Brasil (APIB), 2021. *Luta pela Vida* [Online]. Brasília.

Asociación Interamericana para la Defensa del Ambiente (AIDA), 2021. *Investor Alert: Belo Sun discloses misleading information to investors regarding controversial gold mining project in the Brazilian Amazon* [Online]. Mexico City.

Belo Sun Mining Corp., 2021. *Annual information form for the year ended* [Online].

Boutet, J.-S., 2015. The revival of Québec's iron ore industry: Perspectives on mining, development, and history. *In:* A. Keeling and J. Sandlos, eds., *Mining and communities in Northern Canada: History, politics and memory*. Calgary: University of Calgary Press.

Boutilier, S., 2017. Free, prior, and informed consent and reconciliation in Canada: Proposals to implement articles 19 and 32 of the UN Declaration on the Rights of Indigenous Peoples. *Western Journal of Legal Studies*, 7, 1–21.

Bradshaw, B. and Wright, A., 2013. *Review of IBA literature and analysis of gaps in knowledge: Resources and sustainable development in the Arctic*. Whitehorse: Yukon College. Plymouth: Lakeland University.

Cassell, E., 2013. Some reflections on hydroelectric development and the land rights of the Innu who live in Quebec and those who live in Labrador. *Journal of Human Rights in the Commonwealth*, 1, 3–17.

Conde, M., 2017. Resistance to mining: A review. *Ecological Economics*, 132, 80–90.

Conde, M. and Billon, P. L., 2017. Why do some communities resist mining projects while others do not? *The Extractive Industries and Society*, 4, 681–697.

Emerman, S., 2020. *Evaluation of the tailings dam, cyanide use and water consumption at the proposed Volta Grande Gold Project, Pará, Northern Brazil*. Report written at the request of Amazon Watch, submitted June 1, 2020 [Online].

Eyford, D. R., 2015. A new direction: Advancing Indigenous and treaty rights. *An independent report by the Ministerial Special Representative on Renewing the Comprehensive Land Claims Policy*. Ottawa: Government of Canada.

Fidler, C. and Hitch, M., 2007. Impact and benefit agreements: A contentious issue for environmental and Aboriginal justice. *Environments Journal Volume*, 35, 49–69.

Garzon, B., 2019. The Juruna (Yudjá) People's protocol: A response to a hard-learned lesson. *In:* C. Doyle, A. Whitmore, and H. Tugendhat, eds., *Free prior informed consent protocols as instruments of autonomy: Laying foundations for rights-based engagement*. Köln: Institut für Ökologie und Aktions-Ethnologie (INFOE)/(ENIP).

Gilberthorpe, E. and Hilson, G., 2014. *Natural resource extraction and Indigenous livelihoods: Development challenges in an era of globalization*. Surrey: Ashgate.

Grammond, S., 2003. *Terms of coexistence: Indigenous Peoples and Canadian law*. Toronto: Carswell.

Haida Nation v. British Columbia (Minister of Forests), [2004] 3 S.C.R. 511, 2004 SCC 73.

Hallett, T., 2010. The myth incarnate. *American Sociological Review*, 75, 52–74.

Hallett, T. and Hawbaker, A., 2021. The case for an inhabited institutionalism in organizational research: Interaction, coupling, and change reconsidered. *Theory and Society*, 50, 1–32.

Hallett, T. and Ventresca, M. J., 2006. Inhabited institutions: Social interactions and organizational forms in Gouldner's patterns of industrial bureaucracy. *Theory and Society*, 35, 213–236.

Haugen, H. M., 2016. The right to veto or emphasising adequate decision-making processes? Clarifying the scope of the free, prior and informed consent (FPIC) requirement. *Netherlands Quarterly of Human Rights*, 34, 250–273.

Howlett, C., 2010. Indigenous agency and mining development: A cautious note. *Studies in Political Economy*, 85, 99–123.

Isaac, A., 2022. Understanding conflicting legal traditions: The Wet'suwet'en land struggle and resolving land claims on unceded territory. *Political Science Undergraduate Review*, 7, 25–31.

Kativik Environmental Quality Commission (KEQC), 2012. *Decision regarding the Direct Shipping Iron Ore Project, Project 2A, by Tata Steel Minerals Canada Limited*. Kuujuaq.

Kawawachikamach, 2017. *Position paper on specific claims and comprehensive land claims agreement*. Ottawa: Standing Committee on Indigenous and Northern Affairs.

Leclair, J., Papillon, M., and Forget, H., 2020. Indigenous consultation protocols in Canada: A model of convergence of Indigenous and state legal systems? *Recherches amérindiennes au Québec*, 49, 25–36.

Mikisew Cree First Nation v. Canada (Minister of Canadian Heritage), [2005] 3 S.C.R. 388, 2005 SCC 69.

Ministério Público Federal, Nota Técnica n. 04/2020/6aCC\MPF, 6a Câmara de Coordenação e Revisão do MPF (Populações Indígenas e Comunidades Tradicionais), 9 June 2020. Available from: www.mpf.mp.br/pgr/documentos/6ccr_notapl_mineracao.pdf.

Montambeault, F., Da Silva, L. A. L., and Monteiro Joca, P., 2019. FPIC and Indigenous consultation protocols in Brazil: Comparative trajectories. *Recherches amérindiennes au Québec*, 49 (2), 37–49.

Montambeault, F. and Papillon, M., 2022. Repoliticising Indigenous participation: FPIC protocols in Canada and Brazil. *The International Journal of Human Rights*, 27 (2), 335–358.

Nachet, L., Beckett, C., and MacNeil, K. S., 2022. Framing extractive violence as environmental (in)justice: A cross-perspective from Indigenous lands in Canada and Sweden. *The Extractive Industries and Society*, 12, 1–9.

Natural Resources Canada, 2020. Lands and minerals sector—Indigenous mining agreements. Available from: https://atlas.gc.ca/imaema/en/index.html.

Newman, D. G., 2014. *Revisiting the duty to consult Aboriginal peoples*. Saskatoon: Purich Publishing Limited.

O'Faircheallaigh, C., 2012. International recognition of Indigenous rights, Indigenous control of development and domestic political mobilisation. *Australian Journal of Political Science*, 47, 531–545.

O'Faircheallaigh, C., 2013. Community development agreements in the mining industry: An emerging global phenomenon. *Community Development*, 44, 222–238.

Owen, J. R. and Kemp, D., 2014. 'Free prior and informed consent', social complexity and the mining industry: Establishing a knowledge base. *Resources Policy*, 91–100.

Papillon, M. and Rodon, T., 2016. Proponent-Indigenous agreements and the implementation of the right to free, prior, and informed consent in Canada. *Environmental Impact Assessment Review*, 62, 216–224.

Papillon, M. and Rodon, T., 2019. The transformative potential of Indigenous-driven approaches to implementing free, prior and informed consent: Lessons from two Canadian cases. *International Journal on Minority and Group Rights*, 27, 1–22.

Perrault, M., 2016. Evolution of the balance of power between the Innu and Naskapi Nations in the negotiation process for mining development in Schefferville: From the Iron Ore Company to Tata Steel Minerals Canada. *Cahiers Dialogue*, 1, 88–91.

Povo Juruna, 2017. *Protocolo de Consulta Juruna (Yudjá) da Terra Indígena Paquiçamba da Volta Grande do Rio Xingu*. Brasil: Povo Juruna da TI Paquiçamba.

Ramírez, A. H., 2013. Participación indígena: desarrollo y alcances en torno a la participación ambiental. *Revista Ius et Praxis*, 19 (2), 251–300.

Rodon, T., 2018. Institutional development and resource development: The case of Canada's Indigenous Peoples. *Canadian Journal of Development Studies*, 39, 119–136.

Rodon, T., 2019. *Les apories des politiques autochtones au Canada*. Québec: Presses de l'Université du Québec.

Schertow, J. A., June 14, 2010. Innu block access to mining projects on their territory. *Intercontinental Cry*. Available from: https://intercontinentalcry.org/innu-block-access-to-mining-projects-on-their-territory/

Sosa, I. and Keenan, K., 2001. *Impact benefit agreements between Aboriginal communities and mining companies: Their use in Canada*. Environmental Mining Council of British Columbia Canadian Environmental Law Association. CooperAcción: Acción Solidaria para el Desarrollo.

Taku River Tlingit First Nation v. British Columbia (Project Assessment Director), [2004] 3 S.C.R. 550, 2004 SCC 74.

Thériault, S., Bourgeois, S., and Boirin-Fargues, Z., 2021. Indigenous Peoples' agency within and beyond rights in the mining context: The case of the Schefferville region. *The Extractive Industries and Society*, 12, 1–9.

Ulmer, J. T., 2019. Criminal courts as inhabited institutions: Making sense of difference and similarity in sentencing. *Crime and Justice*, 48, 483–522.

White, G., 2020. *Indigenous empowerment through co-management: Land claims boards, wildlife management and environmental regulation*. Vancouver: UBC Press.

Willow, A. J., 2020. Embrace it, accept it, or fight like hell: Understanding diverse responses to extractive industrial development. *Environment, Development and Sustainability*, 22, 7075–7096.

World Bank, 2021. *Indigenous peoples* [Online]. Available from: www.worldbank.org/en/topic/indigenouspeoples [Accessed June 2022].

Zema, Ana C., 2014. *Direitos e Autonomia Indígena no Brasil (1960–2010): uma análise histórica à luz da teoria do Sistema Mundo e do pensamento decolonial*. Thesis (PhD). Brasília University.

Part II

Braiding Indigenous views in the mining cycle

5 Indigenous Peoples' relationships to large-scale mining in post/colonial contexts

Toward multidisciplinary comparative perspectives

Leah S. Horowitz, Arn Keeling, Francis Lévesque, Thierry Rodon, Stephan Schott, and Sophie Thériault

Introduction

Large-scale mining projects inevitably have widespread impacts on local societies and ecologies. Because of their close relationships to the land, water, and resources therein, and their marginalized social and economic positions, Indigenous Peoples living in current or former settler colonies are particularly vulnerable to mining's impacts, yet have the potential to benefit from its opportunities as well. These impacts and opportunities are shaped by the nature of the mineral and the surrounding environment, the approach of the extractive company, relevant regulatory regimes, socioeconomic conditions, and Indigenous communities' responses, among other factors.

This work updates and further develops ideas presented in an earlier publication (Horowitz *et al.* 2018) that reviews the literature, both published and unpublished, on Indigenous Peoples' relationships to large-scale mining in post/colonial contexts. As all authors are members of MinErAL, a Knowledge Network on Mining Encounters and Indigenous Sustainable Livelihoods (www.mineral.ulaval.ca/en), we focus on the countries covered within that project: Australia, Canada, Finland, Greenland, New Caledonia, Norway, and Sweden. These nations all encompass important—often remotely located—Indigenous populations, as well as economically and politically dominant non-Indigenous groups, and possess mineral resources—often located near Indigenous communities—that attract local and multinational mining companies. Several multinational companies have, or have had, projects in many of these countries, simultaneously or serially. Yet, despite the focus on countries' commonalities, important distinctions exist in terms of legal, political, social, economic, and ecological contexts. While a comprehensive review of all the relevant literature for each country is beyond the scope of this study, we offer an outline of some major arguments and debates in order to conduct comparative analysis and identify future research directions. In providing an interdisciplinary, critical analysis of the relationships between mining companies and Indigenous communities that covers a wide range of developed countries, we aim to generate

DOI: 10.4324/9781003406433-8
This chapter has been made available under a CC-BY-NC-ND 4.0 license.

insights into how large-scale mining projects impact Indigenous Peoples in settler colonies, as well as how Indigenous communities respond to such projects.

In the sections below, we begin by examining, in turn, the various impacts of large-scale mining projects—environmental, economic, and social—and how they specifically affect Indigenous communities. Next, we discuss the legal context of these impacts on Indigenous Peoples in settler colonies. Focusing on the countries listed above, we discuss differences in regimes governing Indigenous Peoples' legal rights and the implications of these rights for negotiations over mining projects. Lastly, we explore ways that Indigenous groups themselves have addressed these impacts by engaging with large-scale mining projects, through both negotiation and resistance. We conclude by summarizing our findings as to what gaps remain in the literature and provide thoughts as to how future research could address those gaps.

Methods

MinErAL is a collaboration between Indigenous and non-Indigenous scholars, and Indigenous communities and organizations. Although the authors of the present article are all non-Indigenous researchers, input from Indigenous colleagues has strongly informed our work, facilitated by a panel organized at the International Congress of Arctic Social Sciences in June 2017.

In reviewing the extensive literature on Indigenous Peoples' relationships to large-scale mining in the countries listed above, we consulted primary academic and legal sources, for example, statutes, regulations, and case law, in English and French, mainly through the major academic and legal databases, including Web of Science, LexisNexis/Quicklaw, Westlaw, HeinOnline, and Legal Track. In consulting the gray literature, we developed a modified version of the plan proposed by Godin *et al.* (2015), incorporating three strategies: a) a customized Google search, b) a search of targeted organizations (advocacy groups, NGOs, government agencies, industry websites, etc.), and c) a systematic examination of each report's bibliographies. While academics, industries, governments, and advocacy groups pursue different goals and thus generate data and analyses that are not always easily comparable, our aim was to identify the complete range of topics that had been identified in the various types of literature.

Environmental impacts, environmental legacies, and extractive development

Modern mineral development entails the large-scale "disassembly" of local environments, to separate target minerals from their geological matrix. As these minerals often represent tiny fractions of the total orebody, large amounts of surface materials (overburden), waste rock, and tailings are created as part of the excavation and mineral separation process. The complex composition of particular ore deposits and mining techniques, combined with the diversity of local ecologies, means that the environmental impacts of mining—and the conflicts they generate—are highly specific to any given development (Voulvoulis *et al.* 2013; Lottermoser 2010).

Mining produces often widespread environmental impacts both directly and indirectly related to extractive processes. These impacts may directly and disproportionately affect the land-based livelihoods and traditional activities of nearby Indigenous communities (Owen *et al.* 2023). At the landscape scale, environmental effects include surface disturbance and deforestation from extractive activities and associated infrastructure, including roads, shipping facilities, exploration and drilling sites, waste impoundment facilities, and power dams (Bridge 2004). Although the overall spatial extent of these disturbances may be small, they nevertheless generate a patchy landscape that can significantly affect local wildlife and vegetation patterns, and persist for long periods (LeClerc and Keeling 2015). Recent comparative research in Sweden and Northern Canada has highlighted Sami, Cree, and Naskapi efforts to document negative effects of mineral development and its associated landscape disturbances on reindeer/caribou habitats and associated Indigenous livelihoods (Österlin *et al.* 2023; Herrmann *et al.* 2014).

Underground mines typically have a small spatial extent but may be situated within leased lands or concessions with restricted access that encompass much wider areas. More dramatic are the landscape changes associated with "open cast" mining, including strip mining (mainly for coal) and open-pit mining. Pioneered in the early twentieth century to permit exploitation of lower-grade deposits, this form of "mass destruction" completely removes the orebody, leaving behind huge voids as permanent features on the landscape (LeCain 2009). These techniques are now used at sites as varied as Australian uranium mines, Canadian Arctic diamond mines, New Caledonian nickel mines, and Scandinavian copper mines. As Jean-Sébastien Boutet (2015, p. 183) documented in Schefferville, Québec (Canada), "the disfigured, deeply altered landscape" of open pits "evokes resentful memories of the company's actions" among Indigenous residents.

Perhaps the greatest environmental concerns for adjacent Indigenous communities arise from mining's substantial and complex waste products. The vast quantities of physical wastes (overburden, waste rock, and tailings) produced may be stored at the surface or deposited in nearby water bodies, including lakes, rivers, or ocean inlets (Lottermoser 2010; Sondergaard *et al.* 2011; Hudson-Edwards *et al.* 2011). Mine tailings have the potential to pollute local watercourses through physical erosion, acid drainage, and the mobilization of heavy metals or trace process chemicals, damaging local fisheries and polluting drinking waters (Kirsch 2014; Byrne *et al.* 2012; Horowitz 2010).

Tailings impoundments (dams) constructed to retain and, in some cases, treat tailings effluent are, themselves, massive landscape features subject to leaks and failures, large and small. When constructed in upstream, often mountainous locations, these failures can be catastrophic, resulting in massive flows that threaten downstream communities with inundation, property damage, and death (Roche *et al.* 2017). Ore processing is also a significant source of pollution and toxicity affecting nearby communities. Smelters are notorious polluters, due to the airborne release of toxic smoke and gasses (containing carbon dioxide, mercury, arsenic, and other contaminants) and the production of slag and ash that may be released or deposited to nearby lands or waters (Lottermoser 2010). These contaminants can affect surrounding plant, animal, and human health, impacting local land use.

At Canada's Giant Mine, arsenic deposition from gold roasting not only poisoned members of the Yellowknives Dene First Nation and domestic animals, but also, at closure, left behind a comprehensively contaminated local landscape and massive toxic site (Beckett 2021; Sandlos and Keeling 2016a).

More recently, the legacies of decades of unregulated mining, and the need for long-term monitoring of post-mining landscapes, have increasingly come into focus. There are likely millions of legacy extractive sites around the globe, ranging from small, relatively innocuous mining disturbances to extensive, complex, degraded, and contaminated landscapes that continue to threaten Indigenous health and land use (O'Faircheallaigh and Lawrence 2019; Keeling and Sandlos 2017; Pepper *et al.* 2014; Worrall *et al.* 2009). Only since the 1970s have some governments in the jurisdictions under consideration begun to require comprehensive closure and rehabilitation plans for extractive developments, as well as some financial securities to ensure they are carried out (Keenan and Holcombe 2021; Dance 2015). Mine remediation may, itself, create controversies with Indigenous communities around restoration goals, the residual health risks, and other transformations of the post-mining landscape (Monosky and Keeling 2021; Beckett and Keeling 2019; Cohen 2017; Kivinen *et al.* 2018; Tsosie 2015).

The intensity of disruption to local land-use practices from industrial development varies considerably across these regions and sites, effects to some extent captured in environmental assessments (see Section 6.1) and local activism. In general, however, recent research demonstrates that, for nearby Indigenous communities, the environmental impacts of mining at all stages are experienced as a form of dispossession and environmental injustice (Nachet *et al.* 2022; Perreault 2013; Boutet 2013; Hall 2012; Horowitz 2011; Keeling and Sandlos 2009). As Mirarr senior traditional owner Yvonne Margarula expressed with respect to Australia's Ranger Mine, "Uranium mining has also taken our country away from us and destroyed it. Billabongs and creeks are gone forever, there are hills of poisonous rock and great holes in the ground with poisonous mud where there used to be nothing but bush" (cited in Graetz 2015a, p. 139). While in some cases Indigenous people may embrace, or at least tolerate, the presence of extractive industries for the economic benefits they bring, these impacts on land and health—whether immediate and acute or long-term and insidious—remain potent sources of concern and conflict.

Positive and negative economic impacts

The economic impacts of extractive industries on Indigenous communities are complex. A number of potential benefits include employment growth, human development, income growth and increased demand for goods and services, diversification of the economic base, increased access to funding, improvement in infrastructure, and enhanced social and health services (Petkova *et al.* 2009). In addition, mining is expected to enhance the small and medium enterprise sector and to generate business income (Evans and Sawyer 2009).

However, more attention to specific flows and distributions of benefits at the household and community level, particularly within Indigenous communities, is needed. While earlier pioneering work on Indigenous employment in mines in Northern Canada (e.g., Hobart 1982; Abele 1989; Gibson and Klinck 2005) and Australia (e.g., Cousins and Nieuwenhuysen 1984; O'Faircheallaigh 2002; Barker and Brereton 2004; Barker 2006) have critically examined challenges such as employment rates, turnover rates, and training, most of the more recent studies on the economic impacts of the mining industry are based on regional or national studies (Eggert 2001; Fessehaie and Morris 2013; Rolfe *et al.* 2011; Sandlos and Keeling 2012; Fleming *et al.* 2015; Langton 2010; Hunter *et al.* 2015; Schott *et al.* 2022).

At the regional or national scale, the resource curse theory points to the economic and sociopolitical risks inherent to an excessive dependence on the extraction and export of natural resources such as minerals (Land *et al.* 2015; Manzano and Rigobon 2001; Sachs and Warner 1995, 2001). In addition, Staples theory (Innis 1930a, 1930b; Innis 1940; Mackintosh 1923) offers a more refined analysis of local or regional linkages and how these could create potential economic development paths. Hirschman (1977) distinguishes between backward, forward, final demand, and fiscal linkages. Each of these linkages is part of a theoretical framework that is used to describe the development path of specific extractive industry operations. Huskey and Southcott (2016) found that although Yukon fiscal resource revenues made up only 10% of the total economic rent between 2000 and 2012, there were substantial other linkages, particularly backward linkages in the form of local spending by mining companies. However, the governance, management, and administration of resource revenues are a major challenge, since these are often used to replace the lack of government social investment in Indigenous communities (Rodon *et al.* 2022).

The scant literature on resource revenue distribution and resource curse effects at the local level indicates mixed results. Levitus's (1991, 2005, 2009) work on the Ranger Mine in Australia showed initial success but an eventual decline in benefits, due to disparity in interests and erosion of stable institutions. For similar reasons, uranium mining in Australia failed to generate significant local development and investment (O'Faircheallaigh 2002). Langton and Mazel (2008) show that little socioeconomic improvement has been achieved in Australian Indigenous communities that neighbor mining operations. Indeed, economic benefits from mining often leak out of local—and, particularly, Indigenous—communities. A recent comparative study on the impact of mining on business development in two Inuit regions of Canada (Nunavik and Nunatsiavut) has shown that 70% of all local Inuit-owned businesses derive less than 10% of their overall revenues from mining in an economy with mining as the predominant sector (Belayneh *et al.* 2018a, 2018b). Local businesses needed to partner up with companies from outside the region to be successful. Similar to the study by Huskey and Southcott, Belayneh *et al.* (2018b) found that final demand linkages leaked out to hubs or centers outside of the mining area. The latter resulted in lost fiscal revenues and lost opportunities and multiplier effects for local consumption and business services and operations. Comparable leakages have been shown for Indigenous Australian communities (e.g., Haslam-McKenzie *et al.* 2009).

Indigenous employment at major mining operations ranges between 4.6% at Golden Grove, Western Australia (Barker and Brereton 2004), and 51% at Voisey's Bay, Canada (Vale 2023). The dynamics of skill development, job quality, turnover rates, and outmigration needs to be examined in more detail. Low-income groups—often Indigenous communities in remote, rural areas—may not possess the skills to secure employment in the mining sector, or the number of jobs may be inadequate. Alternatively, some community members may not be interested in such work, which does not fit with their cultural preferences and priorities, as described in Australia (Taylor 2009) and New Caledonia (Horowitz 2010). These groups' income does not typically escalate enough to compensate for the rising cost of living in a region experiencing a mining boom (Sachs and Warner 2001). This particularly affects Indigenous people who rely on affordable housing, as demonstrated in Western Australia and Alberta, Canada (Haslam-McKenzie *et al.* 2009; Langton 2010), forcing them to relocate or become homeless. Furthermore, the rise of fly-in–fly-out programs promotes block shift work, for example, two weeks in, two weeks out, and causes migration to regional hubs. For instance, in Canada, fly-in–fly-out programs and training centers, in combination with more housing and amenity availabilities, have caused movements from Indigenous communities close to mining sites to communities outside of the region (Jourdren 2014). This was also confirmed in Nunatsiavut (Labrador, Canada). Although 44% of the workforce at Voisey's Bay was Inuit in 2019, only 14.5% actually resided in the region (Schott *et al.* 2022). Often, social services, housing, education, and health services cannot keep up with the rapid expansion of mining centers or mining towns, causing diminishing standards of living for residents (Brueckner *et al.* 2014; Parlee 2015), particularly low-income residents in remote areas that experience housing stress (Hunter *et al.* 2015).

Huskey and Southcott (2016) suggest policy directions for more local sharing in economic rents, including increased local ownership of resource projects, which might be enhanced through impact and benefit agreements (IBAs) and First Nations' land rights. In line with such recommendations, Indigenous communities in many postcolonial situations have been increasing their participation in the mining sector, due to a confluence of legal rights (see below), activism (see below), and a growing business case for inclusion, but there is still a strong need for further increase in Indigenous participation (e.g., in Australia; see Martin *et al.* 2014) and more effective involvement of Indigenous partners in critical assessment of potential benefits versus negative impacts, revenue sharing, and most beneficial resource rent investments (Rodon *et al.* 2022).

Social impacts

Although some researchers have shown that mining activities can have positive economic impacts on Indigenous communities (Langton 2013; Southcott *et al.* 2022), our review indicates that most researchers, NGOs, and government agencies consider mining's negative social impacts to be highly significant (Parlee 2018; Rodon *et al.* 2022; Schweitzer *et al.* 2018).

Academic publications have barely addressed the health impacts of mining on Indigenous communities (Bronson and Noble 2006; Bielawski 2004; Hurtig and San Sebastian 2002; Mactaggart *et al.* 2016; Noble and Bronson 2005; Shandro *et al.* 2011; Keller 2012; Myette and Riva 2021). Impact assessments (IAs) sometimes predict that employment and reduced economic stress will lead to a decrease in the high rates of mental illness (AREVA Resources Canada Inc. 2011; Baffinland Iron Mines Corporation 2012; Hornal & Associates Ltd. 2003). However, evidence suggests that higher incomes can exacerbate preexisting mental health and social issues like drug and alcohol consumption, gambling, prostitution (Gibson and Klinck 2005; Buell 2006; Natural Resources Canada 2003; Fidler and Hitch 2007; Government of the Northwest Territories 2006, 2009), violence, mental illness, absenteeism, poor health, sexually transmitted diseases (Archibald and Crkovich 1999; Buell 2006; Cyzewski *et al.* 2014; Goldenberg *et al.* 2008), and increases in divorce rates (AREVA Resources Canada Inc. 2011) and suicide (Baffinland Iron Mines Corporation 2012; Voisey's Bay Nickel Company Ltd. 1997; Hornal & Associates 2003).

Mining also affects social cohesion (Labrador West Status of Women Council and Femmes francophones de l'Ouest du Labrador 2004; Barrett-Wood, Knotsch *et al.* n.d.; Cyzewski *et al.* 2014; Lévesque and Baril 2020). The upsurge of new workers with different economic, social, and cultural values creates pressures on cultural identity, social integrity, and individual self-esteem, which, in turn, amplifies problems such as alcohol abuse or unsustainable behaviors (Kennett 1999). Moreover, the loss of social norms and structures regulating people's behavior, which is attributed to the speed and extent of changes and to the instability that comes with mining projects, leads to social issues such as anomie and suicide (Parlee and O'Neil 2007) and an increase in crime.

In Northern Canada, family violence and spousal abuse are higher than the national rate (Baffinland Iron Mines Corporation 2012; Hornal & Associates Ltd. 2003; Diakite and Healy 2021; Goelman 2014; Baffinland Iron Mines Corporation 2012; Makivik Corporation 1995). This is problematic, especially since Indigenous women in communities with growing economies are more likely to be the victims of sexual exploitation, violence, and sexually transmitted infections, often through sexual abuse or prostitution (Gibson and Klinck 2005). Familial integrity is also deteriorating, threatened by the demands and stress related to work and the changes in familial roles (Gibson and Klinck 2005; Sosa and Keenan 2001).

The impacts of mining can hinder the capacity of women to contribute to the community's well-being (Kuokkanen 2011), while the lack of childcare services also limits women's opportunities to work within the mining industry or to acquire an education (O'Faircheallaigh 1998). This may explain why women are often excluded from decision-making processes (Natural Resources Canada—Minerals and Metal Sector 2003; Horowitz 2017; Mills *et al.* 2018). Indigenous women are more likely to suffer from poverty (Gibson and Klinck 2005), and households headed by women are more vulnerable to the inflation caused by mining exploitation (Sosa and Keenan 2001).

Social cohesion is not typically discussed in IAs, although various aspects of it are often hinted at. Some IAs mention that the migration of Indigenous or non-Indigenous persons in and out of the communities is to be expected (Baffinland Iron Mines Corporation 2012; Hornal & Associates Ltd. 2003; AREVA Resources Canada Inc. 2011; Hydro-Québec 2004). It is expected that mines will remain dependent on workers from outside the region for many years (AREVA Resources Canada Inc. 2011). A loss of the local Indigenous language is expected because of both the immersion in the English language at work and the increased contact of English speakers with the youth (AREVA Resources Canada Inc. 2011). Only the most recent IAs address this issue, either by providing more services to Inuit in Inuktitut (AREVA Resources Canada Inc. 2011) or by allowing Inuit workers to explain something in Inuktitut to each other, as long as it is then repeated in English for the benefit of other employees (Baffinland Iron Mines Corporation 2012).

Mining can increase intra- and intergenerational inequalities (O'Faircheallaigh 1998; Fidler and Hitch 2007; Irlbacher-Fox and Mills 2007; Davis 2009) and inequalities between communities within the same region (Rodon *et al.* 2022). Communities not located near mines are not a concern of mining companies, nor are regional centers, although these are usually quite impacted by mining development, as an influx of workers can lead to increased demand for healthcare and other services (Gibson and Klinck 2005). In other cases, local communities experience in-migration, which creates pressures on public and private housing (AREVA Resources Canada Inc. 2011; Baffinland Iron Mines Corporation 2012; Hydro-Québec 2004; Hornal & Associates Ltd. 2003). However, mining companies do not contribute to alleviating the housing shortage in impacted communities (Rodon *et al.* 2022).

There are also inequalities between ethnic groups, whereby non-Indigenous workers often earn more than Indigenous workers because of their access to better jobs (Gibson and Klinck 2005; Caron *et al.* 2019).

All of these factors lead to an increase in social stratification and Indigenous people's resentment at being disadvantaged (Duhaime *et al.* 2003). This can also lead to an increase in social tensions (O'Faircheallaigh 1998).

Traditional activities are a source of pride that allow Indigenous Peoples to not only face the hardships related to life on their beloved and respected land but also maintain their identity (Buell 2006; Gilberthorpe and Hilson 2014). Yet, little is known about the real impacts of mining on Indigenous land-use practices (Haley *et al.* 2011a, 2011b; Yamarak and Parton 2023; Southcott and Natcher 2018). Some research indicates that participation in harvesting activities decreases because mining projects take up space and can cause changes and environmental damage to the land (Bjerregaard and Young 1998; Bjerregaard *et al.* 2004; Duhaime *et al.* 2003; Bernauer 2011) and that Indigenous people are forced to adapt their practices to the effects of industrialization (Bernauer 2011; Kruse 2011). Similarly, some IAs predict that less time, overall, will be available for hunting, trapping, and fishing (Hornal & Associates Ltd. 2003). Other researchers, indeed, claim that increased participation in a salary-based economy can lead to decreased participation in the subsistence-based economy (Buell 2006; Kruse 2011).

However, some academic research has shown—and some IAs predict—that mining projects in remote communities do not necessarily lead to a decline in traditional practices (Southcott and Natcher 2018). Local populations may continue to participate in traditional economic activities while taking advantage of the economic diversification offered by mining employment (Boutet *et al.* 2015). Moreover, well-paid jobs allow workers to increase their harvesting activities (Koke 2008; Landry *et al.* 2009; Laneuville 2013; LeClerc and Keeling 2015), largely due to an increased ability to purchase hunting equipment (Hornal & Associates Ltd. 2003). Mining companies even support programs aiming to facilitate the transfer of traditional knowledge and skills (Baffinland Iron Mines Corporation 2012). In any case, it has become evident that subsistence harvesting is becoming increasingly expensive and requires a rapidly growing proportion of income to support hunting and fishing activities (Fergurson 2011; Wenzel 2013).

Legal contexts

Legal regimes governing mining rights and activities, as well as those pertaining to the recognition and protection of Indigenous Peoples' rights, are dense and complex. In the limited scope of this section, we draw a general portrait of the Indigenous Peoples' rights that play a role in structuring the relationships among Indigenous Peoples, the state, and mining companies (O'Faircheallaigh 2016), specifically land rights and participatory rights in mining decision-making processes.

Australia and Canada

In Australia and Canada—both former British settler colonies—judicial recognition of Indigenous Peoples' land rights in common law has exerted significant pressure on governments and third parties to negotiate mining development with affected communities. The 1992 High Court decision in *Mabo v Queensland (No. 2)* spurred the Australian parliament to enact the *Native Title Act 1993* (Cth) (NTA), which, among other provisions, establishes processes and standards through which Aboriginal and Torres Strait Islander People may claim native title where it has not been lawfully extinguished in the past (Bartlett 2014; Strelein 2006; Young 2008). In relation to mineral development more specifically, the NTA provides Aboriginal groups, in addition to the possibility of negotiating Indigenous Land-Use Agreements (Bartlett 2014; Langton and Mazel 2008), a "Right to Negotiate" the terms of mining development projects with the mining proponent and the state government responsible for issuing the mining tenements (Bartlett 2014; Masher 2013). Under the "Right to Negotiate," if an agreement is not reached within the 6 months following the state's notice of the proposed act, either party can refer the matter to arbitration by the National Native Title Tribunal (NNTT), whose decisions can be overturned by the responsible government authority (NTA, sections 25–44, 237). The literature has emphasized that the "Right to Negotiate" regime reinforces the inequalities between Aboriginal peoples and mining companies, the latter benefiting from the strict timeline imposed on negotiations, the absence of an Aboriginal veto, and the NNTT's limited powers and demonstrated favorable stance toward mining development (Bartlett 2014; Masher 2013; O'Faircheallaigh 2016). In comparison, according to the *Aboriginal Land Rights (Northern Territory) Act 1976*, Indigenous Peoples in the Northern Territory may veto mining exploration on their land (unless granting the license is deemed by the Governor General to be in the "national interest"), therefore generating strong incentives for mining companies to negotiate to secure Aboriginal consent (O'Faircheallaigh 2016).

The duty to consult and accommodate in Canadian law has been subject to similar critiques, notably in relation to mining development (Ariss *et al.* 2017; Ariss and Cutfeet 2011; Drake 2015; Lacasse 2017; Thériault 2010, 2015). This duty exists when the government contemplates a conduct—such as permitting mining activities—that might adversely affect established or asserted Aboriginal rights or treaty rights, which are constitutionally entrenched by section 35 of the *Constitution Act* 1982 (*Haida Nation* 2004; *Mikisew* 2005; *Rio Tinto Alcan* 2010). In addition to resource use rights arising from Indigenous traditional activities, "Aboriginal rights" include Aboriginal title, which the Supreme Court of Canada has defined generically as conferring on the title-holders exclusive rights to possess and use the land (including minerals), the right to benefit from the land, and the right to proactively use and manage it (*Tsilhqot'in Nation* 2014, para. 73; *Delgamuukw* 1997, para. 122), provided that the land is not developed "in a way that would substantially deprive future generations of the benefit of the land" (*Tsilhqot'in Nation* 2014, para. 74). The government must seek the consent of the Indigenous People prior to authorizing development projects on lands held under Aboriginal title. However, if consent cannot be secured, the government may nevertheless authorize the project, provided that the procedural duty to consult was upheld and that the infringement can be justified under section 35 (*Ktunaxa Nation* 2017; *Tsilhqot'in Nation* 2014, para. 76; *Haida Nation* 2004, para. 48). Despite recent mining reforms through which the duty to consult has been integrated in some provincial and territorial mining laws and policies, several authors have argued that free entry mining regimes in Canada are fundamentally incompatible with Indigenous Peoples' constitutional rights, especially as they allow mining proponents to register mining claims on lands claimed under Aboriginal title without providing for prior consultation and accommodation (Simons and Collins 2010; Ariss and Cutfeet 2011; Drake 2015; Lacasse 2017; Thériault 2010, 2015).

Indigenous Peoples in Canada may also elect to negotiate their land claims under the federal government's Comprehensive Land Claims Policy. Land claims agreements, which cover most of Canada's northernmost regions, generally provide the Indigenous party some exclusive surface and subsurface rights on limited portions of their traditional territories, harvesting rights on vast areas of public lands, as well as rights to participate in the governance of their lands and resources through co-management boards, environmental assessment regimes, and specific consultation provisions. Some agreements, such as the one negotiated with the Labrador and Nunavut Inuit, also provide for self-government, benefit-sharing, and mandatory IBA negotiations for new development projects on Indigenous lands (Alcantara 2013; Bankes 2013; Rodon 2018). While there is a growing literature analyzing land claims agreements and Indigenous Peoples' agency regarding resource development (Rodon 2017; Samson 2016; Samson and Cassell 2013), legal scholarship comparing the specific terms and the effectiveness of these agreements remains scarce, especially in relation to their provisions pertaining to resource extraction on Indigenous lands.

Norway, Finland, Sweden, and Greenland

Contrary to the *sui generis* rights model prevalent in common law jurisdictions, Sami land and resource rights in Fennoscandia are grounded in general civil law proprietary concepts, including immemorial usage, immemorial prescription, and customary law (Allard 2011, 2013; Allard and Brännström 2021; Ravna and Bankes 2017; Koivurova *et al.* 2015; *Nordmaling* case 2011; Nygaard 2016; *Selbu* case 2001). However, Sami rights in the context of mining development have different levels of legal protection in Norway, Finland, and Sweden (Allard 2011, 2013; Allard and Funderud Skogvang 2015; Koivurova *et al.* 2015).

For instance, in Norway, the 2005 *Finnmark Act* provides the Sami living in the Finnmark area a process through which they can formally establish their land and resource rights "on the basis of prescription or immemorial usage, or on some other basis" (*Finnmark Act*, section 5; Ivsett Johnsen 2016; Ravna 2011; Ravna and Bankes 2017). Moreover, the 2009 *Mineral Act* contains several provisions requiring Sami participation in decision-making processes pertaining to mining projects, with a particular focus on Sami rights and interests in the Finnmark area (Koivurova *et al.* 2015; Nygaard 2016). In the municipalities where their interests are sufficiently represented, the Sami may also benefit from the 2009 modifications to the *Planning and Building Act* (PBA), which enables municipalities to stop a mining project to "protect the natural basis for Sami culture, economic activities, and social life" (PBA, section 3.1; Ivsett Johnsen 2016; Koivurova *et al.* 2015; Nygaard 2016). However, the process established by the *Finnmark Act* has not yet led to the recognition of Sami individual or collective land rights, and both this Act and the PBA have been criticized for providing inadequate and insufficient protection to Sami rights, especially in areas outside Finnmark (Koivurova *et al.* 2015; Nygaard 2016).

In Finland and Sweden, unlike in Norway's Finnmark area, there are still no formal mechanisms to recognize Sami land rights. In Finland, reindeer-herding rights, which are not exclusive to the Sami, have been qualified by some authors as "a weak use right" that "can be trumped by other rights" (Koivurova *et al.* 2015, p. 16). The *Mining Act* provides for several rights to Sami in the Sami Homeland, the Skolt area, and the reindeer husbandry areas. In particular, this Act states that in the Sami Homeland, the permit authority shall, in co-operation with, among other actors, the Sami Parliament, assess the effects of mining activities "on the rights of the Sami as an indigenous people to maintain and develop their own language and culture and traditional livelihoods and shall consider measures required for decreasing and preventing damage." (*Mining Act*, section 38; Koivurova *et al.* 2015; Koivurova and Petrétei 2014). The relative strength of the protection provided by this provision hinges largely on its interpretation and application by state authorities, which have yet to be assessed.

The Swedish *Mineral Act* and Ordinance, as amended in 2014, provide for the participation of "special rights-holders" (including reindeer herders) at different stages of the mining decision-making process (Koivurova *et al.* 2015). Under the Swedish Environmental Code, reindeer herding is a national interest that can be protected against exploitation. However, mining development is also considered a "national interest" under environmental legislation (Lawrence and Klocker Larsen 2017). As emphasized by some authors, mineral extraction tends to prevail over Sami rights and interests (Allard and Curran 2021; Liedholm Johnson and Ericsson 2015).

It is worth noting that international law, including the *International Covenant on Civil and Political Rights*, ILO Convention 169, and the United Nations Declaration on the Rights of Indigenous Peoples (UNDRIP), has played a distinctive role in the development of Sami rights in Norway and, more recently, in Sweden (Allard and Brännström 2021; Allard and Curran 2021; *Fosen* case 2021; *Girjas* case 2020).

The situation in Greenland differs significantly from the other jurisdictions covered in the chapter, since the Inuit, who represent most of the population, exercise considerable political power under the 2009 *Self-Government Act* (Göcke 2012; Mortensen and Barten 2016; Mortensen 2015). The legislative competence over mineral resources was transferred to the Greenlandic Self-Government in 2010. The *Mineral Resources Act* (MRA) provides that "[t]he Greenland Self-Government has the right of use and the right to exploit mineral resources in the subsoil in Greenland." The MRA does not recognize any specific Inuit land rights and interests, following a long-held opinion in Denmark—made explicit in the preamble of the *Self-Government Act*—according to which all Greenlanders are to be considered as one people (Hubbard 2014). Hence, Inuit participation in mining licensing processes occurs primarily through social and environmental impact assessment frameworks, which have been strongly criticized in the literature (Hubbard 2014; Nuttall 2013; Tiainen 2016), and through the negotiation of IBAs among the Greenland Self-Government, the local municipality, and the mining proponent (Mortensen 2015; Tiainen 2016).

Indigenous-led engagements with mining

While the social impacts of mining on Indigenous communities have been examined extensively in the literature, we identified a relative dearth of studies showcasing community members' own perspectives and agency. This section reviews the literature on community-led acts of resistance and negotiation, despite challenges posed by the persistence of industrial capitalism (see Horowitz 2021). We examine, in turn, several sets of relationships: Within and between Indigenous communities; among communities, companies, and governments; and between communities and non-Indigenous allies.

Cohesiveness of Indigenous communities

When diverse Indigenous communities maintain solidarity, despite their differences, they achieve more from engagements with mining companies. For instance, the Labrador Innu and Inuit—despite a history of sometimes-tense relations and overlapping land claims—agreed in advance to recognize each other's interests in contested areas, which included Voisey's Bay, before Inco showed an interest in mining there. The Innu and Inuit then shared information and supported each other in direct actions and litigation, resulting in an extremely strong IBA that achieved nearly all their economic and environmental goals (O'Faircheallaigh 2016). However, rifts between Indigenous communities can jeopardize such outcomes. Around BHP's Ekati Diamond Mine in Canada's Northwest Territories, land claim rivalries inhibited communities from sharing information, which could have allowed each to benefit from terms negotiated by the other (O'Faircheallaigh 2016).

Members of a single community may disagree about how to engage with mining companies. In New Caledonia, a Kanak protest group called Rhéébù Nùù set up roadblocks and burned company equipment in resistance to a mining project, run by Vale, that they felt threatened local marine resources while failing to offer significant employment to the Kanak. While other community members shared these concerns, many disapproved of the protestors' violence (Horowitz 2009). Local customary authorities were especially uncomfortable with these tactics; moreover, while concerned about environmental damage, they supported the project for its promised employment. Ultimately, customary authorities signed an IBA with relatively small environmental and economic benefits, in exchange for halting all direct actions against the company, angering community members who had been excluded from negotiations (Horowitz 2015).

Intra-community disagreements often stem from different positionalities within the community, with overlapping categories of gender, ethnicity, age, socioeconomic status, and culturally specific social hierarchies; those who represent the community in negotiations may not share the interests and concerns of all subgroups. In Australia's Northern Territory, land councils usually represent their communities in negotiations with mining companies. As they often rely on royalty payments from mining, this creates a conflict of interest (Tolazzi 2015). In New Caledonia, customary authorities—all senior men—thought it "normal" that the company chose to negotiate exclusively with them (Horowitz 2017). In Canada, Aboriginal men did not support prioritizing women in employment with the Voisey's Bay Nickel Company (VBNC) (Cox and Mills 2015).

Company–community–government relations

Companies may exploit these different positionalities by excluding marginalized subgroups, such as women and youth, from negotiation processes, as occurred in New Caledonia (Horowitz 2017). However, even formal inclusion in company-organized processes may result in de facto exclusion. For instance, VBNC included women in scoping sessions for its environmental impact statement but ultimately failed to honor women's requests, such as plans for training, retaining, and promoting women (Cox and Mills 2015).

Companies often find ways to short-circuit participatory processes and ignore Indigenous communities' concerns; for example, in Nunavut, Canada, mining projects avoided taking mitigation measures by creating ineffectual stakeholder advisory groups (Bernauer *et al.* 2023). However, other Indigenous communities in Canada and Australia with greater economic and political resources have forged more equitable relationships with companies through strong IBAs (e.g., Vanthuyne and Gauthier 2022) or even equity participation (Kung *et al.* 2022).

Governments, too, exclude Indigenous communities' voices. They often support mining projects despite Indigenous groups' resistance, as with Adani's Carmichael Coal Mine in Queensland, Australia, opposed by Wangan and Jagalingou traditional owners but expedited by both state and federal governments (Jolley and Rickards 2020; Lyons 2019). However, despite multiple constraints, Indigenous communities in Canada are increasingly using legal and extra-legal tools to assert sovereignty over mining activities (Thériault *et al.* 2022).

Alliances

When mining-related engagements grow tense, Indigenous communities may welcome support from non-Indigenous allies. Just as Indigenous groups may have to overcome long-standing differences, Indigenous and non-Indigenous groups may need to set aside unrelated disagreements. For instance, Greenpeace supported Inuit opposition to seismic testing for oil and gas in Nunavut, despite earlier tensions over seal hunting (Zahara *et al.* 2016). The different sources of power of Indigenous and non-Indigenous groups—legal rights, resources, connections, etc.—can complement each other, as when the Gundjehmi Aboriginal Corporation and environmental activists defeated the Jabiluka Mine in Northern Australia (Hintjens 2000). At Voisey's Bay, several nongovernmental organizations and other groups formed a coalition that helped secure media attention and raise public awareness of Innu and Inuit concerns (O'Faircheallaigh 2016). Indigenous groups can also reach out to international institutions that champion Indigenous rights, such as the UN, as Sami leaders (among others) have done (Fjellborg *et al.* 2022).

Nonetheless, alliances are strewn with pitfalls. International organizations may co-opt communities into their own agendas. For example, Indigenous activists in New Caledonia expected the UNESCO World Heritage listing to protect their coral reefs, but UNESCO, which is dependent on state governments, stipulated no restrictions on mining (Horowitz 2016). Even grassroots groups may attempt to manipulate Indigenous-led groups. In Australia, environmentalists often dominate alliances with Aboriginal groups due to their access to financial and informational resources and greater influence within Australian civil society (Tolazzi 2015). Sometimes, though, the tables are turned. In New Caledonia, environmentalists were disappointed when Indigenous activists independently signed an agreement with Vale (Horowitz 2012).

In some cases, Indigenous communities see no alternative to negotiation. Their land rights are restricted to a certain area, often imbued with great cultural, spiritual, and emotional significance. Ultimately, they may face a choice between continuing futile resistance, at great cost and risk, or negotiating some benefits for their communities. Environmentalists, in a moral struggle with no immediate impacts on their own lives, may not understand Indigenous communities' lived realities (Tolazzi 2015).

Conclusions: Gaps in the literature and opportunities for future research

In reviewing the available literature on the relationships of Indigenous Peoples to large-scale mining projects in certain settler colonies, it is strikingly clear that mining activities have particular, and often disproportionate, impacts on Indigenous communities. Indigenous Peoples often rely on natural resources that mining activities disrupt, threaten, or poison, and they have cultural and spiritual relationships to landscapes that may be destroyed or degraded by extractive developments. Meanwhile, the economic benefits of mining tend to leak out of local regions, especially from Indigenous communities that often lack the necessary training to secure employment. Instead, these communities suffer from rising local costs of living, as well as from outmigration to urban areas, compounded by a lack of social services in those areas. These problems, along with an influx of cash and outsiders, can lead, in turn, to an increase in mental health and social issues, including substance abuse, violence, and suicide; a loss of the social cohesion previously provided through shared identities and languages; growing inequities; and changes to traditional land-use practices and community vitality.

Despite the promising opportunities for Indigenous Peoples created by the free, prior, and informed consent (FPIC) principle enshrined in UNDRIP, the implementation of this principle is still being disputed. Indigenous people face institutional opposition to their interpretation of FPIC as a decision-making right. In response, some groups have taken advantage of the legal ambiguity to implement FPIC by putting into place their own decision-making mechanisms (Papillon and Rodon 2020). When they perceive regulatory regimes as inadequate, Indigenous communities may take matters into their own hands through acts of resistance, often in alliance with non-Indigenous groups, which may in turn lead to empowered company–community negotiations.

Despite the compelling nature of these findings, more comparative research is needed on the implications of mining activities for Indigenous Peoples in post/colonial contexts. Below, we identify gaps in this literature and outline opportunities for future research.

Environmental impacts and Indigenous Peoples

The environmental impacts of mineral development are generally well understood and have been for decades, if not centuries. As noted above, the particular effects of these environmental changes on Indigenous Peoples in the settler-colonial contexts of Canada, Australia, New Caledonia, and Scandinavian Samiland are increasingly well documented. The growing recognition of these problems, along with rapidly changing legal, regulatory, and public policy contexts in these jurisdictions, invites further comparative study on how Indigenous communities have encountered, resisted, and adapted to mining-induced environmental change in recent decades (Gilberthorpe and Hilson 2014). For instance, movements for Indigenous rights and sovereignty have, in some cases, achieved greater Indigenous participation in environmental assessment and monitoring. However, the strategies, successes, and failures of including Indigenous communities and knowledge systems bear further investigation (Nygaard 2016; Sandlos and Keeling 2016b). Even as renewed exploration and development returns to previously mined areas, assessments of the regional and cumulative impacts of extractive projects remain crucially underexplored (Keeling *et al.* 2019). In addition, there is new pressure for Indigenous Peoples to open areas as "sacrifice zones" for the extraction of critical minerals (Scott and Smith 2017) to enable a transition to a low-carbon future. Lastly, the "tail-end" of the mining cycle—mine closure and remediation—deserves greater scrutiny, as Indigenous Peoples grapple with the often permanent landscape degradation, pollution, and displacement associated with industrial-scale mining.

Local-scale economic analysis for Indigenous communities

This literature review reveals a lack of detailed analysis of the resource curse, business, employment, and human development at the regional and local scale, and of positive and negative impacts alike on Indigenous communities. This gap in the literature has implications for policies and programs that might address issues relevant for Indigenous and other communities that bear the brunt of resource booms, landscape changes, and drastic developmental change. The potential leakage of revenues and business opportunities, human skill erosion, and in- and outmigration resulting from extractive industries needs to be researched and analyzed in much more detail at the local scale. There is, therefore, an important opportunity for comparative studies within countries and between similar regions that have different institutions, governance regimes, and legal statuses of negotiating local bodies. Further attention should also be given to local business development and long-term diversification in regions that are highly dependent on nonrenewable resource extraction. The stability and diversity of the local business sector will be a crucial factor in the sustainable development of many mining regions. In order to avoid specific regional resource curse impacts, we need to collect more refined local data and prevent negative effects of resource booms for vulnerable, remote Indigenous communities. Future research should involve community and regional data analysis and examine regional regulatory measures, local program development, and governance structures that support sustainable policies and avoid rent capture and corruption at the local or regional level. All of this research needs to feed into local and regional, sustainable development visions and strategies that look well beyond the expected lifetime of a specific resource extraction cycle and that focus on ecological and economic justice as much as rent generation.

Assessing and managing social impacts on Indigenous communities

While our understanding of the social impacts of mining on Indigenous communities in postcolonial contexts has increased considerably, there are still areas where a better understanding would help communities control negative impacts and improve positive impacts. It is obvious that many social impacts are poorly measured or not discussed in IAs, be it by choice or because of the lack of data and resources available. For instance, there are few actual numbers on the employment of Indigenous people, while assessments of project impacts on education are completely absent from IAs. Even if mental health is addressed in most IAs, other health issues are not. Although IAs mention that workers from other regions are likely to migrate to work at the mines, the impact of this situation on Indigenous communities is not considered. IAs also seldom mention impacts on women, although we know from the academic literature that women are disproportionately impacted in many ways by mining development projects. Lastly, different IAs anticipate different (and sometimes opposite) impacts on traditional activities, highlighting the poor quality of information on the subject.

Indigenous rights and mining

As indicated by this overview, Indigenous Peoples' rights, as well as mining laws and regulations, have been examined to varying degrees in most of the jurisdictions we compared in this work. However, legal scholarship at the interface of mining regimes and Indigenous Peoples' rights remains scarce, especially from a comparative perspective. Moreover, the influence of international norms, in particular of FPIC, on the evolution of Indigenous Peoples' rights in the context of mining, while subject to increasing attention from researchers (see, e.g., Papillon and Rodon 2020; Allard 2018; Nygaard 2016; Rice 2020; Szablowski 2010), merits further attention. The literature would therefore benefit from more in-depth comparative analysis aimed at evaluating how different approaches to regulating mining development and protecting Indigenous Peoples' rights shape the relationships between Indigenous Peoples and mining proponents. In turn, it would also be interesting to document, across jurisdictions, how those relationships—and especially negotiated agreements between Indigenous communities and mining proponents—influence policy and law-making regarding Indigenous Peoples' encounters with mining (Le Meur *et al.* 2013). Lastly, from a more general standpoint, it would be useful to analyze and compare, across jurisdictions, rights mobilization strategies by Indigenous Peoples facing mining development on their traditional lands.

Agency, relationships, and strategies in Indigenous-led engagements

In reviewing the literature on Indigenous-led engagements with mining in the countries identified above, several gaps quickly become evident. First, as was also the case for social and economic impacts, we found very little on Sami and Greenland Inuit engagements, especially from an ethnographic perspective. More has been written about such engagements in Australia and Canada, yet much of this focuses on community impacts and how companies can avoid these, with relatively little on Indigenous groups' own agency and how they choose to engage with companies. It would also be useful to see more comparisons of these and other regions to explore how different legal, political, economic, social, and cultural contexts shape Indigenous communities' engagements with companies. Moreover, the literature would benefit from greater research into issues such as intracommunity dynamics (e.g., gender relations) and how these inform engagements, as well as an evaluation of different strategies that communities use, identifying which are more likely to help them achieve their goals.

Acknowledgments

This work was supported by the Social Sciences and Humanities Research Council of Canada, Grant number 895-2016-1015. Parts of this chapter have been reproduced, with permission, from Horowitz *et al.* (2018), Indigenous peoples' relationships to large-scale mining in post/colonial contexts: Toward multidisciplinary comparative perspectives. *The Extractive Industries and Society*, 5 (3), 404–414. We are grateful to the two anonymous reviewers for comments that significantly improved the original article upon which this chapter is based. Any errors are solely our responsibility.

References

Abele, F., 1989. *Gathering strength: Native employment training in the Northwest Territories*. Arctic Institute of North America, University of Calgary.

Aboriginal Land Rights (Northern Territory) Act 1976 (Cth) sch 7.

Alcantara, C., 2013. *Negotiating the deal: Comprehensive land claims agreements in Canada*. Toronto: University of Toronto Press.

Allard, C., 2011. The Nordic countries' law on Sami territorial rights. *Arctic Review on Law and Politics*, 2 (2), 159–183.

Allard, C., 2013. Who holds the reindeer-herding rights in Sweden? A key issue in legislation. *In:* N. Bankes and T. Koivurova, eds., *The proposed Nordic Saami convention: National and international dimensions of Indigenous property rights*. London: Hart Publishing, 207–228.

Allard, C., 2018. The rationale for the duty to consult Indigenous Peoples: Comparative reflections from Nordic and Canadian legal contexts. *Arctic Review on Law and Politics*, 9 (1), 25–43.

Allard, C. and Brännström, M., 2021. Girjas reindeer herding community v. Sweden: Analyzing the merits of the Girjas case. *Arctic Review on Law and Politics*, 12, 56–79.

Allard, C. and Curran, D., 2021. Indigenous influence and engagement in mining permitting in British Columbia, Canada: Lessons for Sweden and Norway? *Environmental Management*. Available from: https://doi.org/10.1007/s00267-021-01536-0.

Allard, C. and Funderud Skogvang, S., eds., 2015. *Indigenous rights in Scandinavia. Autonomous Sami law*. Farnham, UK: Ashgate Publishing.

Archibald, L. and Crkovich, M., 1999. *If gender mattered: A case study of Inuit women, land claims and the Voisey's Bay Nickel Project*. Ottawa: Status of Women Canada.

AREVA Resources Canada Inc., 2011. *Kiggavik Project: Draft environmental impact statement*. Available from: Nunavut Impact Review Board: 120425–09MN003-DEIS Volume 1- Main Document-R1-IMCE.pdf [Accessed 26 October 2023].

Ariss, R. and Cutfeet, J., 2011. Kitchenuhmaykoosib Inninuwug First Nation: Mining, consultation, reconciliation and law. *Indigenous Law Journal*, 10 (1), 1–37.

Ariss, R., MacCallum Fraser, C., and Somani, D., 2017. Crown policies on the duty to consult and accommodate: Towards reconciliation? *McGill Journal of Sustainable Development Law and Policy*, 13 (1), 1–52.

Baffinland Iron Mines Corporation, 2012. *Mary River Project—Final environmental impact assessment*. Available from: Nunavut Impact Review Board: 120213–08MN053-FEIS part1-IT5E.pdf; 120213–08MN053-FEIS part2-IT5E.pdf; 120213–08MN053-FEIS part3-IT5E.pdf; 120213–08MN053-FEIS part4-IT5E.pdf; 120213–08MN053-FEIS part5-IT5E.pdf; 120213–08MN053-FEIS part6-IT5E.pdf; 120213–08MN053-FEIS part7-IT5E.pdf [Accessed 26 October 2023].

Bankes, N., 2013. The forms of recognition of Indigenous property rights in settler states: Modern land claim agreements in Canada. *In:* N. Bankes and T. Koivurova, eds., *The proposed Nordic Saami Convention: National and international dimensions of Indigenous property rights*. London: Hart Publishing, 351–378.

Barker, T., 2006. *Employment outcomes for Aboriginal people: An exploration of experiences and challenges in the Australian minerals industry*. Sustainable Minerals Institute, University of Queensland, Research Paper No. 6.

Barker, T. and Brereton, D., 2004. *Aboriginal employment at the Century Mine*. CSRM, University of Queensland, Research Paper No. 3, April.

Barrett-Wood, Z., Knotsch, C., Davison, C., and Bradshaw, B., n.d. *Translating knowledge on impacts of mining for Aboriginal community health: The issue of rotational, two-week work shifts*. National Aboriginal Health Organization. Poster Presentation.

Bartlett, R. H., 2014. *Native title in Australia*. 3rd ed. Chatswood: LexisNexis Butterworths.

Beckett, C., 2021. Beyond remediation: Containing, confronting and caring for the Giant Mine Monster. *Environment and Planning E: Nature and Space*, 4, 1389–1412.

Beckett, C. and Keeling, A., 2019. Rethinking remediation: Mine reclamation, environmental justice, and relations of care. *Local Environment*, 24, 216–230.

Belayneh, A., Boutet, J.-S., Rodon, T., and Schott, S., 2018a. Mining economies, mining families: Inuit business development and employment in the eastern subarctic. *Northern Review*, forthcoming.

Belayneh, A., Boutet, J.-S., Rodon, T., and Schott, S., 2018b. *The impact of major mining projects on Inuit business development in the Canadian subarctic*. Working Paper. Available from: https://economics.ca/2018/en/program.php.

Bernauer, W., 2011. *Mining and the social economy in Baker Lake, Nunavut*. University of Saskatchewan, Saskatoon. Prepared for the Northern Ontario, Manitoba, and Saskatchewan Regional Node of the Social Economy.

Bernauer, W., Hostetler, G., Greene, E., Tester, F., Harris, R. and Tanguay, L., 2023. Undermining assessment: EIA follow-up, stake-holder advisory groups, and extractive industries in Nunavut, Canada. *Impact Assessment and Project Appraisal*, 41 (2), 87–101.

Bielawski, E., 2004. *Rogue diamonds: Northern riches on Dene land*. Seattle: University of Washington Press.

Bjerregaard, P. and Young, T. K., 1998. *The circumpolar Inuit: Health of a population in transition*. Copenhagen: Munksgaard.

Bjerregaard, P., Young, T. K., Dewailly, É., and Ebbensson, S. O., 2004. Indigenous health in the Arctic: An overview of the circumpolar Inuit population. *Scandinavian Journal of Public Health*, 32 (5), 390–395.

Boutet, J. S., 2013. Opening Ungava to industry: A decentering approach to Indigenous history in subarctic Québec, 1937–1954. *Cultural Geographies*, 21 (1), 1937–1954.

Boutet, J.-S., 2015. The revival of Québec's iron ore industry: Perspectives on mining, development, and history. *In:* A. Keeling and J. Sandlos, eds., *Mining and communities in Northern Canada: History, politics, and memory*. Calgary: University of Calgary Press, 169–206.

Bridge, G., 2004. Contested terrain: Mining and the environment. *Annual Reviews: Environment and Resources*, 29, 205–259.

Bronson, J. and Noble, B. F., 2006. Health determinants in Canadian northern environmental impact assessment. *Polar Record*, 42 (223), 315–324.

Brueckner, M., Durey, A., Mayes, R., and Pforr, C., 2014. Confronting the "resource curse or cure" binary. *In:* M. Brueckner, A. Durey, R. Mayes, and C. Pforr, eds., *Resource curse or cure? On the sustainability of development in Western Australia*. Berlin, Heidelberg: Springer Berlin Heidelberg, 3–23.

Buell, M., 2006. *Resource extraction development and well-being in the North: A scan of the unique challenges of development in Inuit communities*. Ajunnginiq Centre, National Aboriginal Health Organization.

Byrne, P., Wood, P. J., and Reid, I., 2012. The impairment of river systems by metal mine contamination: A review including remediation options. *Critical Reviews in Environmental Science and Technology*, 42 (19), 2017–2077.

Caron, J., Asselin H., and Beaudoin, J.-M., 2019. Attitudes and behaviors of mining sector employers towards the Indigenous workforce. *Resources Policy*, 61 (June), 108–117. Available from: https://doi.org/10.1016/j.resourpol.2019.02.001.

Cohen, T., 2017. Bringing country back? *In:* K. Jalbert, A. Willow, D. Casagrande, and S. Paladino, eds., *ExtrACTION: Impacts, engagements, and alternative futures*. London: Routledge, 137–150.

Constitution Act, 1982, Schedule B to the *Canada Act 1982* (U.K.), 1982, c. 11.

Cousins, D. and Nieuwenhuysen, J., 1984. *Aboriginals and the mining industry: Case studies of the Australian experience*. Sydney: George Allen & Unwin.

Cox, D. and Mills, S., 2015. Gendering environmental assessment: Women's participation and employment outcomes at Voisey's Bay. *Arctic*, 68 (2), 246–260.

Cyzewski, K., Tester, F., Aaruaq, N., and Blangy, S., 2014. *The impact of resource extraction on Inuit women and families in Qamani'tuaq, Nunavut Territory*. A Qualitative Assessment. Pauktuutit Inuit Women of Canada and UBC School of Social Work.

Dance, A., 2015. Northern reclamation in Canada: Contemporary policy and practice for new and legacy mines. *The Northern Review*, 41, 41–80.

Davis, G. A., 2009. Extractive economies, growth, and the poor. *In:* J. P. Richards, ed., *Mining, society, and a sustainable world*. Berlin: Springer, 37–60.

Delgamuukw v. British Columbia, [1997] 3 R.C.S. 1010.

Diakite, N. and Healy G., 2021. *Family violence in Nunavut: A scoping review*. Available from: www.qhrc.ca/wp-content/uploads/2019/02/Scoping-Review-Poster.pdf.

Drake, K., 2015. The trials and tribulations of Ontario's Mining Act: The duty to consult and Anishinaabek law. *McGill Journal of Sustainable Development Law and Policy*, 11 (2), 183–218.

Duhaime, G., Bernard, N., Fréchette, P., Maillé, M.-A., Morin, A., and Caron, A., 2003. *The mining industry and the social stakes of development in the Arctic*. Québec: Chaire de recherche du Canada sur la condition autochtone comparée.

Eggert, R. G., 2001. *Mining and economic sustainability: National economies and local communities*. A Study Prepared for the Mining, Minerals, and Sustainable Development Project, Colorado School of Mines.

Evans, N. and Sawyer, J., 2009. The mining boom: Challenges and opportunities for small business in regional South Australia. *Australasian Journal of Regional Studies*, 15 (3), 355–373.

Fergurson, H., 2011. Inuit food (in)security in Canada: Assessing the implications and effectiveness of policy. *Queen's Policy Review*, 2 (2), 54–79.

Fessehaie, J. and Morris, M., 2013. Value chain dynamics of Chinese copper mining in Zambia: Enclave or linkage development? *The European Journal of Development Research*, 25 (4), 537–556.

Fidler, C. and Hitch, M., 2007. Impact and benefit agreements: A contentious issue for environmental and Aboriginal justice. *Environments Journal*, 35 (2), 49–69.

Fjellborg, D., Beland Lindahl, K., and Zachrisson, A., 2022. What to do when the mining company comes to town? Mapping actions of anti-extraction movements in Sweden, 2009–2019. *Resources Policy*, 75, 102514.

Fleming, D. A., Measham, T. G., and Paredes, D., 2015. Understanding the resource curse (or blessing) across national and regional scales: Theory, empirical challenges and an application. *Australian Journal of Agricultural and Resource Economics*, 59, 624–639.

Fosen case, HR-2021–1975-S (Supreme Court of Norway).

Gibson, G. and Klinck, J., 2005. Canada's resilient North: The impact of mining on Aboriginal communities. *Pimatisiwin: A Journal of Aboriginal and Indigenous Community Health*, 3 (1), 115–139.

Gilberthorpe, E. and Hilson, G. E., 2014. *Natural resource extraction and Indigenous livelihoods: Development challenges in an era of globalisation.* Farnham, Surrey: Ashgate Publishing.

Girjas case, Swedish Supreme Court Case No. T 853–18, decided 23 January 2020.

Göcke, K., 2012. Recognition and enforcement of Indigenous Peoples' land rights in Alaska, the northern regions of Canada, Greenland, and Siberia and the Russian Far East. *The Yearbook of Polar Law*, 4, 279–304.

Godin, K., Stapleton, J., Kirkpatrick, S. I., Hanning, R. M., and Leatherdale, S. T., 2015. Applying systematic review search methods to the grey literature: A case study examining guidelines for school-based breakfast programs in Canada. *Systematic Reviews*, 4, 138.

Goelman, N., 2014. *Mining a better future: Approaching economic growth and sustainable communities in the context of mining in Nunavut.* Master's Thesis. Simon Fraser University, 120.

Goldenberg, S., Shoveller, J., Ostry, A., and Koehoorn, M., 2008. Youth sexual behaviour in a boomtown: Implications for the control of sexually transmitted infections. *Sexually Transmitted Infections*, 84 (3), 220–223.

Government of the Northwest Territories, 2006. *Communities and diamonds.* 2005 Annual Report of the Government of the Northwest Territories under the BHP Billiton, Diavik and De Beers Socio-Economic Agreements. Yellowknife: Government of the Northwest Territories.

Government of the Northwest Territories, 2009. *Communities and diamonds.* 2008 Annual Report of the Government of the Northwest Territories under the BHP Billiton, Diavik and De Beers Socio-Economic Agreements. Yellowknife: Government of the Northwest Territories.

Graetz, G., 2015a. Ranger Uranium Mine and the Mirarr (Part 1), 1970–2000: The risks of "riding roughshod." *Extractive Industries and Society*, 2 (1), 132–141.

Haida Nation v. British Columbia (Minister of Forests), [2004] 3 S.C.R. 511.

Haley, S., Klick, M., Szymoniak, N., and Crow, A., 2011b. Observing trends and assessing data for Arctic mining. *Polar Geography*, 34 (1–2), 37–61.

Haley, S., Szymoniak, N., Crow, A., and Schwoerer, T., 2011a. *Social indicators for Arctic mining.* ISER Working Paper 2011.2. University of Alaska Anchorage: Institute of Social and Economic Research.

Hall, R., 2012. Diamond mining in Canada's Northwest territories: A colonial continuity. *Antipode*, 45 (2), 376–393.

Haslam-McKenzie, F., Phillips, R., Rowley, S., Brereton, D., and Birdsall-Jones, C., 2009. *Housing market dynamics in resource boom towns.* Australian Housing and Urban Research Institute Final Report No. 135, Housing and Urban Research Institute of Western Australia, Perth. Available from: www.ahuri.edu.au/research/final-reports/135.

Herrmann, T. M., Sandström, P., Granqvist, K., D'Astous, M., Vannar, J., Asselin, H., Saganash, N., Mameamskum, J., Guanish, G., Loon, J. B., and Cuciurean, R., 2014. Effects of mining on reindeer/caribou populations and Indigenous livelihoods: Community-based monitoring by Sami reindeer herders in Sweden and First Nations in Canada. *The Polar Journal*, 4 (1), 28–51.

Hintjens, H., 2000. Environmental direct action in Australia: The case of Jabiluka Mine. *Community Development Journal*, 35 (4), 377–390.

Hirschman, A. O., 1977. A generalized linkage approach to development, with special reference to Staples. *Economic Development and Cultural Change*, 25, Supplement, 67–98.
Hobart, C. W., 1982. Industrial employment of rural Indigenes: The case of Canada. *Human Organization*, 41 (1), 54–63.
Hornal, R. & Associates Ltd., 2003. *Socio-economic baseline study of the Kitikmeot communities, Nunavut and Yellowknife, Northwest Territories*. Prepared for Tahera Corporation's Jericho Diamond Project.
Horowitz, L. S., 2009. Environmental violence and crises of legitimacy in New Caledonia. *Political Geography*, 28 (4), 248–258.
Horowitz, L. S., 2010. Twenty years is yesterday: Science, multinational mining, and the political ecology of trust in New Caledonia. *Geoforum*, 41 (4), 617–626.
Horowitz, L. S., 2011. Interpreting industry's impacts: Micropolitical ecologies of divergent community responses. *Development and Change*, 42 (6), 1379–1391.
Horowitz, L. S., 2012. Translation alignment: Actor-network theory, resistance, and the power dynamics of alliance in New Caledonia. *Antipode*, 44 (3), 806–827.
Horowitz, L. S., 2015. Culturally articulated neoliberalisation: Corporate social responsibility and the capture of Indigenous legitimacy in New Caledonia. *Transactions of the Institute of British Geographers*, 40 (1), 88–101.
Horowitz, L. S., 2016. Rhizomic resistance meets arborescent assemblage: UNESCO World Heritage and the disempowerment of Indigenous activism in New Caledonia. *Annals of the American Association of Geographers*, 106 (1), 167–185.
Horowitz, L. S., 2017. "It shocks me, the place of women": Intersectionality and mining companies' retrogradation of Indigenous women in New Caledonia. *Gender, Place and Culture*, 24 (10), 1419–1440.
Horowitz, L. S., 2021. Indigenous rights and the persistence of industrial capitalism: Capturing the law—ideology—power triple-helix. *Progress in Human Geography*, 45 (5), 1192–1217.
Horowitz, L. S., Keeling, A., Lévesque, F., Rodon, T., Schott, S., and Thériault, S., 2018. Indigenous peoples' relationships to large-scale mining in post/colonial contexts: Toward multidisciplinary comparative perspectives. *The Extractive Industries and Society*, 5 (3), 404–414.
Hubbard, R., 2014. Mining in Greenland and free, prior and informed consent: A role for corporations? *Nordic Environmental Law Journal*, 1, 99–118.
Hudson-Edwards, K., Jamieson, H. E., and Lottermoser, B. G., 2011. Mine wastes: Past, present, future. *Elements*, 7 (6), 375–380.
Hunter, B., Howlett, M., and Gray, M., 2015. The economic impact of the mining boom on Indigenous and non-Indigenous Australians. *Asia & The Pacific Policy Studies*, 2 (3), 517–530.
Hurtig, A. and San Sebastian, M., 2002. Geographical differences in cancer incidence in the Amazon Basin of Ecuador in relation to residence near oil fields. *International Journal of Epidemiology*, 31 (5), 1021–1027.
Huskey, L. and Southcott, C., 2016. "That's where my money goes": Resource production and financial flows in the Yukon economy. *The Polar Journal*, 6 (1), 11–29.
Hydro-Québec, 2004. *Projet de l'Eastmain-1-A-Sarcelle-Rupert. Étude d'impact sur l'environnement.*
Innis, Harold A., 1930a. *The fur trade in Canada: An introduction to Canadian economic history*. Toronto: University of Toronto Press, 1956.
Innis, Harold A., 1930b. The importance of Staple products. *In:* W. T. Easterbrook and M. H. Watkins, eds., *Approaches to Canadian economic history*. Toronto: McClelland and Stewart, 1967.
Innis, Harold A., 1940. *The cod fisheries: The history of an international economy*. Toronto: University of Toronto Press.

Irlbacher-Fox, S. and Mills, S. J., 2007. *Devolution and resource revenue sharing in the Canadian North: Achieving fairness across generations*: Walter and Duncan Foundation Discussion Paper.

Ivsett Johnsen, K., 2016. Land-use conflicts between reindeer husbandry and mineral extraction in Finnmark, Norway: Contexted rationalities and the politics of belonging. *Polar Geography*, 39 (1) 58–79.

Jolley, C. and Rickards, L., 2020. Contesting coal and climate change using scale: Emergent topologies in the Adani Mine controversy. *Geographical Research*, 58 (1), 6–23.

Jourdren, M., 2014. *Création d'emplois et contribution des mines au développement de Kangiqsujuaq, Nunavik: un développement local Nickel?* Master's Thesis. Économie du Développement Durable et de l'Environnement, AgroParis Tech et Centre d'écologie fonctionnelle et évolutive de Montpellier.

Keeling, A. and Sandlos, J., 2009. Environmental justice goes underground? Historical notes from Canada's northern mining frontier. *Environmental Justice*, 2 (3), 117–125.

Keeling, A. and Sandlos, J., 2017. Ghost towns and zombie mines: The historical dimensions of mine abandonment, reclamation and redevelopment in the Canadian North. *In:* S. Bocking and B. Martin, eds., *Ice blink: Navigating northern environmental history*. Calgary: University of Calgary Press, 377–420.

Keeling, A., Sandlos, J., Boutet, J-S., Longley, H., and Dance, A., 2019. Knowledge, sustainability, and the environmental legacies of resource development in Northern Canada. *In:* C. Southcott, F. Abele, D. Natcher, and B. Parlee, eds., *Resources and sustainable development in the Arctic*. London: Routledge, 187–203.

Keenan, J. and Holcombe, S., 2021. Mining as a temporary land use: A global stocktake of post-mining transitions and repurposing. *The Extractive Industries and Society*, 8, 100924.

Keller, J., 2012. *Les impacts socio-économiques de l'exploitation minière sur les communautés autochtones de l'Arctique*. Québec: Ministère des Ressources naturelles et de la Faune du Québec et Géologie Québec.

Kennett, S. A., 1999. *Issues and options for a policy on impact and benefits agreements*. Prepared for the Mineral Resources Directorate, Department of Indian Affairs and Northern Development.

Kirsch, S., 2014. *Mining capitalism: The relationship between corporations and their critics*. Berkeley: University of California Press.

Kivinen, S., Vartiainen, K., and Kumpula, T., 2018. People and post-mining environments: PPGIS mapping of landscape values, knowledge needs, and future perspectives in Northern Finland. *Land*, 7, 151.

Koivurova, T., Masloboev, V., Hossain, K., Nygaard, V., Petrétei, A., and Vinogradova, S., 2015. Legal protection of Sami traditional livelihoods from the adverse impacts of mining: A comparison of the level of protection enjoyed by Sami in their four home states. *Arctic Review on Law and Politics*, 6 (1), 11–51.

Koivurova, T. and Petrétei, A., 2014. Enacting a new mining act in Finland—How were Sami rights and interests taken into account? *Nordic Environmental Law Journal*, 1, 119–133.

Koke, P. E., 2008. *The impact of mining development on subsistence practices of Indigenous Peoples: Lessons learned from Northern Quebec and Alaska*. Master's Thesis. University of Northern British Columbia.

Kruse, P., 2011. Developing an Arctic subsistence observation system. *Polar Geography*, 34 (1–2), 9–35.

Ktunaxa Nation v. British Columbia (Forests, Lands and Natural Resource Operations), [2017] 2 S.C.R. 386.

Kung, A., Holcombe, S., Hamago, J. and Kemp, D., 2022. Indigenous co-ownership of mining projects: A preliminary framework for the critical examination of equity participation. *Journal of Energy & Natural Resources Law*, 40 (4), 413–435.

Kuokkanen, R., 2011. Indigenous economies, theories of subsistence, and women: Exploring the social economy model for Indigenous governance. *The American Indian Quarterly*, 35 (2), 215–240.

Labrador West Status of Women Council and Femmes Francophones de l'Ouest du Labrador, 2004. *Effects of mining on women's health in Labrador West*. In Collaboration with MiningWatch Canada and the Steelworkers Humanity Fund, with Generous Assistance from the Lupina Foundation.

Lacasse, J.-P., 2017. Régime minier et droits ancestraux: une conciliation encore difficile. *In:* C. Krolik, ed., *Le droit des ressources naturelles et de l'énergie. Où en sommes-nous? Où allons-nous?* Montreal: LexisNexis, 135–152.

Land, B. C., Chuhan-Pole, P., and Aragona, F. M., 2015. *The local economic impacts of resource abundance: What have we learned?* (No. WPS7263), 1–69. The World Bank. Available from: http://documents.worldbank.org/curated/en/446761467991987706/The-local-economic-impacts-of-resource-abundance-what-have-we-learned.

Landry, V., Bouvier, A.-L., and Waaub, J.-P., 2009. *La planification territoriale autochtone au Canada: Le rôle de l'évaluation environnementale stratégique dans la cogestion adaptative*. Montréal: GEIGER (Groupe d'études interdisciplinaires en géographie et environnement régional).

Laneuville, P., 2013. *Chasse et exploitation minière au Nunavut: une expérience inuit du territoire à Qamani'tuaq*. Master's Thesis. Université Laval.

Langton, M., 2010. The resource curse: New outback principalities and the paradox of plenty. *Griffith Review*, 28, 46–62.

Langton, M., 2013. *The quiet revolution: Indigenous people and the resources boom*. Sydney: ABC Books and Harper Collins Australia.

Langton, M. and Mazel, O., 2008. Poverty in the midst of plenty: Aboriginal People, the "resource curse," and Australia's mining boom. *Journal of Energy & Natural Resource Law*, 26 (1), 31–65.

Lawrence, R. and Klocker Larsen, R., 2017. The politics of planning: Assessing the impacts of mining on Sami lands. *Third World Quarterly*, 38 (5), 1164–1180.

Le Meur, P.-Y., Horowitz, L. S., and Mennesson, T., 2013. Horizontal and vertical diffusion: The cumulative influence of impact and benefit agreements (IBAs) on mining policy production in New Caledonia. *Resources Policy*, 38, 648–656.

LeCain, T. J., 2009. *Mass destruction: The men and giant mines that wired America and scarred the planet*. New Brunswick, NJ: Rutgers University Press.

LeClerc, E. and Keeling, A., 2015. From cutlines to traplines: Post-industrial land use at the Pine Point Mine. *The Extractive Industries and Society*, 2 (1), 7–18.

Lévesque, F. and Baril D., 2020. *Cohabiter avec le navettage aéroporté. Le cas de Val-d'Or en Abitibi-Témiscamingue*. UQAT. 21p.

Levitus, R., 1991. The boundaries of Gagudju Association membership: Anthropology, law and public policy. *In:* J. Connell and R. Howitt, eds., *Mining and indigenous peoples in Australasia*. Sydney: Sydney University Press, 153–168.

Levitus, R., 2005. Land rights and land economies: The Gagudju Association and the mirage of collective self-determination. *In:* D. Austin-Broos and G. Macdonald, eds., *Culture, economy and governance in Aboriginal Australia*. Sydney: University of Sydney Press, 29–39.

Levitus, R., 2009. Aboriginal organizations and development: The structural context. *In:* J. Altman and D. Martin, eds., *Power, culture, economy: Indigenous Australians and mining*, CAEPR Research Monograph No. 30. Canberra: ANU E Press, 73–98.

Liedholm Johnson, E. and Ericsson, M., 2015. State ownership and control of minerals and mines in Sweden and Finland. *Miner Economics*, 28, 23–36.

Lottermoser, B. G., 2010. *Mine wastes: Characterization, treatment and environmental impacts*. 3rd ed. Berlin: Springer.

Lyons, K., 2019. Securing territory for mining when Traditional Owners say 'no': The exceptional case of Wangan and Jagalingou in Australia. *The Extractive Industries and Society*, 6 (3), 756–766.

Mackintosh, William A., 1923. Economic factors in Canadian economic history. *In:* W. T. Easterbrook and M. H. Watkins, eds., *Approaches to Canadian economic history*. Toronto: McClelland and Stewart, 1967.

Mactaggart, F., McDermott, L., Tynan, A., and Gericke, C., 2016. Examining health and well-being outcomes associated with mining activity in rural communities of high-income countries: A systematic review. *The Australian Journal of Rural Health*, 24 (4), 230–237. Available from: https://doi.org/10.1111/ajr.12285.

Makivik Corporation, 1995. *The Raglan Agreement entered into between Makivik Corporation, Qarqalik Landholding Corporation of Salluit, Northern Village Corporation of Salluit, Nunatulik Landholding Corporation of Kangiqsujuaq, Northern Village Corporation of Kangiqsujuaq and Société Minière Raglan du Québec Ltée to which intervened Falconbridge Limited.*

Manzano, O. and Rigobon, R., 2001. *Resource curse or debt overhang?* National Bureau of Economic Research.

Martin, D., Trigger, D., and Parmenter, J., 2014. Mining in Aboriginal Australia: Economic impacts, sustainable livelihoods and cultural difference at Century Mine, Northwest Queensland. *In:* E. Gilberthorpe and G. Hilson, eds., *Natural resource extraction and Indigenous livelihoods: Development challenges in an era of globalization.* Farnham: Ashgate Publishing, 37–56.

Masher, S., 2013. The Australian approach to recognising the land rights of Indigenous peoples: The Native Title Act 1993 (Cth). *In:* N. Bankes and T. Koivurova, eds., *The proposed Nordic Saami Convention: National and international dimensions of Indigenous property rights*. London: Hart Publishing, 323–350.

Mikisew Cree First Nation v. Canada (Minister of Canadian Heritage), [2005] 3 S.C.R. 388.

Mills, S., Martha D., and Cox D., 2018. Gender in research on Northern resource development. *In*: C. Southcott, F. Abele, D. Natcher, and B. Parlee, eds., *Resources and sustainable development in the arctic*. London: Routledge, 251–270.

Monosky, M. and Keeling, A., 2021. Planning for social and community-engaged closure: A comparison of mine closure plans from Canada's territorial and provincial North. *Journal of Environmental Management*, 277, 111324.

Mortensen, B. O., 2015. Arctic mining: The case of Greenland. *The Yearbook of Polar Law*, 7, 102–127.

Mortensen, B. O. and Barten, U., 2016. The Greenland Self-Government Act: The pitfall for the Inuit in Greenland to remain an Indigenous People? *The Yearbook of Polar Law*, 8, 103–128.

Myette, E. and Riva, M., 2021. Surveying the complex social-ecological pathways between resource extraction and Indigenous Peoples' health in Canada: A scoping review with a realist perspective. *The Extractive Industries and Society*, 8 (2), 100901.

Nachet, L., Beckett, C., and MacNeil, K. S., 2022. Framing extractive violence as environmental (in)justice: A cross-perspective from Indigenous lands in Canada and Sweden. *The Extractive Industries and Society*, 12, 100949.

Natural Resources Canada, 2003. *The social dimension of sustainable development and the mining industry*. A Background Paper. Ottawa: Natural Resources Canada, Minerals and Metals Sector.

Noble, B. F. and Bronson, J., 2005. Integrating human health into environmental impact assessment: Case studies of Canada's northern mining resource sector. *Arctic*, 58 (4), 395–405.

Nordmaling case, Case No. T 4028–07, decided on 27 April 2011 (Supreme Court of Sweden).

Nuttall, M. (2013). Zero-tolerance, uranium and Greenland's mining future. *The Polar Journal*, 3 (2), 368–383.

Nygaard, V., 2016. Do Indigenous interests have a say in planning of new mining projects? Experiences from Finnmark, Norway. *Extractive Industries and Society*, 3 (1), 17–24.

O'Faircheallaigh, C., 1998. Resource development and inequality in Indigenous societies: Aboriginal Politics and Public Sector Management. *World Development*, 26 (3), 381–394.

O'Faircheallaigh, C., 2002. *A new approach to policy evaluation: Mining and Indigenous People*. London: Ashgate Press.

O'Faircheallaigh, C., 2016. *Negotiations in the Indigenous world: Aboriginal peoples and the extractive industry in Australia and Canada*. London: Routledge.

O'Faircheallaigh, C. and Lawrence, R., 2019. Mine closure and the Aboriginal estate. *Australian Aboriginal Studies*, 1, 65–82.

Österlin, C., Heikkinen, H. I., Fohringer, C., Lépy, E., and Rosqvist, G., 2023. Cumulative effects on environment and people. *In:* S. Sörlin, ed., *Resource extraction and Arctic communities*. Cambridge: Cambridge University Press, 109–124.

Owen, J. R., Kemp, D., Lechner, A. M., Harris, J., Zhang, R., and Lèbre, É., 2023. Energy transition minerals and their intersection with land-connected peoples. *Nature Sustainability*, 6, 203–211.

Papillon, M. and Rodon, T., 2020. The transformative potential of Indigenous-driven approaches to implementing free, prior and informed consent: Lessons from two Canadian cases. *International Journal on Minority and Group Rights*, 27, 314–335. Available from: https://doi.org/10.1163/15718115-02702009.

Parlee, B., 2015. Avoiding the resource curse: Indigenous communities and Canada's oil sands. *World Development*, 74, 425–436.

Parlee, B., 2018. Resource development and well-being in Northern Canada. *In:* C. Southcott, F. Abele, D. Natcher, and B. Parlee, eds., *Resources and sustainable development in the Arctic*. London: Routledge, 132–155.

Parlee, B. and O'Neil, J., 2007. The Dene way of life: Perspectives on health from Canada's North. *Journal of Canadian Studies*, 41 (3), 112–133.

Pepper, M., Roche, C. P., and Mudd, G. M., 2014. Mining legacies—Understanding life-of-mine across time and space. *In: Proceedings, life-of-mine 2014*. Brisbane, AU. Available from: www.i2massociates.com/downloads/PepperRocheMudd2014Mining-Legacies.pdf [Accessed 6 February 2018].

Perreault, T., 2013. Dispossession by accumulation? Mining, water and the nature of enclosure on the Bolivian Altiplano. *Antipode*, 45 (5), 1050–1069.

Petkova, V., Lockie, S., Rolfe, J., and Ivanova, G., 2009. Mining developments and social impact on communities: Bowen Basin case studies. *Rural Society*, 22 (2), 211–228.

Ravna, O., 2011. The process of identifying land rights in parts of Northern Norway: Does the Finnmark Act prescribe an adequate procedure within the national law? *The Yearbook of Polar Law*, 3 (1), 423–453.

Ravna, O. and Bankes, N., 2017. Recognition of Indigenous land rights in Norway and Canada. *International Journal on Minority and Group Rights*, 24, 70–117.

Rice, R., 2020. The politics of free, prior and informed consent: Indigenous rights and resource governance in Ecuador and Yukon, Canada. *International Journal on Minority and Group Rights*, 27, 336–356.

Rio Tinto Alcan inc. v. Carrier Sekani Tribal Council, [2010] 2 S.C.R. 650.

Roche, C., Thygesen, K., and Baker, E., eds., 2017. *Mine tailings storage: Safety is no accident. Nairobi and Arendal, Norway*. United Nations Environment Programme and GRID-Arendal.

Rodon, T., 2017. Development in Nunavik: How regional and local initiatives redefine sustainable development in Nunavik. *American Review of Canadian Studies*, 47, 176–188.

Rodon, T., 2018. Institutional development and resource development: The case of Canada's Indigenous peoples. *Canadian Journal of Development Studies/Revue canadienne d'études du développement*, 39 (1), 119–136.

Rodon, T., Lemus-Lauzon, I., Séguin, J.-M., and Schott, S., 2022. Resource revenue allocation strategies and Indigenous community sustainable development. *In:* C. Southcott, F. Abele, D. Natcher, and B. Parlee, eds., *Extractive industry and the sustainability of Canada's Arctic communities*. Montreal: McGill-Queen's University Press, 24–52.

Rolfe, J., Gregg, D., Ivanova, G., Lawrence, R., and Rynne, D., 2011. The economic contribution of the resources sector by regional areas in Queensland. *Economic Analysis and Policy*, 41 (1), 15–36.

Sachs, J. D. and Warner, A., 2001. The curse of natural resources. *European Economic Review*, 45 (4–6), 827–838.

Sachs, J. D. and Warner, A. M., 1995. *Natural resource abundance and economic growth*. Working Paper No. 5398. National Bureau of Economic Research. Available from: www.nber.org/papers/w5398.

Samson, C., 2016. Canada's strategy of dispossession: Aboriginal land and rights cessions in comprehensive land claims. *Canadian Journal of Law and Society*, 31 (1), 87–110.

Samson, C. and Cassell, E., 2013. The long reach of frontier justice: Canadian land claims "negotiation" strategies as human rights violations. *The International Journal of Human Rights*, 17 (1), 35–55.

Sandlos, J. and Keeling, A., 2012. Claiming the new North: Development and colonialism at the Pine Point Mine, Northwest Territories, Canada. *Environment and History*, 18 (1), 5–34.

Sandlos, J. and Keeling, A., 2016a. Toxic legacies, slow violence, and environmental injustice at Giant Mine, Northwest Territories. *The Northern Review*, 42, 7–21.

Sandlos, J. and Keeling, A., 2016b. Aboriginal communities, traditional knowledge, and the environmental legacies of extractive development in Canada. *The Extractive Industries and Society*, 3 (2), 278–287.

Schott, S., Belayneh, A., Boutet, J. S., Rodon, T., and Seguin, J. M., 2022. Mining economies, mining families: The impacts of extractive industries on economic and human development in the eastern subarctic. *In:* C. Southcott, F. Abele, D. Natcher, and B. Parlee, eds., *Extractive industry and the sustainability of Canada's Arctic communities*. Montreal: McGill-Queen's University Press, Chapter 3, 53–77. Available from: www.mqup.ca/extractive-industry-and-the-sustainability-of-canada—s-arctic-communities-products-9780228011552.php.

Schweitzer, P., Stammler, F., Ebsen, C., Ivanova, A., and Litvina, I., 2018. Social impacts of non-renewable resource development on Indigenous communities in Alaska, Greenland, and Russia. *In:* C. Southcott, F. Abele, D. Natcher, and B. Parlee, eds., *Resources and sustainable development in the Arctic*. London: Routledge, 42–64.

Scott, D. N. and Smith, A. A., 2017. "Sacrifice zones" in the green energy economy: Toward an environmental justice framework. *McGill Law Journal*, 62 (3), 861.

Selbu case, Rt. 2001 s. 769 (Supreme Court of Norway).

Shandro, J. A., Veiga, M. M., Shoveller, J., Scoble, M., and Koehoorn, M., 2011. Perspectives on community health issues and the mining boom—Bust cycle. *Resources Policy*, 36, 178–186.

Simons, P. and Collins, L., 2010. Participatory rights in the Ontario mining sector: An international human rights perspective. *McGill Journal of Sustainable Development Law and Policy*, 6 (2), 175–216.

Sondergaard, J., Asmund, G., Johansen, P., and Rigét, F., 2011. Long-term response of an Arctic fiord system to lead-zinc mining and submarine disposal of mine waste (Maarmorilik, West Greenland). *Marine Environmental Research*, 71, 331–341.

Sosa, I. and Keenan, K., 2001. *Impact benefit agreements between Aboriginal communities and mining companies: Their use in Canada*. Toronto: Environmental Mining Council of British Columbia, Canadian Environmental Law Association and CooperAcción: Acción Solidaria para el Desarrollo.

Southcott, C., Abele, F., Natcher, D., and Parlee, B., 2022. Extractive industry and the sustainability of Canada's Arctic communities: An introduction. *In:* C. Southcott, F. Abele, D. Natcher, and B. Parlee, eds., *Extractive industry and the sustainability of Canada's Arctic communities*. Montreal: McGill-Queen's University Press, 3–23.

Southcott, C. and Natcher, D., 2018. Extractive industries and Indigenous subsistence economies: A complex and unresolved relationship. *Canadian Journal of Development Studies/Revue canadienne d'études du développement*, 39 (1), 137–154.

Strelein, L., 2006. *Compromised jurisprudence: Native title cases since Mabo*. Canberra: Aboriginal Studies Press.

Szablowski, D., 2010. Operationalizing free, prior, and informed consent in the extractive industry sector? Examining the challenges of a negotiated model of justice. *Canadian Journal of Development Studies*, 30 (1–2), 111–130.

Taylor, J., 2009. Data mining: Indigenous peoples, applied demography and the resource extraction industry. *In:* J. Altman and D. Martin, eds., *Power, culture and economy: Indigenous Australians and mining*. Research Monograph No. 30, CAEPR. Canberra: The Australian National University E Press.

Thériault, S., 2010. Repenser les fondements du régime minier québécois au regard de l'obligation de la Couronne de consulter et d'accommoder les peuples autochtones. *McGill Journal of Sustainable Development Law and Policy*, 6 (2), 1–52, 217–245.

Thériault, S., 2015. Aboriginal Peoples' consultations in the mining sector: A critical appraisal of recent reforms in Québec and Ontario. *In:* M. Papillon and A. Juneau, eds., *Aboriginal multilevel governance*. Canada: The State of the Federation 2013. Montreal: McGill-Queen's University Press, 143–162.

Thériault, S., Bourgeois, S., and Boirin-Fargues, Z., 2022. Indigenous Peoples' agency within and beyond rights in the mining context: The case of the Schefferville region. *The Extractive Industries and Society*, 12, 100979.

Tiainen, H., 2016. Contemplating governance for social sustainability in mining in Greenland. *Resources Policy*, 49, 282–289.

Tolazzi, S., 2015. Channeling Indigenous contestation of uranium mining in Australia: Legislation, negotiation, co-optation. *In:* E. Avril and J. N. Neem, eds., *Democracy, participation and contestation: Civil society, governance and the future of liberal democracy*. London: Routledge, 168–182.

Tsilhqot'in Nation v. British Columbia, [2014] 2 S.C.R. 257.

Tsosie, R., 2015. Indigenous peoples and the ethics of remediation: Redressing the legacy of radioactive contamination for native peoples and native lands. *Santa Clara Journal of International Law*, 13 (1), 203–272.

Vale Website, 2023. Snapshot of Operations. Available from: www.vale.com/voisey-s-bay [Accessed 31 October 2023].

Vanthuyne, K. and Gauthier, M., 2022. Mining the land while sustaining Iiyiyiuituwin: Exercising Indigenous sovereignty through collaboration in Eeyou Istchee. *Canadian Journal of Political Science/Revue canadienne de science politique*, 55 (2), 279–299.

Voisey's Bay Nickel Company Ltd., 1997. Voisey's Bay Mine/Mill Project Environmental Impact Statement. Available from: https://iaac-aeic.gc.ca/archives/evaluations/5EA5DD6D-1/default_lang=En_n=0A571A1A-1_offset=2_toc=show.html [Accessed 8 November 2023].

Voulvoulis, N., Skolout, J. W. F., Oates, C. J., and Plant, J. A., 2013. From chemical risk assessment to environmental resources management: The challenge for mining. *Environmental Science and Pollution Research*, 20 (11), 7815–7826.

Wenzel, G., 2013. Inuit and modern hunter-gatherer subsistence. *Études/Inuit/Studies*, 37 (2), 181–200.

Worrall, R., Neil, D., Brereton, D., and Mulligan, D., 2009. Towards a sustainability criteria and indicators framework for legacy mine land. *Journal of Cleaner Production*, 17, 1426–1443.

Yamarak, L. and Parton, K. A., 2023. Impacts of mining projects in Papua New Guinea on livelihoods and poverty in Indigenous mining communities. *Mineral Economics*, 36, 13–27. Available from: https://doi.org/10.1007/s13563-021-00284-1.

Young, S., 2008. *The trouble with tradition: Native title and cultural change*. Annandale: Federation Press.

Zahara, A., Keeling, A., and Bell, T., 2016. *Social licence to operate: Background and state of knowledge report*. Department of Geography, Memorial University.

6 Environmental assessment as a knowledge infrastructure

Unpacking politics and power in impact evaluations for Indigenous communities

Ella Myette

Introduction

Globally, the mining industry represents an interesting case study of extractive capitalism run rampant, with its unequal distribution of wealth and externalities, problematic corporate–government relations, and massive private profits, often gained at the expense of environmental quality and the public good. In Canada, resource companies have systematically and efficiently expanded their reach into Indigenous lands by establishing new "critical infrastructure" projects (Cowen 2020; Crosby 2021; Pasternak and Dafnos 2018; Spice 2018). These projects have been widely problematized and contested, not only because of their harmful effects on the environment but also because they are seen as infrastructures of settler colonial invasion (Coulthard 2014; Cowen 2020; Hall 2013; Keeling and Sandlos 2015; Pasternak and Dafnos 2018; Spice 2018). Mining projects operating on Indigenous territories fit squarely into a long history of colonial expansion and Indigenous dispossession, and in some cases, corporate entry onto communities' lands has helped advance the extinguishment of Indigenous title and jurisdiction (Coulthard 2014; Hall 2013).

Extraction projects pose a particular risk to Indigenous Peoples' health because they can reactivate trauma, threaten local sources of good health, and physically and discursively separate communities from their lands (Myette and Riva 2021; Richmond and Big-Canoe 2018). Indigenous Peoples have distinct health ontologies that extend beyond biomedical paradigms and draw instead on a broader perspective of health, including emotional, physical, spiritual, and mental health (Richmond and Big-Canoe 2018). Being rooted in place and space and recognizing and honoring connections with the land are of particular importance to maintaining good health for many Indigenous communities (Richmond and Big-Canoe 2018). As such, there are a wide range of direct and indirect pathways through which projects can affect community health (Myette and Riva 2021). However, the impacts of extractive projects on health are evaluated through settler technocratic methods like environmental assessment (EA), a process used by federal and provincial governments to understand how proposed development projects can negatively affect the environment and surrounding communities.

DOI: 10.4324/9781003406433-9

This chapter has been made available under a CC-BY-NC-ND 4.0 license.

Many of the pathways through which extractive projects affect Indigenous health determinants are overlooked in EA, often because standard assessment methodology does not allow for sufficient complexity and nuance (Jones and Bradshaw 2015; Myette and Riva 2021). Indigenous communities are frequently asked to dedicate considerable time and energy to contributing information and feedback during assessments, but there are few avenues for this knowledge to be meaningfully integrated (Simpson 2001). Plus, interacting with government and corporate actors and trying to effectively translate local knowledge into EAs can create intense stress and undue burden on Indigenous communities (Booth and Skelton 2011). Still, EA is unable to account for the full breadth of potential impacts that extractive industries have on Indigenous Peoples' health, or how extraction is situated within a larger history of colonial traumas (Dendena and Corsi 2015; Jones and Bradshaw 2015).

This chapter expands on two earlier works: A scoping review identifying various connections between extractive activities and Indigenous health determinants (Myette and Riva 2021) and an analysis of mining project assessments in Canada (Myette 2022). The scoping review aimed to determine how extractive projects can affect Indigenous health and surveyed myriad pathways through which extractive development can affect a wide range of Indigenous health determinants. The variation in these pathways through space and time showed that a holistic perspective is vital to adequately measure and understand the effects on Indigenous Peoples' health (Myette and Riva 2021). Following this review was a qualitative document analysis of 28 EAs that integrated coding and thematic analysis to investigate if and how Indigenous health determinants are included in the assessment process. This analysis found that assessment methodologies were often at odds with community health ontologies due to their narrow focus and inability to consider the complexity and essentiality of human–nonhuman relations (Myette 2022). Moreover, there was often little regard for communities' recommendations about what should be included in the assessment and how, meaning that community understandings of health were discounted by assessors (Myette 2022). Apart from methodology, there were significant issues with consultation processes that affected the ability of communities to engage, leading to frustration and stress (Myette 2022). Therefore, the primary conclusion of these projects was that EA is largely ineffective at fully or accurately assessing how extractive projects impact Indigenous Peoples' health. This chapter engages with framing techniques from critical infrastructure studies to provide a broader interpretation of these works, describing how EA operates, how this process can affect Indigenous communities, and situating EA policy within Canada's historical and social context.

I argue that not only is EA inappropriate, inaccurate, and insufficient for assessing health effects for Indigenous communities, but also that the supposed problems with this process are not signs of system or design failure. Rather, I characterize EA as a knowledge infrastructure that seeks to enable and support extraction, the material infrastructures that accomplish it, and the colonial and racial dynamics that imbue it. Edwards (2010) defines knowledge infrastructures as "robust networks of people, artifacts, and institutions that generate, share, and maintain specific knowledge about the human and natural worlds" (p. 17). The knowledge infrastructures supporting the mining industry—including EA—play a key role in directing, containing, managing, and reducing the unruly politics and contestation centered at sites of invasion (Barry 2013). Ultimately, I assert that EA prioritizes processual efficiency over substantive engagement and that it largely works to capture and transform criticism in order to expedite the development of material resource infrastructures, regardless of a project's effects.

The EA apparatus

EA legislation in Canada was originally based heavily on an American precedent, the 1969 *National Environmental Policy Act* (NEPA). NEPA was preceded by a decade of rising public concern about the environment in the United States, along with increased support for new legislation designed to limit adverse environmental effects (Caldwell 1998). NEPA's visibility had important implications in Canada, where citizens were becoming increasingly wary of environmental destruction and had "deepening scepticism about Government and corporate reassurances" (Gibson 2002, p. 153). After two decades of legislative development, the *Canadian Environmental Assessment Act* (CEAA) came into force in 1995. CEAA 1995 was overseen by the Canadian Environmental Assessment Agency (the Agency), which sought to "ensure the practices and policies of CEAA are followed by all parties involved, including government" (Paci *et al.* 2002, p. 114). The fundamental principles that the Agency looked to enforce from the Act included facilitating public participation, ensuring that EAs were completed within a set timeframe, and reducing environmental effects (Paci *et al.* 2002). The assessment process included two steps—an initial screening and a more thorough comprehensive study that integrated a regional approach. However, CEAA 1995 was lacking in a few key areas: by retaining "an apparently restrictive definition of 'environment' that omits direct socio-economic and cultural effects" and by designating assessment as a "largely advisory exercise" (Gibson 2002, p. 155). The legislation also contained "only weak provisions for enforcing compliance with the law and with terms and conditions of approvals," limiting its usefulness and strength (Gibson 2002, p. 156).

Stephen Harper's Conservative government began making unilateral changes to CEAA in 2010 in an effort to streamline the process and facilitate development projects. Eventually, a revised version of CEAA 1995 was included in the 2012 Budget Implementation Bill, limiting potential public debate and amendments and obscuring proposed changes (Doelle 2012). This revised version, CEAA 2012, was considerably different from the original legislation, greatly reducing both the number of assessments being conducted and the scope of these assessments. The original two-step process was combined into one environmental impact assessment that was much narrower in focus than either the screening or the comprehensive phases outlined in CEAA 1995 (Doelle 2012). Although the definition of an environmental effect was restrictive in CEAA 1995, there was at least an effort to understand the broader implications of biophysical changes, as per the guidelines of the comprehensive study process. In contrast, CEAA 2012 limited this even further, identifying a very small number of components that would be included in the assessment. This inevitably restricted the ability of EAs to recognize and understand the wide range of impacts that these projects have on Indigenous Peoples, presenting a "critical issue" (Doelle 2012, p. 12).

Under both laws, the primary method used by assessment practitioners for evaluating a project's impacts is through evaluating effects on valued components (VCs). This practice was implemented in Canada in the 1980s to improve the efficiency of and provide more focus on the scope of impact analysis (Olagunju and Gunn 2015). VCs are defined as fundamental elements of the environment with economic, physical, social, or cultural importance that might be affected by a project. During assessment, project proponents and federal authorities identify VCs to undergo analysis by looking at potential environmental impacts, consulting with scientific experts, and receiving feedback from the public (Olagunju and Gunn 2015). As part of this public consultation, Indigenous communities can suggest VCs to be included in the analysis. However, information about this process is largely inaccessible and obscure. Once VCs are finalized, they undergo a residual effects analysis, which identifies expected effects on indicators related to each VC after mitigation measures are implemented (Olagunju and Gunn 2015). These effects are then designated significant or not significant based on a set of criteria. In the sample of projects included in my qualitative analysis, almost none of the indicators (2/652) measured in the residual effects analysis were found to be significant (Myette 2022). Communities also criticized the baseline data, methods, indicators, and mitigation measures included in these analyses. Similarly, the way that significance was determined was controversial, with many communities disagreeing with the criteria used by assessors. These significance criteria often also failed to account for complicated and asymmetrical historical and political contexts (Myette 2022).

While VCs are based on requirements outlined in CEAA 1995 and 2012, there is no legally mandated list of VCs or related indicators that a project must consider. In their study of VC selection for road construction companies, Olagunju and Gunn (2015) found that "most case informants view VC selection as a 'value-ridden' and 'highly subjective' process . . . based on negotiation (as opposed to scientific evidence) conducted with little regard to the specific context of the project" (p. 210). Campbell *et al.* (2020) found similar results in their review of assessments documenting the effects of oil sand development on wildlife. The authors stated that the parameters undergoing analysis were "not comprehensive nor standardized between EIAs, despite a high degree of landscape similarity between projects" and that there was "very little agreement" in the indicators measured across the sample (Campbell *et al.* 2020, p. 129). In my analysis, inconsistency in the EAs created doubt about what was assessed for community members, and proponents' explanations of these analyses were often overly combative, which alienated and frustrated communities (Myette 2022).

The VC selection and measurement process falls especially short for Indigenous communities for several reasons. First, as per CEAA 1995, proponents do not need to take a regional approach in their assessments, meaning that VC measurement tends to be focused exclusively on direct effects stemming from the project site (Ball *et al.* 2013). Not only does this reduce the scope of assessments, it also prevents measurement methods from taking into consideration Indigenous ways of knowing about nature and health, which prioritize balance, interconnection, and relationality (Richmond and Big-Canoe 2018). In EA, the complex interrelations between humans and nonhumans that are visible with an Indigenous perspective are disregarded (LaDuke and Cowen 2020; Myette 2022; Spice 2018). Additionally, Ball *et al.* (2013) explain that assessing the impacts on VCs is "largely 'stressor-based'; that is, focused on identifying project-induced stress and predicting the contribution of that stress to a change in baseline conditions" (p. 470). However, this often fails to account for stressors on communities that exist around the site and that are cumulative over time, including historical and intergenerational traumas that may be reactivated through extraction and the assessment process itself (Myette 2022).

A large part of the external optics of EA is the presence of public participation opportunities and the integration of public feedback. In reality, there are major issues with this process regarding accessibility and timing, namely, due to a lack of formal requirements about how and when to involve communities. While informal guidelines suggested early notification of communities, accessible and appropriate information sharing, co-developing a public participation plan, and flexible and adaptive activities, without formal legal directives, it is difficult to assess how thoroughly these suggestions were implemented during the almost 20-year period when CEAA 1995 was operational. Under CEAA 2012, opportunities for formal public participation became fixed, distilled to five, set periods of between 20 and 30 days for the public and Indigenous communities to give comments and feedback.

While these public comment periods are vital to intervening in, modifying, and improving assessments, they are extremely short. This may inhibit people's ability to participate if they are unable to access, read, and submit their comments in the given timeframe. Limiting comment periods significantly reduces the window for communities to effectively organize and intervene in EAs, which, as Gabrys (2016) writes, "focuses the complexity of civic action toward a relatively reductive if legible set of actions" (p. 203). It is also worth noting that assessment documents are often extremely technical, hundreds of pages long, and generally difficult to read. They are filled with tables and appendices, and the format can be confusing and circular, especially if someone has not been previously exposed to similar reports. These characteristics each pose different accessibility issues, as people may not be able to digest this type and amount of information in such short timeframes. Plus, many communities are increasingly facing multiple, concurrent development proposals, creating a tidal wave of documentation that can be simply unmanageable (Baker and Westman 2018). Together, these conditions make the comment periods not only problematic but also culturally inappropriate for many communities (Baker and Westman 2018).

Understanding infrastructure

According to Edwards *et al.* (2013, p. 15), "knowledge infrastructures do not only provide new maps to known territories—they reshape the geography itself." Although it is difficult to precisely define the term "infrastructure" because of its wide diversity, infrastructures can typically be understood as "extended material assemblages that generate effects and structure social relations, either through engineered (i.e., planned and purposefully crafted) or non-engineered (i.e., unplanned and emergent) activities" (Harvey *et al.* 2016, p. 5). Infrastructures are not neutral, but rather are embedded within and express certain politics (Winner 1980). Many scholars have documented the substantial, and often detrimental, effects that material infrastructures of extraction have on communities (Curley 2021; LaDuke and Cowen 2020; Pasternak and Dafnos 2018; Spice 2018). However, these material infrastructures do not exist alone; rather, they are informed by their social environments, as explained by Filion and Keil (2017):

> We cannot consider physical infrastructures independently of the political, organizational, know-how and financial requirements for their design, construction, operation and maintenance. For all their apparent sturdiness, physical infrastructures are transient relative to the societal conditions essential to their existence . . . Infrastructures cannot be perceived as purely physical artefacts; they must be seen in their broad societal context.
>
> (p. 8)

Material infrastructures of extraction are situated within a web of other interconnecting infrastructures, including financial (Stanley 2016), institutional (Hall 2013), and knowledge infrastructures (Hoogeveen 2016).

I frame EA as a knowledge infrastructure to highlight the power it has over knowledge creation and communication. Infrastructures generally are built and maintained through the "intensely political project of creating and communicating information" (Siemiatycki *et al.* 2020). Operationalizing the term "knowledge infrastructure" in this context helps politicize the ways that knowledge is accumulated, translated, and processed, as well as the infrastructural arrangements that manage, contain, and distribute this information (Edwards 2013). Much of the infrastructural power of EA relates to the types of knowledge prioritized and validated through this process. Important questions about what constitutes data, evidence, expertise, impact, risk, and health are moderated and defined through EA (Myette 2022). And, the answers EA provides to those questions recognize, affirm, and amplify a specific type of knowing grounded in colonial history and racial logics (Myette 2022). Edwards *et al.* (2013) explain that "as knowledge infrastructures shape, generate, and distribute knowledge, they do so differentially, often in ways that encode and reinforce existing interests and relations of power" (p. 14). As such, beyond any effects on people directly participating in this process, EA has other real-world political implications for social relations and hierarchies.

Infrastructural violence

Colonial expansion and racial violence have accompanied and justified Canada's territorial spread since early settlers arrived on the continent (Bernauer 2019; Coulthard 2014; Cowen 2020). Extraction has been an intrinsic part of this development, and extractive industries have both supported and benefited from the settler colonial project (Wolfe 2006; Yellowhead Institute 2019). Understanding these connections helps frame current conversations about the Canadian mining industry within a long history of displacement, dispossession, and violence against Indigenous Peoples. The social and material infrastructures that work to accomplish the settler colonial project perpetrate violence against Indigenous Peoples and other racialized communities (Cowen 2020; Crosby 2021; Wolfe 2006). In particular, the state's push to secure "critical infrastructure" projects from so-called threats from Indigenous Peoples asserting sovereignty or protesting jurisdictional violations has repeatedly been used as justification for physical violence against communities (Crosby 2021; LaDuke and Cowen 2020; Spice 2018; Stanley 2016).

However, there are other, less visible forms of violence against Indigenous communities that also seek to undermine Indigenous jurisdiction and reify the settler colonial project, albeit perhaps more covertly. Nixon (2011) conceptualizes slow violence as a type of violence "that is neither spectacular nor instantaneous, but rather incremental and accretive, its calamitous repercussions playing out across a range of temporal scales" (p. 2). I argue that EA qualifies as one such type of violence, as its infrastructure creates conditions where Indigenous sovereignty can be challenged and colonial knowledge enshrined and legitimized. Dene scholar Glen Coulthard (2014) contends that "in the Canadian context, colonial relations of power are no longer reproduced primarily through overtly coercive or imposed means, but rather through the asymmetrical exchange of mediated forms of state recognition and accommodation" (p. 62). EA represents one example of asymmetrical exchange, as Indigenous communities are asked to participate in demanding and arduous dialogue with companies and government bodies, often without significant ability to disrupt the process or produce a better result (Baker and Westman 2018). This is not to say that these policies and processes act completely unilaterally—communities have carved out spaces for agency and self-determination within these systems through persistent hard work (Yellowhead Institute 2019). However, the playing field for these negotiations remains incredibly uneven, meaning that communities can be subject to extraordinary stress even when engaging in processes where they have more self-determination or influence (Baker and Westman 2018; Booth and Skelton 2011). While this kind of regulatory policy may not seem as insidious as material infrastructures, EA represents a pervasive form of slow violence that accompanies and facilitates physical invasions on Indigenous lands (Ladner 2014; Yellowhead Institute 2019). Together, these different forms of violence equally contribute to Indigenous dispossession and to accomplishing colonial and capitalist goals (Curley 2021; Ladner 2014; Yellowhead Institute 2019).

Considering EA through an infrastructural lens

Popularized by historian of science and technology Geoffrey C. Bowker, "infrastructural inversion" is a method that focuses on "[looking] closely at technologies and arrangements that, by design and by habit, tend to fade into the woodwork" (Bowker and Star 1999, p. 34). Infrastructures are generally invisible, and their role in mediating political acts and practices often recedes into the background (Bowker and Star 1999). Similarly, the critical importance and weight of EA are sometimes overlooked in extractive settings, relative to the physical prominence of other material and technological infrastructures. However, I argue that EA itself represents a site of political significance in the realm of mining development. Infrastructural inversion, as a practice, helps create a critical framework for analyzing infrastructures, their histories, and their functions (Bowker and Star 1999). The methodological exercises related to practicing infrastructural inversion aim to "go backstage," to recognize the choices and processes that have resulted in infrastructures (Goffman 1956). The four framing techniques that I will use to analyze this infrastructure include (1) recognizing the ubiquity, interdependence, and integration of classification schemes and standards; (2) highlighting the material and symbolic nature of these classifications and standards; (3) revising knowledge of the past to seek out different voices and silences; and (4) "uncovering the practical politics of classifying and standardizing" (Bowker and Star 1999, p. 44). For the rest of this chapter, I will focus on implementing these four techniques in order to explore some key issues with EA and how it affects Indigenous communities.

Recognizing ubiquity

The first technique—recognizing ubiquity—is about identifying and unpacking the classifications that inform and support the infrastructure. In this case, that means examining the understandings perpetuated through EA and how they can disregard Indigenous knowledge and reinforce colonial logics. In particular, the foundational classifications of EA, such as "effects," "significance," and "health," warrant interrogation as there is a substantive and political divergence between Indigenous Peoples' and assessment practitioners' ways of knowing (Hoogeveen 2016; Myette 2022). Beyond a lexical difference, Indigenous communities often hold a fundamentally different perspective about how to be in the world—a perspective that encompasses "all our relations" (Hoogeveen 2016). The understanding of health advanced in EA is mostly constrained to the biomedical health model, even though understandings of impact, risk, and health are socially constructed, politically negotiated, and based in community ontologies (Dake 1992). This partiality toward Western biomedicine prevents Indigenous Peoples' holistic visions about health from being properly understood and considered, despite the extraordinary relevance and importance of these ontologies to communities (Myette 2022).

There are also many holes in assessment science that further compound its impracticality for communities. Even the idea that the selection and measurement of VCs are somehow objective feels ironic, given that value is inherently subjective. Barry (2013) explains, "scientists concerned with the problem of environmental impacts do not aim to grasp such issues in all of their complexity; their work is expected to enact impacts in forms that render them amenable to management . . . impacts are abstractions" (p. 118–119). Rather than allowing for a wide scope of analysis, potential impacts on VCs are distilled to establish testable effects, like the amount of a certain toxin in the air or the percentage of habitat lost due to drilling.

In reducing complexity and nuance, processual efficiency is prioritized over substantive engagement with Indigenous communities and values. Not only is this detrimental to communities in that their perspectives are devalued, but it also means that EA recognizes and affirms a kind of scientific knowing characterized by limitations and incuriosity.

Material and symbolic nature

The focus of the second technique is on identifying the material and symbolic value and significance of the infrastructure. EA has a great deal of material force in the realm of extraction, producing information that enables the continued invasion of physical infrastructures on Indigenous lands (Mezzadra and Neilson 2019; Spice 2018). During an assessment, a project's impacts are determined as significant or not, with these results helping to determine whether a project will be approved. Therefore, project approval is based on if and how knowledge about health, risk, and impact is understood and defined in the assessment. Apart from this, assessment processes support an entire side industry of practitioners, including advisors, managers, experts, specialists, and analysts (Baker and Westman 2018). However, these consultants often do not have sufficient education and training in social science practice or enough cross-cultural work experience to produce accurate or sufficient data for impact evaluations (Baker and Westman 2018). Still, these industry professionals bring a secondary level of ethical professionalism and validation to the assessment process through their presence and the information they supply (Barry 2013). The financial support and validation of this industry are other important material outcomes of EA as an infrastructure.

EA also does considerable work in terms of symbolically granting and enforcing legitimacy and credibility. The validity and trustworthiness of EA stem from its supposedly substantive community engagement and integration of community feedback. In recent years, there have been efforts to "ethicize" the mining industry and regain public trust in the assessment process in the face of public criticism (Barry 2013). Increasing involvement of Indigenous communities has been seen as a way to increase corporate accountability. However, assessment strategies can create undue stress on communities, many of whom primarily participate to mitigate potential damage (Booth and Skelton 2011). Still, companies can greatly benefit from engaging with EA infrastructure even in very narrow ways and with the most tokenized and limited forms of community inclusion. Company's brands are able to get a reputation boost and heightened credibility by spreading word of their community engagement activities (Barry 2013; Li 2015). Additionally, selectively opening space for public participation allows companies to shift discussions about responsibility and accountability to frame environmental protection as a "*shared* concern" with the public, rather than the duty of companies and the state (Li 2015, p. 199). As Barry (2013) explains, knowledge infrastructures center certain objects and problems as a matter of public focus, concern, and debate "in the expectation that this will enable the form and intensity of public debate to be contained, by rendering it more rational and informed than it might otherwise have been" (p. 11).

By enveloping community concerns and criticism into EA, companies are able to "manage, channel, and translate" critiques, appropriating them into something that will incidentally benefit the project (Barry 2013, p. 80).

Revising knowledge of the past

The third technique—revising knowledge of the past—asks us to rethink the infrastructure's history and development. For EA, that means recognizing that assessment policy was created within a settler colonial system that is not primarily concerned with Indigenous Peoples' needs, or the needs of racialized communities in general. Not only does this clarify current problems with the infrastructure's design, but it also highlights the persistent inequalities that will continue without substantive systemic action. Historical and contemporary natural resource extraction is intrinsically and closely linked to colonial systems, and particularly to settler colonialism (Bernauer 2019; Coulthard 2014; Hall 2013; Keeling and Sandlos 2015; Wolfe 2006). EA policy was conceived and developed within this colonial social infrastructure, which explains some of its central issues: EA policy is not designed to overturn, or even challenge, the colonial relations or racial logics that are embedded within it or within the infrastructures that surround extractive industries more generally.

Both extractive activities and impact evaluation processes can disproportionately affect Indigenous communities. Projects have inordinate impacts on Indigenous communities' health because they impact a multitude of health determinants simultaneously and because they create cumulative effects that can be compounded by other forms of historical trauma (Myette and Riva 2021). However, EA as an infrastructure itself also disproportionately affects Indigenous communities, continuing a pattern of Crown–Indigenous relations that are often unequal, performative, and harmful. The potential of EA to reactivate stress and trauma to a higher degree for Indigenous communities can be seen as a form of environmental racism—systematic discrimination based on race and furthered by policy. Pulido (2017) coined the term "environmental racism gap" to describe the perpetual inequalities between White and racialized communities resulting from unevenly distributed policies and regulation. In this case, the environmental racism gap constitutes both the differential impacts that mining sites have on Indigenous communities and the heightened stress and marginalization that Indigenous Peoples face in the assessment process compared with non-Indigenous participants. The idea of the "gap" highlights the fact that universal policies and regulations, including EA, do not serve racialized communities because they fail to account for the historic systems that drive contemporary oppression (Pulido 2017). Although EA scientific analyses give an illusion of equality, these methods are not equitable. Until the colonial and racist power structures underlying and supporting EA are changed, this infrastructure can continue to disproportionately affect Indigenous communities.

There are many other impact evaluation practices used alongside EA, ranging from more established practices to ones that are still emerging in the Canadian context. These include social impact assessment, health impact assessment, sustainability assessment, and social life cycle assessment. Dendena and Corsi (2015) explain that "the practice of impact assessment has, in fact, resulted in the flourishing of several methods, aimed to better capture the complexity of reality by introducing different perspectives of analysis" (p. 969). There are a wide range of terms used to describe these other assessment practices to reflect the scope of analysis. Guidelines for these practices emphasize the need for flexibility, accommodation, and community involvement compared with existing EA (Harris-Roxas *et al.* 2012). Still, these practices warrant investigation. In their article analyzing consultation processes on the Athabasca oil sands, Baker and Westman (2018) note that under-trained assessment consultants often "claim to have invented new research methodologies such as integrated assessment, traditional land use assessment, cultural assessment, etc.," to justify their work, putting the usefulness of these practices in question (p. 151).

These alternative assessments largely share the core problems of EA regarding accessibility and transparency. Public participation is still very flawed; like EA, participation is used mostly to legitimize projects rather than to truly interrogate or substantively change proposals (Dendena and Corsi 2015). Also, these assessments often do not take a regional approach, the data included for analysis can be inaccurate or inappropriate, and there can be significant blind spots when identifying impacts due to practitioners' biases (Baker and Westman 2018; Dendena and Corsi 2015; Jones and Bradshaw 2015). Although the language and vocabulary may vary, the foundational logics of these alternative practices are comparable to EA; they are not developed by communities themselves and often do not properly honor or center Indigenous ways of knowing and ontologies (Baker and Westman 2018; Jones and Bradshaw 2015). Health impact assessments (HIAs), in particular, are generally considered to be more interdisciplinary and holistic than other assessment types and to have more integrated stakeholder engagement practices. Still, HIA practitioners often fail to properly incorporate community knowledge and ontologies and instead rely on narrowly scoped assessment methodologies (Mackie 2012). Even when communities lead their own health assessments, HIA scoping and processual guidelines often cannot contain the breadth and scale of mutual relations between Indigenous Peoples and the environment, which have existed since time immemorial and require reciprocal duties and responsibilities (Jones and Bradshaw 2015; Mackie 2012; Richmond and Big-Canoe 2018).

Additionally, this rise in alternative assessments has created a confusing and overlapping impact evaluation landscape where multiple assessments can be conducted for the same project, often simultaneously (Dendena and Corsi 2015). This can be detrimental to communities, who end up wading through more documentation and participating in more consultation processes (Dendena and Corsi 2015). This new evaluation landscape is another manifestation of the environmental racism gap. Alternative assessment strategies, developed within the same colonial infrastructures as EA, do not provide sufficient flexibility or space to fully integrate and build on Indigenous ways of knowing, meaning that these protocols will continue to be limited and inaccurate (Harris-Roxas *et al.* 2012; Mackie 2012). Plus, asking communities to participate in these additional processes alongside EA can exacerbate existing accessibility issues and create more stress. Simply put, unless they change the social infrastructure underwriting impact evaluation, these alternative assessments will maintain the same systemic barriers found in EA.

Uncovering practical politics

Finally, the fourth technique aims to highlight who benefits from the infrastructure and to provide a critical lens for discussing the infrastructure's successes and failures, along with the political consequences of its design. The success of EA hinges on its ability to efficiently and quickly manage information. Conversely, its failure is that the information it includes and manages can be irrelevant and harmful to communities. Plus, the speed and structure of EA create a multitude of barriers for communities trying to represent their interests. I argue, however, that this infrastructural "failure" still has positive outcomes for companies and for the state, who are able to benefit, regardless, from the political implications of performing EA. Here, I borrow Winner's (1980) definition of politics, as "arrangements of power and authority in human associations as well as the activities that take place within those arrangements" (p. 123).

As a knowledge infrastructure, EA works to capture and transform criticism in order to expedite the development of material resource infrastructures. Without the efficient and successful efforts of EA to manage and contain conditions, projects would take a much longer time to reach the construction phase. Mezzadra and Neilson (2019) explain that "there is a complex interplay among technological advances, knowledge production, and financial manipulation that allows capital to prepare the ground for further extraction" (p. 140). I argue that the infrastructure of EA is one means to "prepare the ground" and that the infrastructure's design serves a vital purpose in keeping the supply chain of extraction moving—here, I do not mean the supply chain that delivers raw materials to consumers, but rather the supply chain of information that maintains the flow of new resource projects, delivering benefits to corporations and to the state. EA design seems to be purposefully made to inhibit the potential ability of any person to disrupt or slow it down. This can be seen in the tight legislated timelines for assessment completion, the limited comment periods, and the ignoring of community questions and comments (Myette 2022). If the infrastructural goal of EA is to advance resource development, then the lack of flexibility in this process is a necessary design choice, as mandating substantive engagement would pose a threat to the expediency that this process tries so hard to achieve and maintain. Any moment of pause, hesitation, or uncertainty is a danger to companies and governments, as it puts at risk the money that will be generated from the project.

In the context of mining projects on Indigenous lands in Canada, EA also represents an important political process working parallel to other violent and colonial material infrastructures. As explained earlier, I assert that EA can be considered a form of slow violence because it is a vehicle for colonial knowledge to be legitimized and Indigenous sovereignty to be disputed. Extractive projects have necessitated heightened interactions between the Crown and Indigenous communities, with companies increasingly acting as instigators and mediators in these exchanges (Jones and Bradshaw 2015). EA has become an important public forum for continued dialogue between these actors, but that dialogue can often lead to fraught and unbalanced negotiations regarding ontology, knowledge, and jurisdiction. Coulthard (2014) asserts that

> In situations where colonial rule does not depend solely on the exercise of state violence, its reproduction instead rests on the ability to entice Indigenous Peoples to identify, either implicitly or explicitly, with the profoundly asymmetrical and nonreciprocal forms of recognition either imposed on or granted to them by the settler state and society.
>
> (p. 25)

By enticing communities to participate in processes like EA that grant recognition, albeit asymmetrically, colonial relations can be continually practiced and reproduced (Coulthard 2014). Communities can feel pressured to engage, to try and ensure that their interests are represented, even if the knowledge, values, and comments that they share are recognized in very limited forms or disregarded entirely (Baker and Westman 2018; Booth and Skelton 2011). Not only is this arduous and stressful for communities, but it also means that EA ultimately provides a platform for the colonial state to discursively erase, devalue, and dismiss Indigenous knowledge and ontologies (Booth and Skelton 2011; Ladner 2014; Yellowhead Institute 2019). Nonetheless, the state and project proponents garner credibility and reputation from communities' involvement and from simply participating in these exchanges (Barry 2013; Li 2015).

Conclusion

This chapter discussed the development of EA, how this process operates and impacts Indigenous communities, and how it interacts with and supports material infrastructures related to resource extraction. This discussion drew upon a critical framework based on infrastructure studies and a previous scoping review and qualitative analysis. Together, these works showed that not only are there fundamental issues with assessment that make it often irrelevant and inappropriate for evaluating effects on Indigenous communities, but also the furthering of colonial logics in the public sphere is harmful and contributes to asymmetrical power dynamics between the Crown and Indigenous Peoples. For this reason, I have framed EA as a knowledge infrastructure that reflects and reifies a larger colonial and racist social infrastructure in Canada and as a form of slow violence that supports the continued and ongoing marginalization of Indigenous Peoples.

In light of these findings, it is clear that we need to move past the question of how to make EA more objective and less political. Rather, we need to ask: Given that EA is necessarily political, how do we compensate for the imbalances of power and interest that are invested and encoded in this process? A new EA law, the *Impact Assessment Act* (Bill C-69) was passed in 2019, replacing CEAA 2012. While this new legislation has been lauded as more progressive and inclusive, it carries over many problems from the previous law (Doelle and Sinclair 2019). This suggests that national policy reform is not necessarily a reliable option for liberation, since policies themselves are developed, operated, and implemented by a settler state. Instead, the colonial power structures supporting Canada's current resource governance regime need to be disrupted. Not only does this entail changing policy infrastructures to make them more competent and nuanced, but it also requires upending the judicial practices that facilitate legal land dispossession, including criminalization of land defenders and court injunctions against communities (Yellowhead Institute 2019). In the same vein, Indigenous conceptions of consent and forms of law need to be taken seriously and respected in order to unsettle contemporary forms of resource use and settler hegemony (Coulthard 2014; Yellowhead Institute 2019). Honoring and centering Indigenous justice frameworks and free, prior, and informed consent in community–company–government interactions is vital to strengthening local capacity, as is respecting vetoes. And as researchers and citizens observing this process, it is crucial that we heighten our support of Indigenous communities, question the causal logics of these projects and industries, and contest these regulatory processes and policies when they reify structural oppression.

Acknowledgments

This project was supported by funds from the Social Sciences and Humanities Research Council Partnership Grant—MinErAL. This work was also generously supported by a Social Sciences and Humanities Research Council Canada Graduate Scholarship, a Fonds de Recherche du Québec en Santé Master's Training award, an Eben Hopson Fellowship, and a Lorne Trottier Accelerator Fellowship. I would like to thank my supervisor Mylène Riva and committee member George Wenzel for their support and trust, as well as Darin Barney, who provided thoughtful feedback on an early version of this manuscript.

References

Baker, J. M., and Westman, C. N., 2018. Extracting knowledge: Social science, environmental impact assessment, and Indigenous consultation in the oil sands of Alberta, Canada. *The Extractive Industries and Society*, 5 (1), 144–153.

Ball, M. A., Noble, B. F., and Dubé, M. G., 2013. Valued ecosystem components for watershed cumulative effects: An analysis of environmental impact assessments in the South Saskatchewan River watershed, Canada. *Integrated Environmental Assessment and Management*, 9 (3), 469–479.

Barry, A., 2013. *Material politics: Disputes along the pipeline*. West Sussex, UK: John Wiley & Sons.

Bernauer, W., 2019. The limits to extraction: Mining and colonialism in Nunavut. *Canadian Journal of Development Studies/Revue canadienne d'études du développement*, 40 (3), 404–422.

Booth, A. L., and Skelton, N. W., 2011. "You spoil everything!" Indigenous peoples and the consequences of industrial development in British Columbia. *Environment, Development and Sustainability*, 13 (4), 685–702.

Bowker, G. C., and Star, S. L., 1999. Some tricks of the trade in analyzing classification. *In:* G. C. Bowker and S. L. Star, eds., *Sorting things out: Classification and its consequences*. Cambridge, MA and London: MIT Press, 33–50.

Caldwell, L. K., 1998. *The National Environmental Policy Act: An agenda for the future*. Bloomington, IN: Indiana University Press.

Campbell, M. A., Kopach, B., Komers, P. E., and Ford, A. T., 2020. Quantifying the impacts of oil sands development on wildlife: Perspectives from impact assessments. *Environmental Reviews*, 28 (2), 129–137.

Coulthard, G., 2014. From wards of the state to subjects of recognition? Marx, Indigenous peoples, and the politics of dispossession in Denendeh. *In:* A. Simpson and A. Smith, eds., *Theorizing native studies*. Durham, NC: Duke University Press, 56–98.

Cowen, D., 2020. Following the infrastructures of empire: Notes on cities, settler colonialism, and method. *Urban Geography*, 41 (4), 469–486.

Crosby, A., 2021. The racialized logics of settler colonial policing: Indigenous "communities of concern" and critical infrastructure in Canada. *Settler Colonial Studies*, 1–20.

Curley, A., 2021. Infrastructures as colonial beachheads: The Central Arizona Project and the taking of Navajo resources. *Environment and Planning D: Society and Space*, 39 (3), 387–404.

Dake, K., 1992. Myths of nature: Culture and the social construction of risk. *Journal of Social issues*, 48 (4), 21–37.

Dendena, B., and Corsi, S., 2015. The environmental and social impact assessment: A further step towards an integrated assessment process. *Journal of Cleaner Production,* 108, 965–977.

Doelle, M., 2012. *CEAA 2012: The end of federal EA as we know it?* Rochester, NY: Social Science Research Network.

Doelle, M., and Sinclair, A. J., 2019. The new IAA in Canada: From revolutionary thoughts to reality. *Environmental Impact Assessment Review*, 79, 106292.

Edwards, P. N., 2010. *A vast machine: Computer models, climate data, and the politics of global warming*. Cambridge, MA: MIT Press.

Edwards, P. N., Jackson, S., Chalmers, M., Bowker, G., Borgman, C., Ribes, D., Burton, M., and Calvert, S., 2013. *Knowledge infrastructures: Intellectual frameworks and research challenges*. Ann Arbor, MI: Deep Blue.

Filion, P., and Keil, R., 2017. Contested infrastructures: Tension, inequity and innovation in the global suburb. *Urban Policy and Research*, 35 (1), 7–19.

Gabrys, J., 2016. *Program Earth: Environmental sensing technology and the making of a computational planet* (Vol. 49). Minneapolis, MN: University of Minnesota Press.

Gibson, R. B., 2002. From Wreck Cove to Voisey's Bay: The evolution of federal environmental assessment in Canada. *Impact Assessment and Project Appraisal*, 20 (3), 151–159.

Goffman, E., 1956. *The presentation of self in everyday life*. New York: Doubleday.

Hall, R., 2013. Diamond mining in Canada's Northwest Territories: A colonial continuity. *Antipode*, 45 (2), 376–393.

Harris-Roxas, B., Viliani, F., Bond, A., Cave, B., Divall, M., Furu, P., Harris, P., Soeberg, M., Wernham, A., and Winkler, M., 2012. Health impact assessment: The state of the art. *Impact Assessment and Project Appraisal*, 30 (1), 43–52.

Harvey, P., Jensen, C. B., and Morita, A., 2016. Introduction: Infrastructural complications. *In:* P. Harvey, C. B. Jensen, and A. Morita, eds., *Infrastructures and social complexity*. London, UK: Routledge, 19–40.

Hoogeveen, D., 2016. Fish-hood: Environmental assessment, critical Indigenous studies, and posthumanism at Fish Lake (Teztan Biny), Tsilhqot'in territory. *Environment and Planning D: Society and Space*, 34 (2), 355–370.

Jones, J., and Bradshaw, B., 2015. Addressing historical impacts through impact and benefit agreements and health impact assessment: Why it matters for Indigenous well-being. *Northern Review*, (41), 81–109.

Keeling, A., and Sandlos, J., 2015. *Mining and communities in Northern Canada: History, politics, and memory* (Vol. 3). Calgary, AB: University of Calgary Press.
Ladner, K. L., 2014. Political genocide: Killing nations through legislation and slow-moving poison. *In:* A. L. Hinton, A. Woolford, and J. Benvenuto, eds., *Colonial genocide in Indigenous North America*. Durham, NC: Duke University Press, 226–245.
LaDuke, W., and Cowen, D., 2020. Beyond Wiindigo infrastructure. *South Atlantic Quarterly*, 119 (2), 243–268.
Li, F., 2015. *Unearthing conflict: Corporate mining, activism, and expertise in Peru*. Durham, NC: Duke University Press.
Mackie, J. H., 2012. *"How Do You Measure the Loss of a Lake?": Assessing community relevance of health impact assessment frameworks to the Tl'azt'en Nation of northern-interior British Columbia*. Prince George, BC: University of Northern British Columbia.
Mezzadra, S., and Neilson, B., 2019. *The politics of operations: Excavating contemporary capitalism*. Durham, NC: Duke University Press.
Myette, E., 2022. *Unearthing the discursive politics of mining on Indigenous lands: Knowledge, health, contestation, and power in contemporary Canadian regulatory infrastructures*. Unpublished Master's Thesis. McGill University.
Myette, E., and Riva, M., 2021. Surveying the complex social-ecological pathways between resource extraction and Indigenous Peoples' health in Canada: A scoping review with a realist perspective. *The Extractive Industries and Society*, 8 (2), 100901.
Nixon, R., 2011. *Slow violence and the environmentalism of the poor*. Cambridge, MA: Harvard University Press.
Olagunju, A. O., and Gunn, J. A., 2015. Selection of valued ecosystem components in cumulative effects assessment: Lessons from Canadian road construction projects. *Impact Assessment and Project Appraisal*, 33 (3), 207–219.
Paci, C., Tobin, A., and Robb, P., 2002. Reconsidering the Canadian Environmental Impact Assessment Act: A place for traditional environmental knowledge. *Environmental Impact Assessment Review*, 22 (2), 111–127.
Pasternak, S., and Dafnos, T., 2018. How does a settler state secure the circuitry of capital? *Environment and Planning D: Society and Space*, 36 (4), 739–757.
Pulido, L., 2017. Geographies of race and ethnicity II: Environmental racism, racial capitalism and state-sanctioned violence. *Progress in Human Geography*, 41 (4), 524–533.
Richmond, C., and Big-Canoe, K., 2018. The geographies of Indigenous health. *In*: V. A. Crooks, G. J. Andrews, and J. Pearce, eds., *Routledge handbook of health geography*. London, UK: Routledge, 179–188.
Siemiatycki, M., Enright, T., and Valverde, M., 2020. The gendered production of infrastructure. *Progress in Human Geography*, 44 (2), 297–314.
Simpson, L., 2001. Aboriginal peoples and knowledge: Decolonizing our processes. *The Canadian Journal of Native Studies*, 21 (1), 137–148.
Spice, A., 2018. Fighting invasive infrastructures: Indigenous relations against pipelines. *Environment and Society*, 9 (1), 40–56.
Stanley, A., 2016. Resilient settler colonialism: "Responsible resource development," "flow-through" financing, and the risk management of Indigenous sovereignty in Canada. *Environment and Planning A: Economy and Space*, 48 (12), 2422–2442.
Winner, L., 1980. Do artifacts have politics? *Daedalus*, 121–136.
Wolfe, P., 2006. Settler colonialism and the elimination of the Native. *Journal of Genocide Research*, 8 (4), 387–409.
Yellowhead Institute, 2019. Land back: A Yellowhead Institute Red Paper. Available from: https://redpaper.yellowheadinstitute.org/wp-content/uploads/2019/10/red-paper-report-final.pdf

7 Realizing Indigenous rights

Effective implementation of agreements between Indigenous Peoples and the extractive industry

Ciaran O'Faircheallaigh and Thierry Rodon

Introduction

In Australia and Canada, and increasingly in the Global South, negotiated agreements constitute a critical mechanism through which Indigenous legal rights are given concrete expression. In Canada, court cases on Aboriginal rights have made it clear that First Peoples need to be consulted about developments that may impinge on their rights and that, in some cases, their consent may be necessary (Papillon and Rodon 2017). This has created uncertainty for extractive industries and led many mining companies to negotiate agreements, referred to as impact and benefit agreements (IBAs), which provide the consent of Indigenous communities for development on their lands, and so reduce legal uncertainty. In addition, the negotiation of IBAs is mandatory under the most recent comprehensive land claim settlements ("modern treaties") (Bradshaw and McElroy 2014; Prno *et al.* 2010; Papillon and Rodon 2017). The drivers in Australia are somewhat different. Australia, unlike Canada, introduced national legislation—the *Native Title Act 1993*—providing for the recognition of inherent Indigenous rights in land ("native title"). Native title has now been found to exist in nearly half of Australia's land mass, including most major mineral-producing regions. Aboriginal holders of native title are granted a "right to negotiate," which requires mining companies to seek their consent through the negotiation of agreements (O'Faircheallaigh 2016). In both countries, the adoption of "corporate social responsibility" by major mining companies has added to the momentum for the signing of agreements. The end result is that there are now hundreds of IBAs in Australia and Canada.

In the past, extractive industries located on Indigenous lands imposed huge costs and generated few benefits for affected Indigenous Peoples. Costs included environmental degradation, destruction of Indigenous heritage, loss of livelihoods, negative cultural and social impacts associated with large-scale in-migration of non-Indigenous populations, and loss of self-esteem due to the inability of Indigenous Peoples to influence the exploitation of their ancestral lands. In addition, potential benefits associated with project-generated employment, training, business development, and revenue streams were monopolized by non-Indigenous governments, businesses, workers, and shareholders. Negotiated agreements have the potential to change this situation by 1) providing measures to protect the

DOI: 10.4324/9781003406433-10

This chapter has been made available under a CC-BY-NC-ND 4.0 license.

environment and cultural heritage in addition to those available in general legislation; 2) generating revenues for Indigenous governments and affected communities; 3) allowing Indigenous Peoples access to income-earning opportunities; and 4) restoring to them a measure of influence over how development occurs.

The signing of an agreement does not of itself allow these benefits to be realized. Numerous activities must be undertaken and maintained over time to ensure that agreements are effectively implemented. These activities include identifying, recruiting, training, and retaining Indigenous mine workers; establishing sustainable Indigenous businesses to provide goods and services to projects; ensuring that financial entitlements are correctly calculated and effectively managing the revenues they generate; monitoring exploration and mining activity to ensure that cultural heritage and environmental protection provisions are being observed; educating non-Indigenous workers and managers regarding their obligations under agreements; and reviewing agreement provisions to ensure they are effective.

There is considerable empirical evidence that highlights the challenges associated with these activities. For example, a review of large samples of agreements and case studies of agreement implementation for individual projects show that implementation failure is common (Cree Nation of Mistissini *et al.* 2008; Crooke *et al.* 2004; O'Faircheallaigh 2002a, 2002b). The result is that Indigenous employment and business development targets may not be met; project revenues accruing to Indigenous entities may not be used effectively and, in some cases, may be wasted or misappropriated; cross-cultural education may not continue beyond the first few years of an agreement; and measures to protect the environment and cultural heritage may fail to deliver the hoped-for protection or fall entirely into disuse. Such outcomes not only result in a failure to minimize project costs and maximize benefits but also often lead to recriminations and conflict within Indigenous communities.

At a broad level, implementation failure can be attributed to weaknesses in agreements themselves and to capacity limitations within signatory Indigenous communities and mining companies. However, there is little detailed research on how the content of agreements shapes implementation patterns and very little, indeed, that documents the experience of Indigenous communities in seeking to grapple with implementation challenges. Yet, an understanding of these two areas is essential if agreement implementation is to be improved.

Implementation is a major challenge in the field of public policy in general. It is important to also draw on experience from the wider domain to identify causes of implementation failure and potential strategies for dealing with them.

The goal of this chapter is to identify obstacles to the implementation of agreements between mining companies and communities; analyze how the content of agreements shapes implementation models; describe the experience of Indigenous communities seeking to meet the challenges of implementation; and identify ways of addressing those challenges.

In the next section, we review public policy literature on implementation. We then discuss the methodology we used in undertaking our case studies of agreement implementation. The first case study involves the Innu nation of Matimekush-Lac

John in Québec, Canada, and focuses broadly on implementation of agreement provisions that involve the participation of both the First Nation and the signatory mining companies. Royalty payments under IBAs tend to be managed by the recipient communities independently of mining companies and involve specific and complex issues that are the subject of a separate and large literature, often under the theme of the "resource curse." Management of mining royalties is the focus of our second case, the Ely Bauxite Mine Beneficiaries Trust, based in Northern Queensland, Australia. In combination, the case studies allow us to address the full range of implementation issues that arise in relation to IBAs. We conclude with a discussion of ways in which implementation of agreements can be improved so as to increase the benefits that agreements bring to First Peoples.

Public policy literature on implementation

There is a large literature on the difficulties associated with implementation of public policies. A key finding of that literature, and the genesis for much research in the field, is that implementation does not happen automatically and, indeed, is often challenging. This realization was delayed by a conviction held by many public administration scholars in the decades after 1945 that implementation should be straightforward and, indeed, automatic. In their view, democratically elected politicians had a mandate to bring about change and/or to deliver societal benefits; politicians delivered policy directions to bureaucrats, who faithfully implemented their directions. Empirical research in the United States on the outcomes of the "Great Society" programs launched by President Lyndon B. Johnson in 1964–1965 revealed a fundamental flaw with this model of public administration. Either implementation did not happen at all or what was implemented was a far cry from what was intended by policy-makers in Washington (Pressman and Wildavsky 1973). This discovery led to a large body of research designed to understand implementation processes, the causes of implementation failure, and how to remedy that failure. No consensus emerged on these issues, but the literature does generate insights that are relevant to the implementation of IBAs (for a summary, see O'Faircheallaigh 2002a, pp. 24–45).

One issue on which there is broad agreement is that problems with implementation can arise because inputs required for its achievement (funding, personnel, and the attention of decision-makers) are not provided. This problem clearly occurs with IBAs. Two decades ago, O'Faircheallaigh (2002b) documented that many IBAs fail to provide dedicated funding or personnel focused specifically on implementation, with the result that key tasks for the delivery of employment and training, business development, and environmental provisions of agreements cannot be put into effect.

The causes of implementation failure are not only an absence of resources. A key discovery by public policy scholars was that actors involved in the "implementation chain," other than policy-makers, affected policy outcomes. "Street-level bureaucrats" (Lipsky 1980), that is, those responsible for implementing policy on the ground and for the interface with the recipients or targets of policy, hold significant and independent sources of power. These include the exercise of discretion in applying general policy in individual cases; control over flows of information; shared values with coworkers, especially strong among professional coworkers such as doctors, lawyers, and engineers, which can be used to maintain internal cohesion and resist external control; and, emerging from all of these, considerable power of obstruction in the face of policies they do not support. The opportunity for street-level bureaucrats to shape implementation is increased by the tendency of public policies to be expressed in broad terms or to lack clarity, making it inevitable and necessary to interpret them to allow their implementation (O'Faircheallaigh 2002a, p. 34). Obvious parallels may exist in relation to IBAs, including the fact that, particularly in Canada, IBAs are often negotiated by lawyers who are not involved in implementation (CRDDN 2020). Implementation is left to operational staff and community representatives who may not be aware of why specific provisions were negotiated or of the trade-offs that lie behind them. This is compounded by the fact that IBA provisions are often broadly framed and lack specifics, for instance those involving commitments to "maximize employment opportunities for Indigenous community members."

Another important finding was that the targets or recipients of policy also influence its implementation. In many areas, policies depend on the cooperation and support of these recipients, which may not be forthcoming. Clients of a service may subvert policy goals by their behavior. For example, recipients of cashless debit cards that can only be used to buy essential items such as food and clothing, and that are designed to stop recipients spending welfare payments on alcohol and gambling, may sell food and clothing they purchase with their cards, and spend the proceeds on the "forbidden" items. This discussion is also relevant to IBAs. The behavior of "recipients" of IBA provisions will influence their outcome, as where politically astute community members succeed in diverting funds intended for long-term community investment to other uses that benefit only themselves.

The public policy literature also emphasizes that successful implementation requires not only availability of the necessary inputs, such as money and personnel, and outputs such as training courses, but also a valid theory of causality linking outputs with desired outcomes (O'Faircheallaigh 2002a, pp. 26–27). If policy design is based on a causal theory that turns out to be invalid, desired outcomes will not be achieved even if all required inputs and outputs are present. For example, low levels of Aboriginal employment in a mining project may be attributed to a skills deficit on the part of potential recruits. If other factors, for example, fear of racist behavior by non-Indigenous miners, or cultural opposition to mining, are actually causing Aboriginal recruits not to take up employment in mining, delivery of training courses that would otherwise be effective may have no impact. The public policy literature suggests that a focus on causal theory is important for another reason, which is that unanticipated factors in the external environment can either facilitate or retard policy implementation. For example, a scheme to encourage diversification by primary producers engaged in monoculture may prove ineffective because of short-term price increases for their traditional crops. Conversely, policy goals may be achieved not because policy outputs are produced and causal theory is valid but because unrelated events bring about the desired change (O'Faircheallaigh 2002a, pp. 225–227). In an IBA context, a sharp decline in economic conditions in other sectors might lead to increased Aboriginal employment in a mining project, an outcome that could be wrongly attributed to a training program designed to address a supposed skills deficit.

In summary, the public policy literature illustrates that a variety of factors can explain the failure or success of implementation efforts and the policy goals to which they relate. These include the presence or absence of required inputs and outputs; the behavior and influence of individuals responsible for implementation, and of the recipients of services or benefits; the validity of causal theories underlying policy initiatives; and the influence—positive and negative—of unanticipated events in the external environment. The appropriate response to implementation failure depends on its causes. If an input or output is missing, the appropriate response is to provide it. If service delivery staff and policy recipients are undermining implementation, the appropriate response may be to create incentives for them to change their behavior. If causal assumptions are faulty, the policy approach may have to be radically altered. If unanticipated events hinder implementation, but appropriate outputs are being produced and the causal theory appears valid, the appropriate response may be to maintain the status quo and wait for these external events to pass. All of these considerations can apply equally to implementation of IBAs.

Methodology

This research has been undertaken in a fully collaborative manner with Indigenous organizations and governments that have considerable experience of agreement implementation and are working to minimize implementation failure. Their collaboration has allowed access to information sources and practical knowledge that would otherwise not be available to researchers. Research agreements have been signed with the Ely Bauxite Mine Beneficiaries Trust ("Ely Trust") in Northern Queensland, Australia, and the Innu Nation of Matimekush-Lac John in Northern Québec, Canada. These agreements address participation in project design, access to research materials and sites, confidentiality, contribution to community-based research, access to preliminary results, exchange of feedback, and knowledge mobilization.

For the Canadian case study led by Rodon, two workshops were conducted under Chatham Rules with mining companies and Indigenous leaders involved in IBA implementation, in September 2020 and December 2022. In addition, five phone or in-person interviews were conducted from October 2019 to January 2023 with key informants from Matimekush and Tata Steel Minerals Canada Ltd (hereafter Tata Steel) who were involved in implementation of the IBA. Most asked to remain anonymous, so we refer to them only by interview number. Rodon also had access to the IBAs signed by the community of Uashat mak Mani-Utenam with Tata Steel, which are available online, were negotiated at the same time as the Matimekush IBA, and are very similar. Matimekush and Ushat mak Mani-Utenam have a long-shared history. The interviews allowed the researcher to gather information on the specificities of the Matimekush IBA.

For the Australian case study, O'Faircheallaigh drew on his personal experience in helping negotiate the IBA that led to the establishment of the Ely Trust in the discussion of the early history of the Ely Agreement. In early 2018, he was asked by the directors of the trust to assist in planning for the expected cessation of mining in 2027. He attended every meeting of trust directors between 2018 and 2021 and was given access to all board papers and other trust documentation. He also attended numerous meetings of trust members in the communities where its beneficiaries lived. His engagement with the trust was conducted on a basis agreed informally with trust directors until August 2020, when a research agreement was signed between the Ely Trust and Griffith University. Preparation of the case study was governed by this agreement, which addresses ownership of intellectual property, treatment of confidential information, and provision of research findings to the trust.

Innu Nation of Matimekush-Lac John in Québec, Canada, and Tata Steel

The Innu Nation of Matimekush-Lac John is located at the 55th parallel north in Québec, near the Labrador border and the city of Schefferville, a former mining town that was closed in 1984, along with the IOC mine. The Innu of Matimekush are part of the Innu Nation, with nine communities in Québec and two in Labrabor. The Schefferville area was used by Innu families from the community of Uashat mak Mani-Utenam to set up winter camps for trapping and hunting, but with the opening of the mine, some families decided to take up residence permanently in the region. That led the federal government to create a reserve in Lac John in 1956 and then in Matimekush in 1968. Matimekush-Lac John is a reserve with 839 registered members, although only 750 reside in the community. There is also the Naskapi community of Kawawachikamach, 40 km away, that relocated from Fort Mackenzie, 250 km to the north, for the promise of a better future with employment and lodging at the new mine.

This region can only be reached by train or aircraft. The town of Schefferville, which used to have 3,000 inhabitants, now has only 120 (Rodon *et al.* 2022a). A new mine operated by Tata Steel opened 20 km from Matimekush in 2013. However, it is located in Labrador, and all the employees are working on a fly-in–fly-out schedule, with no contribution to Schefferville (Rodon *et al.* 2022a). Legally, Matimekush-Lac John is an "Indian Reserve," meaning it is a federal land reserved for Indigenous people. It is governed under the *Indian Act* by a chief and a band council elected every 3 years. The band council and the services it provides to the population are funded by the government of Canada.

Matimekush has signed four active IBAs with four different mining companies: Tata Steel, Minerai de fer du Québec, ArcelorMittal, and Rio Tinto/IOC. Tata Steel and Minerai de fer du Québec are new mines that negotiated their IBAs prior to production, while ArcelorMittal and Rio Tinto are older mines that decided to sign IBAs to avoid litigation, as the jurisprudence on Aboriginal rights has evolved significantly since the beginning of their operations. Only Tata Steel is close to Matimekush and directly impacts the communities; the other three mines are located more than 250 km south of Matimekush on land used by only a few families from Matimekush and Uashat mak Mani-Utenam.

In examining the IBAs we were able to consult, it is obvious that these agreements use a pro forma model that evolves during the negotiation process. All IBAs have chapters on royalties to be paid to the community, Indigenous training and employment targets, business opportunities for Indigenous enterprises, environmental monitoring, implementation, dispute resolution, and confidentiality. In addition, all the IBAs have a consent clause that precludes Indigenous signatories from taking any action against the projects.

Each IBA calls for an implementation committee composed of mining companies and Innu representatives. In the case of Matimekush, many of these meetings are attended by a consultant working for the community. These committees generally meet twice a year and deal mainly with Indigenous employment and with subcontracting opportunities for Innu businesses. These were the issues most discussed during the two workshops conducted as part of the project, suggesting that they are the most difficult to implement (CRDDN 2020, 2022).

In the case of employment, discrimination and racism in the workplace were two points mentioned by Indigenous participants at the Montréal workshop (CRDDN 2022). The issue of taxation is also a barrier to mining employment since, in Canada, Indigenous people working on reserves are exempted from taxation. Many Innu are, therefore, reluctant to work directly for a mining company since it means their wages will be taxed (CRDDN 2022). Lastly, the minimum high-school education required by mining companies for most of their jobs makes it difficult for the Innu to apply, since most of them did not graduate from high school (Matimekush interviewee 2 2022; CRDDN 2022).

Another issue is relocation. While Tata Steel is close to the community, the other three mines are located 250 km south of Matimekush, and there is no road connecting them to the community. Furthermore, both ArcelorMittal and Rio Tinto/IOC expect their workers to relocate to their mining towns, and very few Innu are willing to move to a mining town far from their communities. This is an example of factors in the external environment affecting implementation. Lastly, Indigenous communities are often quite small. For example, with an active population of only 450 people available for work (Statistics Canada 2017), Matimekush has difficulty providing many workers for four mines.

Taking advantage of subcontracting opportunities is also problematic, since Indigenous communities have few private enterprises. Moreover, in some cases, bidding on contracts is difficult either because of the large size of the contract or because the online bidding process is cumbersome (CRDDN 2020, 2022).

Challenges of implementing the IBA with Tata Steel

Tata Steel is a partnership between Tata Steel (82%) and the government of Québec (18%). It was the first mine to open since the closure of the IOC mine in 1984. The mining operation started in 2013, after Tata Steel signed IBAs with the Naskapi

Nation of Kawawachikamach (2010) and with Ushat mak Mani-Utenam and the Innu Nation of Matimekush-Lac John (2011). Each of the three Indigenous nations has its own IBA, but the content of each is fairly similar.

The Matimekush chief, when confronted with this new project, decided to submit the IBA to a referendum asking the Innu population to give him a clear mandate to either go ahead or to stop the project (Matimekush interviewee 1 and interviewee 2). The prior experience of the community with mining had been fairly negative. At the time of IOC, the Innu from Matimekush benefited from employment at the mine; however, they were used as seasonal unskilled labor without the benefits of full-time employment (Boutet 2015). The closure of the mine in 1984 was also quite traumatic, since most of Schefferville's housing and community infrastructure was destroyed by IOC at that time. Only a few buildings, including the arena, were saved, thanks to the protests of a group of Innu women who wanted to preserve it for the community's youth (Boutet 2015; Wolfe 1992). In the end, there was no remediation of the IOC mining site, and there are still huge mining pits and tailing piles around the community. The referendum was held in 2011 just before the signing of the IBA with Tata Steel, and 52% of Innu voters supported the signing (Matimekush interviewee 1 2019, and Matimekush interviewee 2 2022). It was mostly the young people who were in favor, with the older people retaining bitter memories of mining (Nachet 2019; Matimekush interviewee 1 2019). The support of young people can be explained by high unemployment in the community, which had a 33% unemployment rate in 2016 (Statistics Canada 2017).

Matimekush interviewee 1 (2019) considers that Tata has respected neither the IBA with Matimekush nor the general spirit of good cooperation with the community. First, the level of employment and business contracts have been disappointing. Second, he feels that the relationship between the community and Tata is so bad that it would be better for the company (which is in severe financial trouble) to be bought out. According to Matimekush interviewee 1 (2019), Tata Steel is a "bad payer" and "is not transparent" with the community. And lastly, the company has a poor environmental management record, as illustrated by the leakage of iron-contaminated water from a retention pond in the lakes and rivers surrounding the mine in the summer of 2019. Following this incident, it took 4 years, after being fined by the Québec government, for Tata Steel to repair one of the facilities that caused the red water spill (Jung 2023).

A major problem is the difficulty for Tata Steel to meet the employment targets that were set at 23 employees in the first year of the IBA, with a final target of 35. At the time of writing, only ten Innu from Matimekush were employed at Tata. While Tata did have some training programs, they were mostly in the construction phase, and only for heavy equipment drivers. Most of the mining jobs require a completed high school education, something rarely achieved in Matimekush, which is one of the significant barriers to the employment of Innu from the community (Matimekush interviewee 2 2022; CRDDN 2022).

Another issue is the work schedule. Tata Steel mine operates on a fly-in–fly-out schedule of 28 days' work/28 days' break. This schedule is also applied to the Innu workers from Matimekush, but they go home every day instead of staying at the fly-in–fly-out housing units. This does not work very well for the Innu women employees who need to take care of their children:

> They have children. They have to get up at 6 a.m., take the bus, go there and come back here at 6 p.m., exhausted, make supper at night, go to sleep at 8 p.m. Wake up at 6 a.m. They couldn't do it. It was too hard for them, and for others. Plus, the daycare was always full.
>
> (Matimekush interviewee 2 2022)

The issue of taxation is also compounding problems. To circumvent this issue, the Innu workers were subcontracted through the band council, allowing Innu employees not to be taxed. However, this arrangement proved unworkable. First, the band council was told by Canada Revenue Agency that it was not allowed to hire Innu mine workers for tax exemption purposes. Furthermore, Tata Steel always paid the band council very late. These delays created tensions and led the chief and council to blockade the mine access in 2018. After negotiations, the blockade was lifted, but the parties decided to end this arrangement, and now the Innu employees work directly with Tata Steel, and their income is taxed. However, the parties are still trying to find a way to avoid the Innu employees being taxed as this would make working at Tata more attractive (Matimekush interviewee 1 2019; CRDDN 2022).

There are no targets in the IBA for contracts for Innu businesses, only a general priority for business contracts offered by the mine. To benefit from contracts, an Indigenous consortium called Naskinnuk was created. The consortium consists of the Innu Nation of Matimekush-Lac John, the Naskapi Nation of Kawawachikamach, and Fédération des cooperatives du Nouveau-Québec, a coop owned by its 14-member co-ops in the Inuit communities along the Hudson and Ungava coasts of Northern Québec, or Nunavik. This consortium obtained the exclusive contract to provide fuel for the mine. The contract could have been very profitable; however, Naskinnuk accumulated CA$28 million in unpaid fuel bills from Tata (Matimekush interviewee 1 2019). This outstanding debt was settled just before the mine closure (Matimekush interviewee 1 2019). Such long payment delays were also experienced by the other contractors working with Tata Steel (Jung 2022a).

Between 2010 and 2019, the level of royalties was disappointing for Matimekush (interviewee 1) since these are linked to the total iron ore produced. However, in 5 years of operation, Tata Steel only managed to extract a total of 8–10 million tons, thus producing around 2–3 million tons of iron ore per year. By comparison, in the good years of IOC, 12–14 million tons of iron ore per year could easily be produced. Matimekush interviewee 1 (2019) regrets that they did not negotiate an alternative royalty calculation that was not as dependent on Tata's iron ore output. The royalties are managed by the band council and are primarily used to maintain community services (arena, housing, road paving, etc.) as the band council has been in a very precarious financial position (Matimekush interviewee 1 2019).

There is a chapter in the IBA on environmental monitoring that calls for the creation of an environmental management committee that meets regularly. However, the IBA does not provide any funds for an Innu environmental coordinator, so participation in the committee has to be covered by the band council. At the time of writing, the band council has appointed consultants to participate in the committee on its behalf. In the opinion of interviewee 2, the environmental chapter is too weak:

> What I deplore about Tata Steel is that there was no agreement on [financing] the environmental issue. They should have funded guardians of the territory to monitor what was going on and tell us if there were mistakes or things that were done poorly.
>
> (Matimekush interviewee 2 2022)

Chief Mckenzie, who had signed the IBA, lost the 2016 election, and a new chief, Tsauni Ambroise, was elected. The change of leadership affected the relationship with Tata Steel. Chief Ambroise was less inclined to engage with Tata Steel since he was opposed to mining and organized a mine blockade in 2018. Chief Mckenzie was reelected in 2019, but the relationship with Tata Steel was, by then, quite strained.

Tata Steel also had leadership problems. The relationship between Tata Steel and Matimekush was managed by a vice president for intergovernmental relations, an Innu from Matimekush, and a community affairs manager. The vice president resigned in 2020, and the community affairs manager has been on leave for the past 2 years. Since their departure, the relationship with the community has been managed by the president's office via a Tata Steel executive from India with little knowledge about Indigenous issues. This has led to further tensions between the company and the Innu of Matimekush. Recently, a new community affairs manager, a former Matimekush high school teacher, was appointed and plans to restore a better relationship with the community.

In the fall of 2022, Tata Steel finally agreed to pay all its dues to the band council but, shortly after that, decided to suspend mining activities for a few months, blaming the low price of iron ore. However, Tata Steel has committed to keeping all its Innu employees on the payroll during the temporary closure of the Schefferville mine (Jung 2022a) and has announced that it will resume operations in February 2023.

In this case study, most of the elements discussed in the public policy literature on implementation were present. First, the absence of required inputs had a major impact. For example, resources were not provided for an environmental coordinator or for the training required to prepare Innu people to work in the mine, and key positions at Tata (VP intergovernmental relations and community relations manager) were left unfilled, with the result that some of the human resources needed for implementation were lacking. In many instances, Tata Steel was very late in paying its subcontractors, to the point that some Indigenous companies were forced to file for bankruptcy (Jung 2022b). This did not happen in the case of Naskinnuk, but only because one of the partners had sufficient funds to keep the company up and running. The behavior of the individuals responsible for implementation and that of the recipients of services or benefits was also a factor. For example, the change in leadership in Matimekush has affected implementation, as has the deterioration in the relationship between Tata and Matimekush. The tension between Tata's front-line employees and Tata leadership is evident as well. This was the first foray into Canada for Tata Steel, and the company leadership had poor knowledge of Indigenous issues in Canada. Lastly, factors in the external environment, in particular, low levels of formal education among the Innu and the location of the community in relation to some of the mines, adversely affected the implementation of employment goals. Compounding the problem, the IBA did little to address these barriers or to deal with matters such as work schedules.

Ely Bauxite Mine Beneficiaries Trust

One specific aspect of implementation involves the use of monies payable to Indigenous communities under mining agreements. Effective implementation is especially important in this area for several reasons. Revenues from agreements have the potential to create significant benefits, given that they can be allocated to uses that are a high priority for the recipient community. Unlike other economic benefits such as employment and training, and business contracts, they are almost certain to materialize while a mine is operating. However, revenues also have the potential to cause considerable harm, for example, if their inequitable or wasteful allocation leads to social conflict or if their distribution to individuals leads to an increase in destructive activity such as substance abuse. Their potential to create either significant benefits or serious harm means that it is especially important to ensure that agreement provisions designed to ensure the effective management of royalties are implemented.

There is an extensive literature on the use of revenues derived from extractive industries by state authorities. This literature suggests that while it is not impossible to achieve positive outcomes from resource revenues, formidable obstacles arise in seeking to do so. These include the tendency for their arrival to generate rent-seeking behavior that militates against effective and equitable application of revenues; their instability, associated with fluctuating market conditions; and the tendency of their recipients to regard them as "windfall" gains that need not be subject to the fiscal discipline applied to more conventional and reliable sources of state revenues. In the Indigenous context, the economic disadvantage faced by many Indigenous communities heightens pressures to apply revenues to meet immediate needs for consumer goods and basic services, making it difficult to retain a portion of income to even out revenue fluctuations and to generate a capital base that can ensure an ongoing flow of income after mining ends (for a fuller discussion of these issues, see O'Faircheallaigh 2018 and Rodon *et al.* 2022b).

The literature indicates that the quality of institutions that receive revenues plays a key role in determining whether these challenges can be met. There is limited guidance on what constitutes "quality" in this context and very little, indeed, on how high-quality institutions can be developed and maintained. One specific gap is the lack of recognition of the role leadership can play in creating and maintaining "institutional quality" in this context (O'Faircheallaigh 2018, pp. 104, 115). These limitations are even more pronounced in the Indigenous context. This part of our chapter uses a case study of revenue flows generated by a specific agreement—for the Ely Bauxite Mine in Australia's North—to shed light on these issues.

History of the Ely Agreement

A brief history of the *Ely Bauxite Mining Project Agreement*, 1997 (the "Ely Agreement") provides important context for the analysis of the use of Agreement revenues through the Ely Bauxite Mine Beneficiaries Trust ("Ely Trust"). In 1965, the Queensland government awarded the Canadian aluminum company, Alcan, the Ely lease, covering an area of some 140,000 hectares in Western Cape York in far north Queensland. Alcan identified bauxite deposits on the lease but decided not to develop them at that time. Alcan needed bauxite because it was involved in building an alumina refinery to process bauxite in Gladstone in Central Queensland (alumina is smelted into aluminum, which is used in thousands of different ways). The refinery in Gladstone started production in 1967 and was a partnership between four aluminum producers: Alcan, Comalco, Kaiser Aluminum, and Pechiney. The partnership involved each company providing a share of the bauxite that was needed for the refinery in Gladstone. Comalco and Alcan agreed that rather than Alcan building its own mine in Western Cape York, Comalco would sell bauxite to Alcan from its existing Weipa mining lease, which was adjacent to the Ely lease, to meet Alcan's commitment to supply bauxite to the Gladstone refinery. Over the years, Comalco steadily increased the price of the bauxite it supplied to Alcan. The long-term contract that Alcan had signed with Comalco was due to end in January 2000. Alcan announced in 1995 that it would develop its bauxite deposits on the Ely lease, providing it with a much cheaper source of bauxite to meet its commitments to supply the Gladstone refinery.

At that time, there was no legal requirement for Alcan to negotiate a mining agreement with the traditional owners of the Ely lease or the affected Aboriginal communities because Alcan was granted the Ely mining lease in 1965, before the recognition of inherent Indigenous rights in land in Australia's legal system. However, Noel Pearson, then head of the regional Aboriginal land organization—the Cape York Land Council (CYLC)—persuaded Alcan that a negotiated agreement with traditional owners and the CYLC would facilitate the speedy development of the Ely project, a key consideration for Alcan, as it needed to have its own mine on stream before its contract with Comalco expired in January 2000. Otherwise, Alcan would have to sign another contract with Comalco. Alcan agreed to negotiate with the Aboriginal community to see if a mutually beneficial agreement could be negotiated.

At the same time, the Aboriginal community in Western Cape York was involved in negotiating an agreement with Comalco, whose mining operations were much larger than Alcan's planned project. By 1997, these negotiations were deadlocked, and a particular sticking point was Comalco's refusal to accept the Aboriginal community's demand for payments based on the value of bauxite it extracted. Senior leaders of the traditional owners and the CYLC approached Alcan with an offer to quickly conclude an agreement if Alcan would agree to certain conditions, including a royalty payment, that Comalco was resisting. The leaders felt that if Alcan accepted these conditions, it would be very difficult for Comalco, which was also mining bauxite on Aboriginal land just 30 km from Ely, to reject them. This strategy was ultimately successful (O'Faircheallaigh 2016), but it did mean that the Ely Agreement had to be negotiated more quickly than would otherwise have been the case. Most of the preparation and negotiation occurred in the space of just 6 months in 1997, and the Ely Agreement was signed in September of that year. The agreement makes very little mention of how financial payments from Alcan to the Aboriginal community should be allocated or used. It states that payment compensation shall be made to the Ely Trust, a trust to be established by the parties to the agreement, other than Alcan. These parties are the six traditional owner groups that signed the agreement, the communities of Mapoon, New Mapoon and Napranum, and the CYLC. The purpose of the trust is to allow the Aboriginal community, which is the term that includes traditional owners and the three communities, to receive and deal with the financial compensation.

What happened next is critical in terms of the history of the Ely Agreement. Comalco realized that Alcan was serious about building its own mine, so Comalco approached Alcan and offered it a much better deal if Alcan would continue to buy bauxite from it. In 1999, Comalco and Alcan signed what was called a Bauxite Mining Exchange Agreement. Under this agreement, Comalco would mine the bauxite for Alcan from Comalco's lease, and Comalco would later be allowed to mine the same amount of bauxite from the Ely lease. In effect, Comalco would lend Alcan the bauxite, and Alcan would later repay the bauxite from the Ely lease. The result was that Alcan did not go ahead with building the Ely mine. If the Ely mine had been built, the history of the Ely Agreement and its implementation would have been very different.

The Ely Trust

Under the Bauxite Mining Exchange Agreement, when Comalco took out bauxite for Alcan, it would pay the royalty that Alcan was supposed to pay under the Ely Agreement into an Ely Trust that was provided for under the agreement. While the negotiations with Alcan were wrapping up, Aboriginal Elders told CYLC that no money should be spent until CYLC and the Aboriginal community had had an opportunity to discuss how it should be managed. At that time, it was assumed that the Ely mine would be built and that everybody involved would be coming back and talking to each other about many issues involved in implementing the agreement. When the project did not go ahead, this did not happen.

Alcan made an initial payment of AU$250,000 on the signing of the agreement in 1997. There was no trust at the time, and that money was put into an account managed by CYLC as an interim measure. Rumors soon began to circulate in the Aboriginal community that CYLC and its CEO, in particular, were using the $250,000 for their own purposes. In an unfortunate coincidence, CYLC replaced the CEO's leased vehicle, and a rumor started that some of the Ely money had been spent to buy the CEO a new vehicle. Despite the CEO traveling to the Ely Agreement communities and presenting a bank statement showing that the $250,000 was intact and earning interest, this failed to quell the rumors.

At the same time, there was discussion among the traditional owners that some money should be paid out to the Elders who had been involved in the negotiations and in the struggle over the decades, including those who were forcibly relocated when the community of Mapoon was closed by the Queensland government in 1963 to make way for a second port planned for Comalco's bauxite mine. It was felt that older people had endured a hard life, had struggled to achieve recognition of Aboriginal rights, had fought to achieve the Ely Agreement, but would not benefit from the jobs or business contracts the agreement was expected to generate. The CEO of CYLC, tired of the rumors that the money was being misused, agreed to distribute the $250,000 in equal payments to traditional owners over 55 years of age.

In 2000, royalties started to be paid on the bauxite that was being mined for Alcan by Comalco, and the younger people in the community argued that it was now their turn to get a payment. It was stated that this should only happen with the first royalty, after which a part of the payments would be used for community projects and long-term investments. The first royalty was distributed to people who were younger than 55 years. However, the idea was now established that money should be handed out to individuals. At the same time, the work required to develop long-term policies and allocation mechanisms for the Ely Trust was not undertaken, work that would very probably have occurred had the mine gone ahead.

Payments in the vicinity of AU$1 million per year were made to the Ely Trust by Rio Tinto for the first 15 years or so. The trust appears not to have operated for a number of those years, with no distributions to beneficiaries. A number of people, including some of the lawyers and accountants hired by the trust, bear responsibility for this situation. More recently, the trust began operating efficiently again but, given its limited funds, was not able to do more than resume small payments of about $600 per annum to its Aboriginal beneficiaries, make some small loans from its business development fund, and contribute a modest amount of money to educational bursaries.

In late 2017, Rio Tinto informed the trust that, due to changes in its mine plan, bauxite production (and, subsequently, royalty income) would jump by a factor of 10, but that mining would end in 2027. The trust's directors, all traditional owners for the mine lease area, set about devising a strategy to ensure that some benefits could continue to flow after mining ended and that, in the interim, significant benefits would be created for the three Aboriginal communities that are beneficiaries of the trust. The strategy involved allocating about 40% of the annual income to a long-term investment fund designed to create a capital base that could generate ongoing income after 2027. The sole criteria for this fund involved profit maximization, and professional investment advice was secured on a competitive basis to help achieve the highest possible return and to ensure that the trust's affairs were conducted in a tax-effective manner. A further 25% of income was set aside for "community projects," to be allocated based on applications from community organizations and designed to create broad benefits for community members and avoid duplicating basic services that the government should provide. Funding already available under the Ely Agreement to encourage business development would be supplemented, as a second part of the strategy, to make sure that benefits could continue to flow after mining ended, while existing payments to individuals would be maintained. This strategy was discussed and endorsed after two rounds of community meetings during 2018.

Directors decided that the trust should discontinue individual payments in 2019 after Rio Tinto revised its forecast of future royalty payments downward due to declining world demand and prices for bauxite which, along with a rapid rise in the number of beneficiaries claiming the payments, would make it impossible to maintain the trust's other initiatives. The proposal to discontinue payments was also discussed in community meetings and, despite opposition from some recipients, was strongly endorsed because of support for the strategy of maintaining long-term investment and community benefits.

The Ely Trust's investment strategy has been successful to date, with almost AU$20 million now held in the long-term investment fund, and forecasts indicating that this amount will rise to $40 million by 2027. In 2020–2021, the trust allocated $1.5 million to community projects. Some of these involve items that the government will not provide under any circumstances. One example involved the purchase of a hearse for the beneficiary community of Mapoon, where the need to hire a hearse from the regional center of Weipa had generated considerable cost and stress for grieving families. Another involved the allocation of funds to purchase a second dialysis chair for the Weipa hospital. In this case the government, using a standard funding formula, had built a dialysis facility which it will maintain and had staff on an ongoing basis. By funding an additional dialysis chair, the trust was able, at marginal cost, to increase the capacity of the dialysis facility and reduce the need for patients to travel for treatment away from their families and their homelands.

Conclusion

In their work on IBAs in Canada's North, Caine and Krogman (2010) ask whether IBAs are "powerful or just plain power-full." Focusing on the power relationship between the proponent and Indigenous people, the authors argue that, under some conditions, IBAs may provide more direct engagement with industry and a sharing of benefits from resource development. However, depending on negotiation processes and agreement outcomes, IBAs can also stifle the capacity of Indigenous people to benefit from resource development. Outcomes are determined not only by the provisions of agreements but also by how well or poorly these provisions are implemented. Our case studies illustrate a range of obstacles to the successful implementation of IBAs, many of which are also highlighted in the literature on public policy implementation. Identifying and explaining these obstacles, which limit the capacity of Indigenous people to benefit from resource projects, provide a basis for achieving more successful implementation and, therefore, more favorable outcomes.

A key issue was that the human and financial resources needed for implementation were frequently absent, as occurred in the Tata–Matimekush IBA. Funding was not provided for training programs and environmental monitoring, Tata failed to fill key managerial positions for extended periods of time, and major delays occurred in paying Innu subcontractors. In the case of the Ely Agreement, a major problem was the lack of clarity in the agreement regarding the purposes and governance of the trust set up to receive royalties.

Both IBAs illustrate the risk that behavior of potential beneficiaries of agreements may undermine their purpose. This occurred in Ely, as beneficiaries of the trust deflected resources into individual payments, an outcome not planned or contemplated when the agreement was signed. In the case of Matimekush, the lack of commitment by a new chief, as well as by Tata Steel, to the IBA contributed to a deterioration of the relationship. The Tata Steel–Matimekush experience illustrates the broader issue that development and maintenance of positive relations between the parties are critical for essential implementation. In summary, adequate resources, clear goals, and a sustained focus on maintaining positive relations between the parties involved are critical to the effective implementation of IBAs.

Turning to the effective implementation of financial provisions, the Ely case illustrates that, as suggested in the "resource curse" literature, governance is a key factor, with accountability and transparency being especially important. The Board of the Ely Trust made extensive use of community meetings to tell beneficiaries about its planned strategies and to win support, for example, for the decision to reallocate individual payments to community projects and long-term investment. Another factor is the capacity to undertake and implement strategic planning, to clearly demarcate priorities both across time (current benefits versus building a future capital base) and across alternative uses for funds, as, for example, in the Ely Trust's support for dialysis services. The capacity to obtain and effectively deploy expert financial advice is also important. The Ely Trust has shown an ability to use such expertise in developing strategies that avoid high-risk investment and so preserve its capital, but at the same time to develop a diversified portfolio of shares and property that allow it to build a capital base that will be sufficient to maintain income once mining operations end.

As noted in reviewing the general literature on the resource curse, insufficient attention has been paid to the role of leadership, whose importance is evident in the examples discussed here. While leadership may be difficult to define, its presence or absence is crucial. For instance, leadership by the Ely Trust directors was critical in responding effectively to the changed circumstances created by Rio Tinto's revision of its mining plans. From a less positive perspective, the absence of leadership in Tata and conflicting approaches to the IBA among Matimekush's leaders created significant barriers to effective implementation.

Acknowledgments

This research was supported by MinErAL funded by the Social Sciences and Humanities Research Council of Canada (grant number 895-2016-1015). The authors would like to thank the anonymous reviewers for their advice to improve the manuscript. This research was made possible thanks to the anonymous respondents and our Indigenous partners, the Innu Nation of Matimekush-Lac John, and the Ely Bauxite Mine Beneficiaries Trust. We would also like to thank the respondents from the mining industry who shared their insights with the authors. Any errors are solely the responsibility of the authors.

References

Boutet, J. S., 2015. The revival of Québec's iron ore industry: Perspectives on mining, development, and history. *In:* A. Keeling and J. Sandlos, eds., *Mining and communities in Northern Canada: History, politics, and memory*. Calgary: University of Calgary Press, 169–206.

Bradshaw, B., and McElroy, C., 2014. Company—community agreements in the mining sector. *In: Socially responsible investment in the 21st century: Does it make a difference for society? (Critical studies on corporate responsibility, governance and sustainability, Vol. 7)*. Bingley: Emerald Group Publishing Limited, 173–193.

Caine, K. J., and Krogman, N., 2010. Powerful or just plain power-full? A power analysis of impact and benefit agreements in Canada's North. *Organization & Environment*, 23 (1), 76–98.

Chaire de recherche INQ sur le développement durable du Nord (CRDDN), 2020. *Seminar on the implementation of impact and benefit agreements*. Mushuau Nipi. Available from: www.mineral.ulaval.ca/en/seminar-implementation-impact-benefit-agreements-summary-discussions [Accessed 3 February 2023].

Chaire de recherche INQ sur le développement durable du Nord (CRDDN), 2022. Seminar on the implementation of impact and benefit agreements. Available from: www.mineral.ulaval.ca/sites/mineral.ulaval.ca/files/report_seminar_ibas_montreal_2022.pdf [Accessed 3 February 2023].

Cree Nation of Mistissini, Cree Regional Authority, and Inmet Mining Corporation, 2008. *Implementing the Troilus Agreement—A joint study of Cree employment and service contracts in the mining sector: Final report*. Montreal: Roquet Vincent and Associates Inc.

Crooke, P., Harvey, B., and Langton, M., 2004. Implementing and monitoring Indigenous land use agreements in the minerals industry: The Western Cape Communities Co-Existence Agreement. *In*: M. Langton *et al.*, eds., *Settling with Indigenous people*. Sydney: Federation Press, 95–114.

Jung, D., 2022a. La minière Tata Steel accusée d'être une "mauvaise payeuse." *Radio-Canada*, 16 August. Available from: https://ici.radio-canada.ca/rci/fr/nouvelle/1905803/schefferville-miniere-tata-steel-accusee-etre-mauvaise-payeuse [Accessed 3 February 2023].

Jung, D., 2022b. Fermeture temporaire: Tata Steel affirme que les Innus garderont leurs emplois. *Radio-Canada*, 17 November. Available from: https://ici.radio-canada.ca/espaces-autochtones/1933453/fermeture-temporaire-mine-tata-steel-innus-schefferville?depuisRecherche=true [Accessed 3 February 2023].

Jung, D., 2023. Eaux rouges de Tata Steel: un bassin est réparé, mais une inspection reste à faire. *Radio-Canada*, 18 January. Available from: https://ici.radio-canada.ca/espaces-autochtones/1949305/schefferville-mine-environnement-eaux-rouges-tata-steel-bassin-inspection-faire [Accessed 3 February 2023].

Lipsky, M., 1980. *Street level bureaucracy: Dilemma of the individual in public services*. New York: Sage.

Nachet, L., 2019. Journal de bord du séminaire sur les villes en transition du Québec-Labrador. Available from: www.mineral.ulaval.ca/sites/mineral.ulaval.ca/files/field_diary_pierre_nachet.pdf [Accessed 3 February 2023].

O'Faircheallaigh, C., 2002a. *A new model of policy evaluation: Mining and Indigenous people*. Aldershot: Ashgate Press.

O'Faircheallaigh, C., 2002b. Implementation: The forgotten dimension of agreement-making in Australia and Canada. *Indigenous Law Bulletin*, 5, 14–17.

O'Faircheallaigh, C., 2016. *Negotiations in the Indigenous world: Aboriginal peoples and extractive industry in Australia and Canada.* New York: Routledge.

O'Faircheallaigh, C., 2018. Using revenues from Indigenous impact and benefit agreements: Building theoretical insights. *Canadian Journal of Development Studies*, 39 (1), 101–118.

Papillon, M., and Rodon, T., 2017. Proponent–Indigenous agreements and the implementation of the right to free, prior, and informed consent in Canada. *Environmental Impact Assessment Review*, 62, 216–224.

Pressman, J. L., and Wildavsky, A., 1973. *Implementation: How great expectations in Washington are dashed in Oakland; or, why it's amazing that federal programs work at all, this being a saga of the Economic Development Administration as told by two sympathetic observers who seek to build morals on a foundation.* Berkeley, CA: University of California Press.

Prno, J., Bradshaw, B., and Lapierre, D., 2010. Impact and benefit agreements: Are they working? *Proceedings of the Canadian Institute of Mining, Metallurgy, and Petroleum conference*, 11 May. Vancouver, BC: CIM.

Rodon, T., Keeling, A., and Boutet, J. S., 2022a. Schefferville revisited: The rise and fall (and rise again) of iron mining in Québec-Labrador. *The Extractive Industries and Society*, 12.

Rodon, T., Lemus-Lauzon, I., Séguin, J. M., and Schott, S., 2022b. Resource revenue allocation strategies and indigenous community sustainable development. *In:* C. Southcott, F. Abele, D. Natcher, and B. Parlee, eds., *Extractive industry and the sustainability of Canada's Arctic communities*. Montreal: McGill-Queen's University Press, 24–52.

Statistics Canada, 2017. *Matimekosh, IRI [Census subdivision], Quebec and Quebec [Province] (table). Census Profile. 2016 Census* (Catalogue No. 98-316-X2016001). Ottawa: Statistics Canada. 29 November 2017. Available from: https://www12.statcan.gc.ca/census-recensement/2016/dp-pd/prof/index.cfm?Lang=E [Accessed 3 February 2023].

Wolfe, J., 1992. Schefferville: The crisis in the Quebec–Labrador mining region (B. Mine closure in single industry towns and the problem of residual activities). *In*: C. Neil, M. Tykkläinen, and J. Bradbury, eds., *Coping with closure: An international comparison of mine town experiences*. London: Routledge, 192–207.

8 Comparative perspectives on the social aspects of mine closure and mine site transition in Canada and Australia

Sarah Holcombe, Sandy Worden, and Arn Keeling

Introduction

Since Bainton and Holcombe (2018) published their critical review of the social aspects of mine closure, there has been an upsurge in scholarly and gray literature examining this dimension of closure planning (O'Faircheallaigh and Lawrence 2019; Everingham *et al.* 2022). International standards and guiding materials have been published as governments, the mining industry, and host communities grapple with the challenges of closure. The *Integrated Mine Closure Good Practice Guide* (2019) drafted by the International Council on Mining and Metals (ICMM), for example, promotes the integration of closure planning across the mining lifecycle. The guide emphasizes the importance of applying socially and culturally appropriate remediation techniques, and the integral role that communities should play in the development of closure goals, visions, success criteria, and post-mining land uses (PMLUs). Elsewhere, the importance of early and long-term community engagement in remediation planning is increasingly recognized as a means of maximizing benefits and minimizing adverse impacts of closure (Cowan *et al.* 2011; Everingham *et al.* 2020). Community-engaged closure planning creates opportunities to center the needs and expertise of the people who will inherit the land that is left behind, and opens up space for co-learning between mining companies and host communities. Furthermore, actively including Indigenous interests, priorities, and expertise in closure planning offers an opportunity to remedy past exclusion and adverse impacts (Beckett *et al.* 2020).

Despite this more nuanced understanding of the social aspects of mine closure in the literature (Bainton and Holcombe 2018), mine closure regulations in many jurisdictions continue to overlook the need for socioeconomic plans and clear objectives for social outcomes (Kabir *et al.* 2015; Monosky and Keeling 2021a). This deficiency inhibits mitigation of the often devastating socioeconomic effects of closure and the integration of community objectives for mine site transitions into planning processes. In this chapter, we compare mine closure regulatory frameworks in three jurisdictions—Nunavik (Northern Québec) in Canada, and Queensland and the Northern Territory (NT) in Australia—through a lens that also considers Indigenous rights and interest holders. The jurisdictions have broadly similar mining histories, structures of mineral governance, and legacies of colonial

DOI: 10.4324/9781003406433-11

This chapter has been made available under a CC-BY-NC-ND 4.0 license.

relations with Indigenous peoples. We highlight the contextual factors that shape the regulations, including governance structures, Indigenous rights, negotiated agreements, and the scale and scope of mineral development activities. In doing so, we assess the following:

- How provincial/state and regional authorities regulate and assess mine closure and reclamation
- The extent to which these regulations and practices address the social aspects of closure and mine site transitions (economic impacts, cultural impacts, and PMLUs)
- The barriers and opportunities for local and Indigenous community participation
- The emerging practices of community engagement in mine closure and transition planning.

Section 2 outlines the national and subnational contexts. The body of the chapter (Sections 3 and 4) is structured around the thematic accountability framework used by Bainton and Holcombe (2018), which identified two overarching dimensions: 1) Procedural fairness (the administrative elements of mine closure) and 2) social risk (the things that mining companies are accountable for). The themes within the procedural dimension are integration and sustainability; stakeholder engagement; baselines, risks, and impact assessments; governance processes and the state; and local-level agreements with communities and affected landholders. Within the social risk dimension, the themes are housing and town normalization; service and infrastructure provision; economic linkages and transitions; Indigenous engagement in PMLU; and cultural heritage management. These themes have been reordered, adapted, and, in some cases, combined to reflect our jurisdictional contexts, the focus on Indigenous rights and interest holders, and available data. Table 8.1 provides a synoptic summary of the key actors, regulations, and findings from our researcher, and our conclusions are presented in Section 5.

National and regional contexts

Canada

Mineral development in Canada is largely regulated by subnational (provincial and territorial) governments, although in the territorial north these powers have only gradually and recently been "devolved" to territorial authorities. National laws and policies around both Indigenous rights and environmental assessment, therefore, intersect in complex ways with the patchwork of subnational resource laws and policies, federal constitutional authority (and court decisions), and both historical and modern treaties between the Crown and Indigenous Peoples (Thériault, Bourgeois, and Boirin-Fargues 2022). For instance, the (belated) implementation of a national *United Nations Declaration on the Rights of Indigenous Peoples* (UNDRIP) *Act* (2021) is complicated by these jurisdictional divisions and overlaps: "While some provinces have endorsed UNDRIP in policy statements, few have taken concrete

Table 8.1 Overview of the three jurisdictional regulations and procedures related to mine closure and reclamation

Mine Closure and Reclamation Governance	*Nunavik (Canada)*	*Queensland (Australia)*	*Northern Territory (Australia)*
Relevant acts and policies: National	• *United Nations Declaration on the Rights of Indigenous Peoples* (UNDRIP) *Act* (2021) • *Impact Assessment Act* (2019) • Collaborative Modern Treaty Implementation Policy (2023)	• *Environmental Protection and Biodiversity Conservation Act* (1999) • *Native Title Act* (1993)	
Relevant acts, policies, and guidelines: Subnational	• *Québec Mining Act* (2013) • Québec Guidelines for Preparing Mine Closure Plans (2017) • Nunavik Mining Policy (2014) • Plan Nord (2011)	• *Mineral Resources Act* (1989) • *Mineral and Energy Resources (Financial Provisioning) Act* (2018) • *Strong and Sustainable Resource Communities (SSRC) Act* 2017 • *Environment Protection Act* (1994) • *Aboriginal Cultural Heritage Act* (2003) • *State Development and Public Works Organisation Act* (1971) • Guidelines for Progressive Rehabilitation and Closure Plans (2023) • Social Impact Assessment (SIA) Guideline (2018)	• *Mining Management Act* 2001 • *Mineral Titles Act* 2010 • *Environmental Protection Act* (2019) • Only mine closure guidelines provided online: Voluntary federal LPSDP (2016)

(*Continued*)

Table 8.1 (Continued)

Mine Closure and Reclamation Governance	*Nunavik (Canada)*	*Queensland (Australia)*	*Northern Territory (Australia)*
Key regulatory bodies/agencies	• Ministère des Ressources naturelles et des Forêts (formerly MERN) • Ministère de l'Environnement et de la Lutte Contre les Changements Climatiques (MELCC) • Société du Plan Nord • Makivik Corporation • Kativik Environmental Quality Commission	• Department of Resources • Department of Environment and Science • Office of the Coordinator-General, Department of State Development, Infrastructure, Local Government, and Planning	• NT Department of Industry, Tourism, and Trade • NT Environment Protection Authority
Key closure governance challenges	• Multilevel and evolving governance systems • Two impact and benefit agreements (IBAs) and a modern land claims agreement • Mix of public government and Inuit organizations • Several regional governance bodies with overlapping mandates • Regional bodies responsible for environmental and social impact assessments (ESIAs) • Québec has final decision-making power for all aspects of mineral development, including closure	• A suite of new closure and PMLU regulations since 2016 • Multiple Indigenous land-use agreements (ILUAs) through which closure discussions occur • Vast areas without prescribed body corporates (PBCs) or land councils representing Indigenous interests • Queensland government has final decision-making power	• *Mining Management Act* 2001 under review (very dated) • Four regional Indigenous governance bodies (land councils) • Multiple ILUAs through which closure discussions occur • No coordination with the *Commonwealth ALRA Act* (1976) and NT laws • NT government has final decision-making power

steps to establish a policy or legislative framework for its implementation" (Papillon and Rodon 2020, p. 323). These policies overlay a complex and contested recent history of Indigenous engagement in mineral development and regulation, largely driven by their participation (however limited) in impact assessment processes and bilateral negotiations, with industry actors aimed at ensuring some form of consent for (and benefit from) mining development on Indigenous lands.

Although requirements for both remediation and Indigenous consultation have improved in recent decades, questions remain about how successfully Indigenous priorities are incorporated into both environmental assessment and remediation planning (Boerchers *et al.* 2018; Beckett *et al.* 2020). Historically, mine operators in Canada suffered minimal repercussions for not complying with already lax closure requirements, resulting in a plethora of abandoned sites and legacy impacts, including toxic contaminants, stunted economies, social dislocation, and disruptions to Indigenous access to traditional lands (Dance 2015; Rodon and Lévesque 2015; Sandlos and Keeling 2016). Dance *et al.* (2022) report that remediation efforts for both new and legacy mines within northern territories and regions remain complicated by jurisdictional overlap, as well as a lack of any clear remediation objectives. These policy and capacity shortcomings result in highly variable closure planning strategies and practices for currently operating mines. Closure plans across Northern Canada suffer from vague descriptions of community engagement, inconsistent or nonexistent application of Indigenous and local community knowledge, and a lack of any meaningful acknowledgment of the socioeconomic aspects of closure (Monosky and Keeling 2021a; Squires *et al.* 2022).

Regional context: Nunavik, Québec

The northern region of Nunavik covers one-third of the province of Québec, from the 55th parallel to the Hudson Strait coast (Fig. 8.1). It is also one of four regions within Inuit Nunangat, the circumpolar Inuit homelands. Today, Nunavik is home to more than 13,000 people, 11,800 (85%) of whom are Inuit (Statistics Canada 2017). The population resides in 14 coastal communities, with the largest being the administrative center of Kuujjuaq. Due to its Arctic climate, small population, and lack of road connections, Nunavik remains remote from major settlements in the south of Canada.

Natural resource development in Nunavik has been closely tied to the development of local and regional government systems. Contentious hydroelectric dam developments in Northern Québec in the 1970s pushed the province into negotiations with Indigenous groups, leading to the *James Bay and Northern Québec Agreement* (JBNQA) signed with Cree and Inuit in 1975. This treaty created Nunavik's current land regime, which provides Inuit residents with special land rights, albeit to only a small portion of their territory. In spite of long interest in its mineral resources, only three mines have operated in the region—the now abandoned Asbestos Hill Mine (1972–1984) and two currently operating nickel mines, Raglan Mine (1997–present, owned by Glencore) and Nunavik Nickel (2012–present, operated by Canadian Royalties). Notably, Asbestos Hill predated the creation of

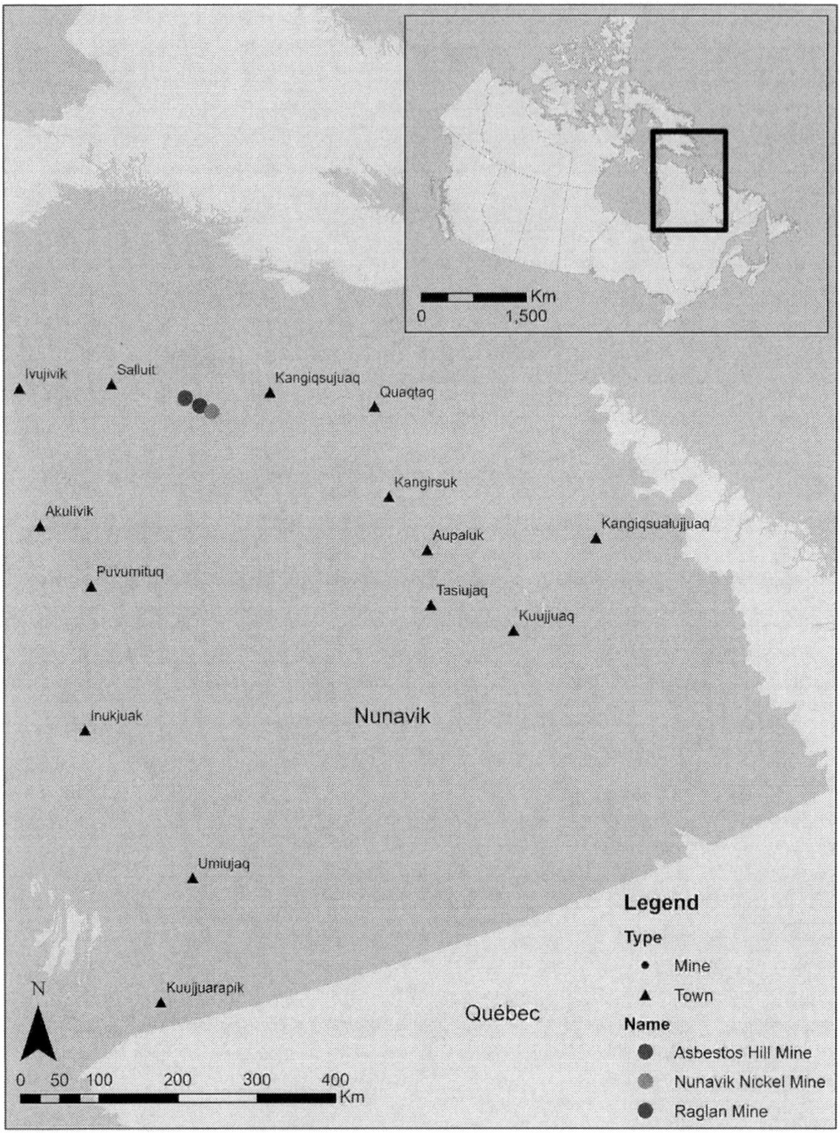

Figure 8.1 Map of Nunavik mines and communities

Map by Isabella Richmond

Nunavik's contemporary governance landscape, as well as modern environmental and mining regulations. Weak provincial regulations resulted in widespread contamination at the mine site and the nearby Deception Bay port, with virtually no remediation undertaken (Poirier and Brooke 2000; Carney 2016). After decades of inactivity, the site was finally put on the provincial list of abandoned sites in 2019 (MERN 2020).

Mining and mine closure in Nunavik are governed within a relatively new and complex system that includes multiple regional authorities with different but sometimes overlapping mandates, two impact and benefit agreements (IBAs), and a land claims agreement that provides some special rights to Nunavimmiut on some of their territory. At the regional level, the JBNQA created regional current governance bodies, like Makivik Corporation, the Kativik Regional Government, the Kativik Environmental Quality Commission (KEQC), and the Kativik Environmental Advisory Committee, which are a mix of "public" government and Inuit organizations (Rodon 2014; Fabbi *et al.* 2017; Nungak 2017). These bodies have varying roles in regulating mining and ensuring Inuit participation in decision-making. Only the KEQC has a defined role in closure planning through the socioeconomic and environmental impact assessment (SEIA) process and its regular review of closure plans. Makivik Corporation may sit on committees with mining companies where closure is discussed, but these engagements originate more from negotiated agreements than from government regulations.[1]

These regional governance bodies are nested within the provincial regulatory system of Québec, which has its own complicated set of ministries, operating procedures, and political agendas related to resource development. Provincially, Ministère de l'Environnement et de la Lutte Contre les Changements Climatiques (MELCC) and Ministère de l'Énergie et des Ressources Naturelles (MERN)[2] play the greatest role in regulating mine closure. Québec has the final decision-making authority for all aspects of mineral development in Nunavik, including mine closure. MELCC and MERN are responsible for governing how mine closure happens, what mine closure plans (MCPs) must contain, and what the standards are for remediation. Revised and updated closure plans must be submitted to the provincial government for review and approval every 5 years leading up to the final closure.

Australia

In Australia, mining is also regulated primarily by subnational (state and territory) governments, and there is considerable diversity in legislation, regulation, and guidance at this level. Federal legislation only exists where mining intersects with matters of national environmental and cultural significance, such as biodiversity (*Environmental Protection and Biodiversity Conservation Act* 1999 [EPBC Act]). The EPBC Act also includes the regulation of uranium mining, which is assessed and approved by federal and state or territory governments. The federal government has developed a series of 17 leading practice sustainable development handbooks for the mining industry,[3] which include handbooks on "mine closure" and "working with Indigenous communities." Although they do offer some good practice advice, these voluntary guidelines are more than a decade old, and engagement with them is inconsistent across Australia. *The Native Title Act* 1993 (Cth) is the other national legislation that impacts mining through the provision of a consultation and negotiation mechanism for Indigenous rights-holders, as discussed below.

Echoing the Canadian experience, Australian regulations have largely failed to hold proponents accountable for remediating environmental and social impacts from mining, resulting in an estimated 50,000 abandoned mines across the country (Unger *et al.* 2015). This legacy continues to pose significant environmental, safety, and cultural heritage risks, with subnational governments being hit with a rehabilitation liability of up to AU$1 billion (Cooper 2019). While regulatory reforms in Queensland and New South Wales have sought to address issues of financial assurance and progressive rehabilitation, Young *et al.* (2019) note that clear and transparent relinquishment processes do not feature in Australia's mining regulations, hampering relinquishment. In fact, only 25 mines have been officially relinquished (LPSDP 2016), none of which are large, open-cut mines.

The *Native Title Act* provides one of the few levers that Indigenous people have to negotiate with the mining industry across all Australian jurisdictions. Native title is the recognition of Indigenous Australians' rights and interests in land and waters according to their own traditional laws and customs. It may include the exclusive possession, use, and occupation of customary lands, or, more often, it relates to nonexclusive rights (AIATSIS 2016) that "coexist" with the rights and interests of others, such as pastoral or mining leaseholders. In the case of large-scale mining, coexistence is largely incompatible with the cultural priorities of Traditional Owners, as access to country is usually severely limited under mine safety regulations, and cultural heritage sites are routinely destroyed by mining and associated infrastructure, as demonstrated by the recent parliamentary inquiry into the destruction of the 46,000-year-old Juukan Gorge cave sites (Australian Government 2021). The *Native Title Act* is not a right to protect sites, only a right to negotiate with proponents. Furthermore, native title rights do not encompass subsurface rights to the minerals or petroleum, which are owned by the state or territory (AIATSIS 2016; Howlett and Lawrence 2019). The Act has also established an Indigenous governance landscape across the majority of Australia, with representative organizations known as prescribed body corporates (PBCs). PBCs act as the formal bodies through which industry engages and negotiates with Indigenous rights-holders, following the successful determination of native title. In 2023, there were 249 registered PBCs across Australia (AIATSIS 2023).

Subnational Australian mineral resource contexts: Queensland and the NT

Queensland is Australia's most northeastern state, spanning 1,729,742 km^2 (Geoscience Australia 2022). It has a population of 5,156,140 people, approximately 5% of whom are Indigenous (ABS 2021). Native title has been recognized over 25% of Queensland's land and waters (AIATSIS 2016). Figure 8.2 highlights Indigenous land tenures in Queensland and the Northern Territory. Mining is Queensland's third most important industry, adding AU$86.5 billion in nominal gross value to the economy in 2022/2023, after the health care and social assistance industry (Queensland Treasury 2024). Coal is by far the most valuable mineral export commodity, followed by zinc, copper, bauxite, and lead (QDR 2022a, 2022b).

In Queensland, mining is authorized under the *Mineral Resources Act* 1989 (MR Act), and mining leases are granted and administered by the Department of Resources. A proponent must obtain an environmental authority issued by the

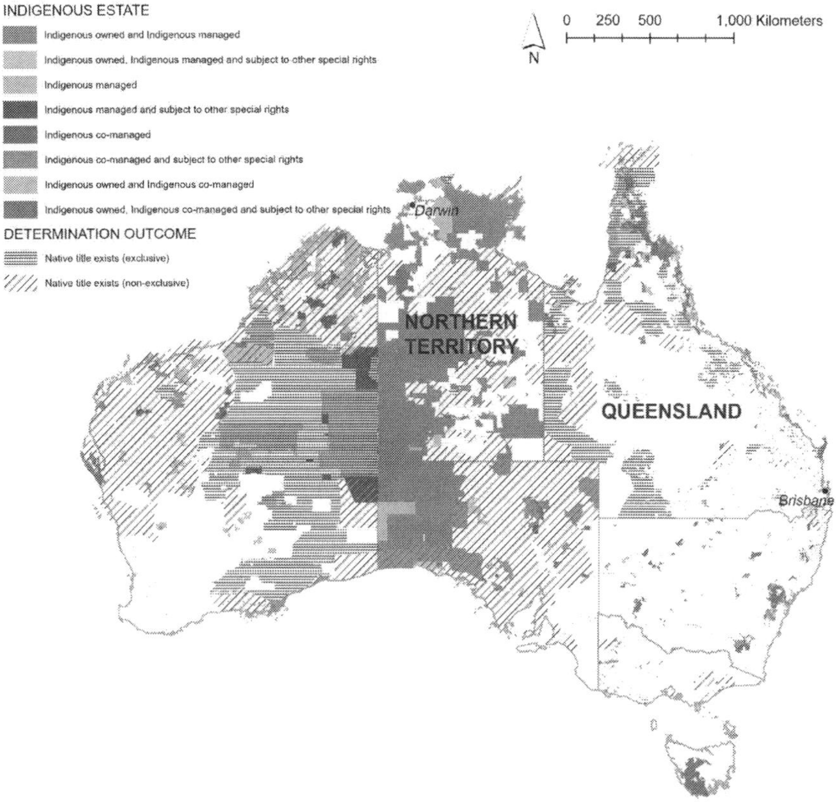

Figure 8.2 Australia map highlighting NT and Queensland Indigenous land tenures

Map by Julia Loginova

Department of Environment and Science under the state's *Environmental Protection Act* 1994 (EP Act) to undertake activities on a mining lease. The EP Act requires that all areas disturbed within the relevant mining tenure must be rehabilitated to a PMLU or managed as a nonuse management area (NUMA). Any undisturbed land within the relevant mining tenure must also be identified as a PMLU. In 2021, the government appointed a mine rehabilitation commissioner to develop "leading practice advice," which at this stage does not include any engagement with Indigenous rights and interest holders (see, for instance, Purtill and Littleboy 2023).[4] In 2023, the Queensland government developed its Guidelines for Progressive Rehabilitation and Closure Plans (Queensland Government 2023).[5] According to Purtill and Littleboy (2023), the period from 2016 to present has seen the most significant reforms to mine rehabilitation policy in Queensland's history.

There are 711 active mines in Queensland (Werner *et al.* 2020) located within four key mining regions—the North West Minerals Province (minerals), North Queensland (minerals), the Bowen Basin (coal), and the Clarence Moreton/Surat (coal and gas). Werner *et al.* (2020, p. 6) report there are around 2,820 abandoned or "neglected" mines, although estimates are as high as 15,000 (Cooper 2019). Queensland's mine closure context is further complicated by the fact that no mine has ever been fully rehabilitated (Werner *et al.* 2020), and no large, open-cut mine has been officially relinquished (Vivoda *et al.* 2019).

The NT is Australia's largest territory, located between Queensland and Western Australia (WA). The NT has a small and dispersed population of approximately 250,000, 30% of whom are Indigenous Peoples[6] and who experience high levels of disadvantage on the major indicators of education, health, housing, and employment. More than 70% of the Indigenous population lives in remote or very remote areas with significant historical deficits in civil society and public infrastructure. Globally, the NT is one of the few jurisdictions with legislation recognizing inalienable collective Indigenous rights to land, under the *Aboriginal Land Rights Act (NT)* 1976 (ALRA), which includes a form of free, prior, and informed consent (FPIC) over development on this land.[7] Under the ALRA, more than 50% of the NT is now recognized as Aboriginal freehold title, while much of the remaining land is subject to native title rights and interests, including 85% of the coastline. However, with few mainstream economic opportunities in remote communities, the NT government has shown a strong preference for mining and other extractive industries, such as coal seam gas, as the primary development path.

The largest economic sector in the NT is government and community services, contributing 23% of gross state product in 2019, followed by mining, which accounts for 19% of gross state product (NTG DEPWS 2020). Unlike Queensland and WA, the NT does not have distinct mining regions, which has resulted in a smaller industry and mining footprint. In 2020, there were 189 authorized mining projects (NTG DEPWS 2020). Like Queensland, no mine in the NT has been relinquished without liability to the state. The NT is the only jurisdiction in Australia with a profit-based royalty system, and because many extractive industry operations are capital and infrastructure-intensive, this often means that no royalties are paid by industry. This system also assumes that any industry-funded infrastructure or development has flow-on and broader benefits to communities and the region in a "trickle-down effect," both during and beyond life-of-mine (Holcombe 2021, p. 207).

Closure requirements in the NT are addressed via the *Mineral Titles Act* 2010 (MTA) and the *Mining Management Act* 2001 (MMA). Importantly, however, the *Environment Protection Act* 2019 (EPA) only assesses the initial potentially significant impacts of a mining proposal and conditions required for an environmental approval, as the MMA currently regulates for ongoing environmental management and closure. The MMA is regarded by many as outdated legislation (ALEC 2021; EDO 2021). The Hawke Review of the Northern Territory Environmental Assessment and Approval Process (2015) notes that the NT MMA "inherently presumes that the mining activity will in all cases trump the potential environmental impact" (Hawke 2015, p. 35) and that it is a largely non-risk-based approach to regulating mining activities. The NT government is exploring regulatory reforms to address these issues. What is lacking in the proposed reforms is consideration of the social aspects of mine closure, specifically how engagement with and participation of Indigenous rights and interest holders is going to occur, and harmonization of legislation that pertains to their interests (see also ALEC 2021). Indigenous Peoples remain positioned as stakeholders rather than rights-holders. Similar to the Nunavik context, the regional Indigenous representative body may sit on committees with mining companies where closure is discussed, but these engagements originate from negotiated agreements rather than government regulations. It is the NT government that decides what the mine closure remediation standards are and what the MCPs should contain. The exception to this practice is the closure of uranium mines, which also has federal regulatory oversight.

Procedural dimensions

Regulation and socioeconomic impacts of mine closure

Bainton and Holcombe (2018) advocate for the integration of closure planning throughout the mine lifecycle, with the social impacts of closure highlighted at each stage. They note, however, that "successful integration [in transition planning] is often undermined by a failure to start early" (Bainton and Holcombe 2018, p. 472). These shortcomings are evident in the Canadian and Australian jurisdictions and relate to both entrenched closure planning practices and regulatory shortcomings. In the Nunavik context, Monosky and Keeling (2021b) report that statements about whether and how companies should assess and mitigate negative socioeconomic impacts do not exist, are ill-defined, or are not related to closure specifically. The province of Québec's current guidelines for preparing MCPs state that "reclamation techniques may affect wildlife, plants, and the social environment" and that "the reclamation of accumulation areas must attain technical, environmental, and social objectives" (MERN 2017, p. 27). However, this document does not define "social objectives" and provides no guidance on how to measure success in meeting them. Unsurprisingly, then, MCP documents for active mines in Nunavik (like those elsewhere in Northern Canada) largely lack any meaningful acknowledgment of the socioeconomic aspects of closure (Monosky and Keeling 2021a).

Recognition of these aspects of mine closure is also largely absent in Queensland's regulatory framework. The mining reforms introduced in 2019 offer little in the social domain beyond a requirement for community consultation during the development of a progressive rehabilitation and closure plan (PRCP) and in the identification of PMLUs (Queensland Government 2019a, 2019b). Community consultation guidance for PRCPs acknowledges the cultural interest "Indigenous communities, Traditional Owners including native title holders" may have, but it positions Indigenous Peoples as stakeholders rather than rights-holders (Queensland Government 2019a, p. 1). The gap in addressing socioeconomic impacts is reflected in the MCPs produced by proponents. In our review of more than 10 MCPs and draft plans for Queensland operations that were produced prior to the requirement for a PRCP, none addressed the socioeconomic impacts of closure. The social dimensions of closure are limited to brief discussions about community and/or stakeholder engagement processes to be utilized, and some include a list of potential stakeholders to be consulted, including Indigenous Peoples. Socioeconomic impacts of mining are covered in sections 9, 10, and 11 of the *Strong and Sustainable Resource Communities (SSRC) Act* 2017 and the Social Impact Assessment (SIA) Guideline (Queensland Coordinator-General 2018). However, these regulatory instruments are focused on the early stages of the mining lifecycle (project development) and do not specifically deal with closure.

In the NT, like Queensland, neither the current nor the proposed regulatory reforms address the socioeconomic aspects of closure beyond that of community consultation for PMLUs. The government has not published any guidance on these aspects. Likewise, the ESIA process is focused on the front end of the operation. Under the current regulatory arrangements, once a mine is approved, the Department of Industry, Tourism, and Trade (DITT) becomes the "one-stop shop" for all environmental and resource regulation, which presents a risk of regulatory capture.

In our review of NT MCPs, we found considerable diversity in the coverage of socioeconomic and cultural dimensions, from minimal coverage to significant detail (see NTG 2023). Ranger Uranium Mine's MCP represents the detailed case. The mine has always been contentious due to early opposition to its development from Mirrar Traditional Owners and to its ecologically sensitive location (Howe 2020) adjacent to the World Heritage-listed Kakadu National Park. Lawrence (2022) notes that particular scientific knowledges have been privileged throughout Ranger's rehabilitation process. Now that the mine is in closure, the aim is for it to be reabsorbed into the national park. The complexity of this aspiration is reflected in the MCP, which is reviewed annually. In addition to meeting corporate closure and PMLU requirements, the MCP needs to address federal government closure criteria. It also contains cultural closure criteria (ERA 2022) and an analysis of future occupancy intentions, including for customary harvest use. Notwithstanding the level of detail in the MCP, Lawrence (2022) contends that Ranger will continue to threaten the environment and Mirrar long after closure. MCPs for other mines on Aboriginal land in the NT are also starting to include cultural criteria, for example, operations owned by major mining companies Rio Tinto and Newmont. These MCPs, however, do not consider broader Indigenous socioeconomic issues, such as what happens when the royalties stop flowing and how impacts to employment, housing, etc., will be addressed. In addition, the extent to which cultural criteria are translated from closure planning into practice is unclear. At the other end of the coverage spectrum (i.e., the Nolans rare-earth and Jervois base metal project), reference is made to the EIS and the socioeconomic impacts identified, but little to no consideration is given to engaging Traditional Owners in closure and PMLU planning (see NTG 2023). Expanding coverage of socioeconomic and cultural dimensions in these MCPs will require regulatory guidance, which is currently absent.

SIA and mine closure

Environmental assessment and SIA processes, Bainton and Holcombe (2018) note, offer at least a *potential* avenue for the recognition of social impacts of closure and strategies for their mitigation. In particular, SIA provides potential approaches for identifying social risks related to mine closure. It also presents opportunities for stakeholder engagement and community participation in setting (and evaluating) post-mining objectives. Yet, frameworks for including remediation planning in SIA are unevenly applied across jurisdictions, and it is unclear how updates to remediation plans are connected to the mitigations and objectives outlined in these assessments (Beckett *et al.* 2020). There is a persistent tendency of impact assessment processes to focus on the construction and operational phases of development, and mainly on the environmental aspects of closure and reclamation. Where mine closure and remediation are considered in these processes, these may be "conceptual" closure plans that bear scant resemblance to the social or environmental conditions at closure; social indicators and expectations established through impact assessment may be long forgotten by the time remediation is being enacted (Devlin and Tubino 2012).

In Nunavik, the KEQC, through the ESIA process, plays the most direct role in regulating mine closure in the region. The ESIA process allows for regional governments and communities to communicate their needs and expectations for a new mine. A preliminary closure plan is also provided to the KEQC as part of the ESIA. The KEQC can review, approve, reject, and set additional conditions for it. But despite the important role that it plays in ensuring companies are acting in the best interests of Nunavimmiut, the province can overturn any of the KEQC's decisions at any point, meaning the province holds more formal power (Rodon 2018). The territory's operating mines—Raglan Mine and Nunavik Nickel—each proceeded through ESIA (and regular follow-up reviews of closure plans), but the resulting closure plans do not address any closure-related social objectives. Either the ESIA process is not adequately accounting for important aspects of closure or there are no effective follow-up mechanisms to ensure that the impacts identified in the ESIA make it into MCPs (Monosky and Keeling 2021b).

In both Queensland and the NT, ESIAs are triggered when proponents of large-scale mining projects apply for an environmental authority. The ESIA assesses and reports on a project's social, economic, and environmental impacts and the mitigation measures identified, as is standard. The focus in this process, however, is on the early stages of the mining lifecycle, and there is little evidence that attention is given to the socioeconomic impacts of closure. The Queensland closure plans reviewed, for example, do not refer back to the socioeconomic impacts identified in the SIAs. Through the SIA consultation process, potentially impacted individuals and communities can share their aspirations for the new mine, how they may be affected by its development and operation, and their perspectives on monitoring requirements. The regulator then assesses whether the benefits of the project outweigh its costs. There is no obligation for local and Indigenous community views to be addressed by the proponent or the state government even where native title has been determined over the mining lease area. Similar to Nunavik, project development decision-making power in Queensland rests with the state. This is not the case in the NT, where Traditional Owners have veto rights for developments on Aboriginal land. However, if Traditional Owners provide their consent for mine development on their land, then, under the subsequent ESIA process, they are considered just another stakeholder group as on any other land tenure. ESIAs are rarely led by Indigenous Peoples, and they rarely meet their expectations (Holcombe 2021; Roche *et al.* 2021).

While the Queensland SIA guideline includes the principle that SIAs should be lifecycle-focused (i.e., they should also address closure), there is little evidence that this occurs in practice. The SSRC Act does not mention mine closure or the mining lifecycle at all. To better assess the socioeconomic impacts of closure, some proponents in Queensland and the NT are starting to use SIAs in their mine closure planning, for example, Rio Tinto, for Gove. However, these SIA methods need to be examined through the lens of Indigenous rights and interest holders. The use of SIAs in closure planning is not a regulatory requirement, and few examples have been published in the scholarly literature or elsewhere. One exception is Kabir's (2021) case study of Rio Tinto Coal Australia's 2005 SIA for the closure of the Blair Athol Coal Mine in Central Queensland. Kabir contends that the SIA supported a collaborative approach to closure planning, enabled the company to integrate community and employee concerns early in the planning process and mitigate socioeconomic risks, and contributed to the development of an MCP that was "acceptable" to the local community (Kabir 2021, p. 24).

Community engagement and negotiated agreements

The Blair Athol Mine SIA example highlights the issue of community engagement for closure planning. Drawing on a number of case studies and reviews, Bainton and Holcombe (2018, p. 5) argue that "a failure to engage stakeholders in a clear and stable process constitutes a primary failure of mine closure processes." Engagement with stakeholders and rightsholders is crucial to identifying, addressing, and mitigating the potential social impacts of mine site transitions. To be effective, such engagement should occur at the earliest stages of project planning and evaluation and continue throughout the mine life, rather than being initiated during "winding-down" phases (Devlin and Tubino 2012; Xavier *et al.* 2015; Everingham *et al.* 2018).

In Queensland, it is a regulatory requirement for proponents to engage with impacted local and Indigenous communities on mine rehabilitation and closure and provide them with the opportunity to comment. Proponents must document how they plan to undertake ongoing consultation on mine rehabilitation and the extent to which each proposed PMLU is consistent with the outcome of that community consultation (Queensland Government 2019a). Methods for collaborative decision-making about closure, PMLUs, or post-mining transition are not articulated. A key weakness in the regulatory framework is that closure planning is undertaken on a mine-by-mine basis, with no consideration of cumulative impacts in mining regions. This narrow approach can result in closure implications primarily being considered from the proponent's perspective, that is, a focus on achieving relinquishment.

A further weakness across Australia is the limited incorporation of Indigenous knowledges into closure planning (Bond and Kelly 2020). As discussed in Chapter 5, Indigenous communities whose customary lands and practices are directly affected by environmental impacts from large-scale extractive developments can have particularly complicated relationships with proponents (Kabir *et al.* 2015). The rights and interests of these communities in the restoration of post-mining impacts are increasingly being recognized through formal consultation and consent requirements and benefit-sharing agreements. Yet, as Monosky and Keeling (2021a, p. 2) note, "it remains unclear how these novel arrangements apply to the long-term social and environmental challenges that emerge at the end of the mining cycle." Such agreements may not include mine closure and reclamation provisions nor create governance mechanisms that require ongoing consultation and engagement. Negotiated agreements also have complicated relationships with state-driven project review and closure governance, such as ESIA or closure-specific regulations.

The intricate dynamics of community engagement and negotiated agreements are apparent in the case of Nunavik. Both provincial and regional governments require community engagement between mining companies and Indigenous communities, although this is largely limited to the early stages of a project's life. The *Québec Mining Act* (2013) requires mining companies to establish a joint company–community committee that "must be established within 30 days after the lease is issued and must be maintained until all the work provided for in the rehabilitation and restoration plan has been completed" (Government of Quebec 2020). However, specific guidance is not provided on how broadly communities should be engaged or what issues should be addressed.

In 1995, Falconbridge, Raglan Mine's original owners, signed the first IBA between a community and mining company in Canada. Canadian Royalties (Nunavik Nickel) followed suit with its IBA in 2008.[8] The two IBAs differ in their details but generally contain provisions for preferential hiring of Inuit, preferential contracting for Nunavik businesses, training programs, profit sharing, and requirements for environmental protections (Bird and Nixon 2004; Rodon 2018). Each agreement establishes a company–community implementation committee for the duration of the mine's life. While the IBAs do not focus on closure, the Raglan IBA parties recognized the need for greater consultation on closure and reclamation and subsequently established the Raglan Closure Plan Subcommittee. Through the subcommittee, community members are learning about closure activities and contributing their knowledge and priorities to the MCP. The subcommittee is not a regulatory requirement.[9]

The focus of agreements in Australia is similar. In their analysis of Indigenous land-use agreements (ILUAs) across the country's major mining regions, O'Faircheallaigh and Lawrence (2019) found that many of them failed to address mine closure in any substantive way. Indeed, 30 of the 50 agreements make no reference to closure, except in relation to the termination of financial benefits if production is suspended or a mine closes. While it is a regulatory requirement to negotiate ILUAs, there is no requirement that the agreements address closure and its associated legacies.

Risk dimensions

Economic and employment transitions

The impacts of industrial cycles and mine closure on local, resource-dependent communities are well-documented in the closure literature. "Coping with closure," in many regions, has meant grappling with the individual and collective displacement caused by the economic shocks of mass job losses and attendant socioeconomic impacts (Neil *et al.* 1992; Keeling 2010; Rodon *et al.* 2022). This is particularly evident in rural and remote regions, where resource towns emerged alongside industrial developments. Closure can result in outmigration and community economic decline, even collapse. The potential for post-mining repurposing of mined lands and mining infrastructure is highlighted by some observers (and industry advocates) as a potential buffer for these problems (Kivinen 2017; Keenan and Holcombe 2022), but longer-term economic transitions may be overshadowed by environmental concerns, as well as the challenges of community survival faced by remote communities. As Bainton and Holcombe note in their review, local and regional governments facing the crisis of closure may have few planning levers at their disposal, particularly as the financial capacities of regions and companies are undermined by the loss of revenues.

In Nunavik (and elsewhere in Northern Québec), the experience of closure as an economic trauma is not unfamiliar. For instance, the closure of iron ore mines near Schefferville (just south of the Nunavik boundary) in 1982 led to the near collapse of the community and mass outmigration of workers. It became a paradigmatic case of community-wide closure impacts. While the town did not entirely disappear (being reoccupied, somewhat ironically, by previously marginalized local Indigenous residents), the lack of planning for economic transition was evident (Rodon *et al.* 2022). Within Nunavik, local impact of the closure of the Asbestos Hill Mine (1972–1984) was buffered by the fact that this remote site (near the current Raglan Mine) was mainly staffed by fly-in, fly-out (FIFO) workers from the south of Québec (Carney 2016). There were very few Inuit employees, and the mine contributed no direct revenue, which limited the impact of closure on local Nunavik economies.

In spite of these (and other) historic examples, provincial mine closure policy neglects economic transition planning. But the socioeconomic impacts of closure, while largely absent in policy, are relatively well known to current regional actors in Nunavik. Indeed, there appears to be a disconnect between what regional authorities know about closure impacts and what is accounted for in policy. Many regional government participants interviewed for Monosky and Keeling's (2021b) study recognize that the communities have benefited from mining activities but remain skeptical that these benefits are contributing to community sustainability or that they will last long after closure. Concerns about the abrupt decline or end of employment, tax and business revenue, and profit-sharing arrangements came up just as frequently in the interviews as did issues of tailings stability, water quality, and landscape changes. Participants expressed concern about the impact that mine closure may have on the region's ability to maintain infrastructure and services, and how unemployment could affect the health and well-being of families and communities. Another issue identified was the possibility that Inuit employees (whose numbers are growing) would lose their jobs without transferable skills or employment opportunities in other industries in the region or community. This region's very remote location, with its harsh climate, complicates planning for post-mining economic development.

Economic transitions for resident Indigenous populations in Queensland and the NT are likely to have very different impacts from non-Indigenous populations. In the NT, a key reason for this is the persistent lack of direct employment of Indigenous landowners in mines operating on their land, despite local employment and procurement targets set out in ILUAs. This has been the case even after pre-employment training programs for Traditional Owners. For instance, at Rio Tinto's Gove Mine in the NT, Pearson and Daff (2013, p. 53) found that "relatively few Indigenous people demonstrate[d] a preference to work in the minerals extraction industry at Nhulunbuy" (aka Gove). This is also the case for the Ranger and McArthur River mines. Further research is required to identify causal factors. One possibility is that FPIC had not been granted. Gove and Ranger predated the ALRA, and the McArthur River Mine is on a pastoral lease, so FPIC was not sought.[10] However, at Newmont's Tanami Mine, where the Warlpiri Traditional Owners provided their consent for the mine on their land, TO employment numbers have been persistently low for this long-life mine (Kung *et al.* 2019).

At Weipa, in Queensland's western Cape York Peninsula, where Rio Tinto operates a bauxite mine, there is some innovative local employment activity of Wik-Waya Traditional Owners in the supply chain side of progressive rehabilitation. A comprehensive ILUA was negotiated here (Crooke *et al.* 2006), and over the past several years, a local TO group has established a native seed supply business. At Amrun Mine, an extension of Rio Tinto's bauxite operation, the Wik-Waya Traditional Owners are actively engaged in cultural heritage management, seed collection, and rehabilitation activities on their land. Through this program, they have been closely involved throughout the mining lifecycle, from planning and development to operations and, eventually, closure (Barnes *et al.* 2020). Indigenous involvement in mining restoration supply chains is an emerging and evolving sociopolitical process with potentially far-reaching implications for mining practices, plans, and policies (Urzedo *et al.* 2022).

Long-term planning for multiple mine closures and economic transition across mining regions in Queensland represents a policy vacuum. The government is prioritizing investment in low-emissions technology, renewable energy, and "new economy minerals" (Queensland Government 2022, p. 3) over the development of a policy regime for regional post-mining transition planning. The government's growth strategy is part of its decarbonization agenda.[11] The policy vacuum is evident in the 68-page Queensland Resources Industry Development Plan (Queensland Government 2022), which, apart from a brief note about ongoing regulatory reforms, fails to address mine closure or post-mining transitions. Within this decarbonization policy regime, questions remain about the impacts of potential mine closures on regional economies and employment opportunities. As the Schefferville experience shows, the lack of long-term transition planning can be devastating. For the mining-dependent regions of the North West Minerals Province and Bowen Basin, this is problematic. Local governments will face significant revenue losses, and without well-paying jobs in the industry, most employees are expected to migrate out of the regions.

Housing, services, and infrastructure

Mine closure and community decline can manifest in deteriorating infrastructure, withdrawal of community services such as schools and hospitals, and, ultimately, community abandonment ("ghost towns"). To some extent, the history and legacies of closure-induced town collapse have motivated the industry-wide transition to FIFO or drive-in, drive-out (DIDO) arrangements (Storey 2016). While decreasing the exposure of employees and residents to the social and economic risks of closure, FIFO installations also include infrastructure of potential interest for remote regions and/or Indigenous communities considering post-extraction economic activities. The long-standing issue of town closure has become less of a concern in the past two decades with the increase in FIFO, while it also has an additional set of impacts on local Indigenous Peoples (Haslam McKenzie and Hoath 2016; Jensen and Sandström 2020).

The FIFO context is particularly germane to the Nunavik situation. Both operating mines are staffed entirely by commuting labor, including workers drawn from Nunavik Inuit communities (which are located distant from the mine sites). Housing and community infrastructure, therefore, are only indirectly related to the mines themselves. Nevertheless, mine infrastructure and transportation facilities, including haul roads, a deepwater port at Deception Bay, power-generation installations, and buildings such as residences at the site, represent important potential assets. The Raglan Agreement IBA discusses the right of first refusal regarding surplus equipment and property, which "gives Inuit Parties a right of first refusal prior to the removal or demolition of facilities that Raglan considers to be surplus to requirements." In other words, Inuit parties are given the option to obtain or purchase any assets after closure. However, current closure plans, following regulatory guidance, propose the complete dismantling and removal of buildings, roads, and power infrastructure at the mine, in spite of clear interest by local communities in maintaining or acquiring some or all of these facilities. In a regional context, transportation and energy infrastructure have the potential to provide significant economic and social benefit for local communities, but only if their repurposing is effectively planned for.

This inconsistency between MCPs and negotiated agreements is similar in the NT context. The town of Jabiru was established to house Ranger Mine workers, and Nhulunbuy to house workers from Gove. But with town transfers to the state (Jabiru) and to Aboriginal Traditional Owners (Nhulunbuy) in their infancy, and with little published research, we are yet to witness how these transfers will unfold. As occurred in Schefferville, it is likely that these towns will be reoccupied by previously marginalized local Indigenous residents who do not work at the mines.

Queensland's SSRC Act prohibits large resource operations from having a 100% FIFO workforce. Consequently, mines have a mix of FIFO, DIDO, and local housing arrangements. Few mining towns in the state will need to undergo a process of normalization once mines close, as most of them are under the jurisdiction of local governments. One exception is Weipa, which is governed by Rio Tinto through the Weipa Town Authority. As mines close, local governments will, however, need to grapple with significant impacts on their revenues and the cost of maintaining stranded assets.

Indigenous participation and cultural heritage

As Bainton and Holcombe note, Indigenous engagement in mine closure planning aims to recognize rights, integrate Indigenous/traditional knowledge, and address concerns for cultural heritage values in mine reclamation and PMLUs. In connection with Canadian closure planning, Monosky and Keeling (2021b, p. 2) argue, "For mine closure to be successful in a northern context it must incorporate community expertise, emerge from the values and priorities of the Indigenous peoples whose lands mines are operating on, and account for a wider scope of social, economic, and cultural impacts." For Nunavimmiut, the key avenues for involvement in mine closure and PMLU planning are through regional Inuit organizations (through ESIA and subsequent review processes) or through IBA implementation committees (for signatory communities and organizations) (Monosky and Keeling 2021a). Nevertheless, neither the Raglan nor the Nunavik Nickel Mine in Nunavik explicitly incorporates Inuit knowledge in its closure plans, in part because it is not mandated by Québec's mine closure regulation. This shortcoming is despite regional actors having a wealth of important environmental and cultural knowledge that could benefit closure planning. In the Nunavik context, where cultural heritage is tied closely to the preservation of language and land-based activities, as well as specific cultural sites (Makivik Corporation 2014), incorporation of Indigenous knowledge would seem essential to PMLU planning. For instance, the continuation of hunting and fishing and ensuring the safety of travel routes and harvesting areas are central concerns for Inuit when assessing the long-term impacts of closed and remediated sites.

Likewise, in Australia, without regulatory levers or contractual guidance in ILUAs, inconsistencies arise in how Indigenous agreement beneficiaries are engaged in cultural heritage management at closure. There are some emerging good practices, for instance, at the Ranger Mine, where "cultural closure criteria" (Smith 2009) and "cultural reconnection committees" (Brady *et al.* 2021) are specified in the MCP. Cultural criteria are also integrated into the PMLU, closure objectives, and closure criteria documents (ERA 2022). This approach, however, is rare. Even on Aboriginal land that has preemptive laws to protect cultural heritage, the usual industry approach is ad hoc and inconsistent. As mines are built, they usually destroy sacred sites, places of significance, and hunting grounds, resulting in cumulative impacts over the life-of-mine. These impacts are felt on the landscape and in relationships, including the perceived complicity of some Traditional

Owners in the destruction of a sacred site which, in turn, can be a cause for continuing tension and ill-feeling within the Indigenous community (Lewis and Scambary 2016). Reclamation and restoration of the environment present an opportunity for Traditional Owners to restore social relationships and for industry to remedy its past adverse impacts.

A return to original ecosystems is a mine closure objective for some Traditional Owners in Australia. Early evidence from the closure of the Ranger Mine suggests that Traditional Owners want a productive site to enable a return to customary harvest activities. They do not want any evidence of mining to remain. Cohen (2017) reports that at Weipa (Queensland), the Alngith Traditional Owners also sought rehabilitation outcomes that resembled prior ecosystems. They wanted their land to support multiple livelihood aspirations, ranging from economic ventures, for example, establishing cattle operations, to fine-scaled enhancements of specific locations for recreational or other economic purposes, for example, fishing or swimming sites. In some instances, these enhancements were intended to increase populations of introduced plant and animal species, like mango and wild pig, that serve as local food sources. Other examples aimed to increase the local density of native bush foods.

Conclusions

In spite of the growing recognition of the need to integrate social criteria and objectives into closure planning, this review highlights the significant gaps in regulation and practice that remain. There are important similarities between the Canadian and Australian contexts. Both jurisdictions give very limited regulatory attention to the social aspects of mine closure. Negotiated agreements, likewise, take an ad hoc approach to addressing closure impacts, which limits opportunities for Indigenous rights and interest holders to manage closure risks and advance initiatives that support their values and aspirations. While IBAs and ILUAs incorporate development opportunities for the operational stage of a mine's life, in the Australian context this is largely missing for the closure stage. In Nunavik, there is inconsistency between agreement expectations and regulatory requirements.

In both Australia and Canada, there are no "public policy framework[s] that guide [agreements] negotiation, terms of reference, or implementation" (Cameron and Levitan 2014, p. 26; Papillon and Rodon 2020).[12] Any potential benefits from IBAs and ILUAs hinge on their successful implementation. This includes the capacity of the company, the Indigenous communities, and their representative bodies to uphold the conditions of the agreements. Given the significant level of disadvantage that requires addressing and, thus, what these agreements are expected to achieve, this public policy gap is both surprising and a major gap (Holcombe 2021). Addressing this gap is also a missed opportunity in the current revision of the NT mining legislation.

The focus on the front end of the mining lifecycle is clearly apparent not only in the IBAs and ILUAs but also in the ESIA process, which is rarely applied to mine closure (Beckett *et al.* 2020; Getty and Morrison-Saunders 2020; Morrison-Saunders *et al.* 2016). Although this is changing in Australia, ESIAs are largely scoped from a proponent's perspective, rather than those of impacted Indigenous groups or local communities. It is challenging, however, to track the extent of change, as few ESIA reports are publicly available. Like agreements, ESIA processes tend to focus on the impacts of mine development and operations. Closure is only considered in conceptual or general terms. Yet, both operational and closure stages have social and environmental impacts that extend beyond life-of-mine and require advanced assessment, planning, and mitigation. While normative industry frameworks, such as the ICMM's integrated mine closure planning guide, call for closure planning across all stages of the mining lifecycle, there is little evidence that this is being implemented with any consistency in these jurisdictions.

In the absence of adequate regulation and policy, management of social aspects of closure tends to comprise ad hoc negotiations between companies, governments, and mining-affected communities. There is a clear need for more effective and prescriptive closure planning regulations and guidelines at the subnational level, as well as processes for local and Indigenous community participation in closure governance. Regulatory requirements should not supersede negotiated agreements; rather, they should help provide clarity and support planning for PMLUs and socio-economic transition. The objectives of relinquishment should also meet reclamation and environmental justice concerns of mined lands. Regulations should further stipulate the long-term environmental monitoring and protection required to ensure the resumption or continuation of Indigenous land use and ownership.

Acknowledgments

We would like to thank to the editors and anonymous reviewers for their input. Some of the Queensland data came from a study that was commissioned and funded by the Queensland Resources Council and the ICMM. The Nunavik sections of this chapter are adapted with permission from Monosky, M. and Keeling, A. 2021b. "Social considerations in mine closure: Exploring policy and practice in Nunavik, Quebec." *The Northern Review*, 52 (December) 29–61. Research for this section was generously supported by grants from the National Sciences and Engineering Research Council of Canada network, Towards Environmentally Responsible Resource Extraction (TERRE-NET), and the Social Sciences and Humanities Research Council of Canada.

Notes

1 For greater detail on these regional organizational roles and responsibilities, see Monosky and Keeling (2021b).
2 Recently renamed Ministère des Ressources naturelles et des Forêts.
3 See them here: www.industry.gov.au/publications/leading-practice-handbooks-sustainable-mining
4 www.qmrc.qld.gov.au/about/what-we-do
5 There is one mention of Indigenous peoples' interests in these guidelines, and that is in relation to consideration of PMLU options: "Any regulatory constraints on, or objectives for, the proposed land use (e.g., Indigenous Land Use Agreements, site management

plan [contaminated land], local and regional land use plans, endangered species, or registered heritage places)" (2023, p. 21).

6 Within the broader Australian population, Indigenous peoples make up only approximately 3.5% of the total.

7 The ALRA is Commonwealth (federal government) legislation. The right to FPIC is a qualified one, however, as a "yes" to exploration is also a "yes" to mining, while a "no" to exploration is only for 5 years, when the company can return to ask again. There have been some strong campaigns by TOs that "no means no" and that the company is not welcome to return.

8 The communities of Salluit and Kangiqsujuaq are signatories of the Raglan Mine IBA, called the *Raglan Agreement*. Salluit, Kangiqsujuaq, and Puvirnituq are signatories of the Nunavik Nickel IBA, called the *Nunavik Nickel Agreement*. Note that neither the federal nor provincial governments are parties to these negotiated agreements.

9 Author Keeling is an invited member of this closure plan subcommittee, providing input, guidance, and research support on the social aspects of closure.

10 The Ranger Mine consultation and consent process was compromised at the time, as the ALRA (1976) NT was being introduced and this mine was seen as a test case (Howe 2020).

11 The Queensland government has set a 50% renewable energy target by 2030, is investing AU$1.16 billion in clean energy transition (via the Powering Queensland Plan), is funding nine programs aimed at reducing greenhouse gas emissions, and is supporting low-emissions infrastructure, such as the H2-Hub™ Gladstone Project. The hub is a proposed multi-billion-dollar chemical complex producing green hydrogen and green ammonia (Miles 2022). In addition, there were more than 80 renewable energy projects slated for the state in 2021 (QDNRM 2021).

12 In Canada, IBA "toolkits" and guidebooks have been produced by researchers and negotiators experienced in the field. See "The IBA Community Toolkit," available from https://gordonfoundation.ca/initiatives/iba-toolkit/, and "Impact and Benefit Guidebook," available from www.sfu.ca/rem-dev/planning/research/iba/handbook.html.

References

Arid Lands Environment Centre (ALEC), 2021. ALEC's submission on 'regulation of mining activities: Environmental regulatory reform'. Available from: https://assets.nationbuilder.com/aridlands/pages/57/attachments/original/1648529894/ALEC%E2%80%99s_submission_on_%E2%80%98Regulation_of_mining_activities__environmental_regulatory_reform%E2%80%99.pdf?1648529894

Australian Bureau of Statistics (ABS), 2022. Census all persons QuickStats. Queensland, 2021. Available from: https://abs.gov.au/census/find-census-data/quickstats/2021/3

Australian Government, 2021. *Parliament of the Commonwealth of Australia. A way forward: Final report into the destruction of Indigenous heritage sites at Juukan Gorge.* Joint Standing Committee on Northern Australia. Available from: https://parlinfo.aph.gov.au/parlInfo/download/committees/reportjnt/024757/toc_pdf/AWayForward.pdf;fileType=application%2Fpdf

Australian Institute of Aboriginal and Torres Strait Islander Studies (AIATSIS), 2016. Native title information handbook: Queensland. Available from: https://aiatsis.gov.au/sites/default/files/research_pub/native_title_information_handbook_2016_qld_2.pdf

Australian Institute of Aboriginal and Torres Strait Islander Studies (AIATSIS), 2023. Find a PBC. Available from: https://nativetitle.org.au/find/pbc

Bainton, N., and Holcombe, S., 2018. A critical review of the social aspects of mine closure. *Resources Policy*, 59, 468–478.

Barnes, R., Holcombe, S., and Parmenter, J., 2020. *Indigenous groups, land rehabilitation and mine closure: Exploring the Australian terrain.* Brisbane: Centre for Social Responsibility in Mining, University of Queensland.

Beckett, C., Dowdell, E., Monosky, M., and Keeling, A., 2020. *Integrating socio-economic objectives for mine closure and remediation into impact assessment in Canada*. SSHRC Knowledge Synthesis Report, 48 pp. Available from: http://research.library.mun.ca/id/eprint/14487.

Bird, F., and Nixon, R., 2004. The Raglan mine and Nunavik inuit. *In:* F. Bird and S. W. Herman, eds., *International businesses and the challenges of poverty in the developing world: Case studies on global responsibilities and practices*. London: Palgrave Macmillan.

Boerchers, M., Sinclair, A. J., Gibson, R. B., and Halden, N. M., 2018. Sustainability is finding the next mine: The complicated relationships among legacies, sustainability, and EA. *Environmental Impact Assessment Review*, 71, 84–93.

Bond, C., and Kelly, L., 2020. Returning land to country: Indigenous engagement in mined land closure and rehabilitation. *Australian Journal of Management*, 46 (1), 174–192.

Brady, C., Christopherson, P., and O'Brian, J., 2021. Incorporating Indigenous knowledge in mine closure: Ranger Uranium Mine. *The Royal Society of Victoria*, 133, 18–22.

Cameron, E., and Levitan, T., 2014. Impact and benefit agreement and the neoliberalization of resource governance and Indigenous-state relations in Northern Canada. *Studies in Political Economy*, 93, 25–52.

Carney, J., 2016. Seeking closure: Legacies of the asbestos hill mine in Nunavik. Available from: www.chairedeveloppementnord.ulaval.ca/sites/chairedeveloppementnord.ulaval.ca/files/final_report_jeanette_carney.pdf

Cohen, T., 2017. Bringing country back? Indigenous aspirations and ecological values in Australian mine-site rehabilitation. *In:* K. Jalbert, A. Willow, D. Casagrande, and Stephanie Paladino, eds., *ExtrACTION: Impacts, engagements, and alternative futures*. Abingdon: Routledge, 137–150.

Cooper, S., (2019). Maximising post-mining land use: Queensland Government reforms. *In:* A. B. Fourie and M. Tibbett, eds., *Proceedings of the 13th international conference on mine closure*. Perth: Australian Centre for Geomechanics, 969–982.

Cowan, W. R., Mackasey, W. O., and Robertson, J. G., 2011. *Policy framework in Canada for mine closure and management of long-term liabilities*. Sudbury, ON: National Orphaned and Abandoned Mines Initiative.

Crooke, P., Harvey, B., and Langton, M., 2006. Implementing and monitoring Indigenous land use agreements in the minerals industry: The Western Cape Communities Co-existence Agreement. *In*: K. Shain, L. Palmer, M. Langton, and O. Mazel, eds., *Settling with indigenous people: Modern treaty and agreement-making*. Annandale: Federation Press.

Dance, A., 2015. Northern reclamation in Canada: Contemporary policy and practice for new and legacy mines. *The Northern Review*, 41, 41–80.

Dance, A., Monosky, M., Keeling, A., and Sandlos, J., 2022. Mine remediation policy and practice in Northern Canada. *In:* C. Southcott, F. Abele, D. Natcher, and B. Parlee, eds., *Extractive industry and the sustainability of Canada's Arctic communities*. Montreal: McGill-Queen's University Press.

Devlin, J., and Tubino, D. I., 2012. Contention, participation, and mobilization in environmental assessment follow-up: The Itabira experience. *Sustainability: Science, Practice and Policy*, 8 (1), 106–115.

Environmental Defenders Office (EDO), 2021. Submission on the Regulation of mining activities: Environmental regulatory reform consultation paper. Available from: https://depws.nt.gov.au/__data/assets/pdf_file/0004/984946/edo-submission-01-mar-21-on-mining-consultation-paper.PDF

Energy Resources Australia (ERA), 2022. Post-mining land use, closure objectives and closure criteria, Chapter 8. Available from: www.energyres.com.au/uploads/Mine-Closure-Plan/Chapter_8_Post_closure_land_use__closure_objectives_and_closure_criteria_2022.pdf

Everingham, J., Rolfe, J., Lechner, A. M., Kinnear, S., and Akbar, D., 2018. A proposal for engaging a stakeholder panel in planning post-mining land uses in Australia's coal-rich tropical savannahs. *Land Use Policy*, 79, 397–406.

Everingham, J., Mackenzie, S., Svobodova, K., and Witt, K., 2020. *Participatory processes, mine closure and social transitions*. Brisbane: Centre for Social Responsibility in Mining, University of Queensland.

Everingham, J. A., Svobodova, K., Lèbre, É., Owen, J. R., and Worden, S., 2022. Comparative capacity of global mining regions to transition to a post-mining future. *The Extractive Industries and Society*, 11, 101136.

Fabbi, N. C., Rodon, T., and Finke, E. W., 2017. Makippugut (we are standing up): Public policy and self-determination in Nunavik. *American Review of Canadian Studies*, 47 (2), 117–126.

Geoscience Australia, 2022. Area of Australia—states and territories. Available from: www.ga.gov.au/scientific-topics/national-location-information/dimensions/area-of-australia-states-and-territories

Getty, R., and Morrison-Saunders, A., 2020. Evaluating the effectiveness of integrating the environmental impact assessment and mine closure planning processes. *Environmental Impact Assessment Review*, 82 (1), 106366.

Government of Quebec, 2020. Mining Act M-13.1. Available from: http://legisquebec.gouv.qc.ca/en/ShowDoc/cs/M-13.1

Hawke, A., 2015. Review of the Northern Territory environmental assessment and approval processes. Available from: https://depws.nt.gov.au/__data/assets/pdf_file/0011/262919/hawke-review-of-the-northern-territory-environmental-assessment-and-approval-process.pdf

Haslam McKenzie, F. M., and Hoath, A., 2016. Aboriginal mine workers: Opportunities and challenges of long-distance commuting. *In:* F. M. Haslam McKenzie, ed., *Labour force mobility in the Australian resources industry*. Singapore: Springer.

Holcombe, S., 2021. The State's selective absence: Extractive capitalism, mining juniors and Indigenous interests in the Northern Territory. *In:* N. Bainton and E. E. Skrzypek, eds., *Absent presence: Resource extraction and the state in Papua New Guinea and Australia*. Canberra: ANU Press.

Howe, K., 2020. The Ranger Uranium Mine agreement revisited: Space-times of agreement making in Aboriginal Australia. *Transformations*, 33, 97–115.

Howlett, C., and Lawrence, R., 2019. Accumulating minerals and dispossessing Indigenous Australians: Native title recognition as settler-colonialism. *Antipode*, 51 (3), 818–837.

International Council on Mining and Metals (ICMM), 2019. Integrated mine closure. Good practice guide, 2nd ed. Available from: www.icmm.com/en-gb/guidance/environmental-stewardship/2019/integrated-mine-closure

Jensen, T., and Sandström, J., 2020. Fly-in/fly-out and the fragmentation of communities: A case study of a uranium mine on Indigenous land. *Journal of Rural Studies*, (78), 78–86.

Kabir, Z., 2021. The role of social impact assessment (SIA) in the development of a mine closure plan in regional Australia. *Journal of Environmental Assessment Policy and Management*, 23 (3&3).

Kabir, S. M. Z., Rabbi, F., Chowdhury, M. B., and Akbar, D., 2015. A review of mine closure planning and practice in Canada and Australia. *World Review of Business Research*, 5 (3), 140–159.

Kivinen, S., 2017. Sustainable post-mining land use: Are closed metal mines abandoned or re-used space? *Sustainability*, 9 (10), 1705.

Keeling, A., 2010. Born in an atomic test tube: Landscapes of cyclonic development at Uranium City, Saskatchewan. *Canadian Geographer*, 54, 228–252.

Keenan, J., and Holcombe, S., 2022. Mining as a temporary land-use: A global snapshot of mine site transitions and repurposing. *The Extractive Industries and Society*, 8 (3), 100924.

Kung, A., Rogers, P., Skrzypek, E., and Barnes, R., 2019. *Newmont Tanami Operations social impact assessment*. Brisbane: Centre for Social Responsibility in Mining, Sustainable Minerals Institute, the University of Queensland.

Lawrence, R., 2022. Rehabilitating Ranger uranium mine: Scientific uncertainty, deep futures and the production of ignorance. *Environmental Politics*, 31 (1), 49–69.

Lewis, G., and Scambary, B., 2016. Sacred bodies and ore bodies: Conflicting commodification of landscape by Indigenous peoples and miners in Australia's Northern Territory. *In:* P. McGrath, ed., *The right to protect sites: Indigenous heritage management in the era of native title*, 221–253. Available from: https://aiatsis.gov.au/publication/35045

LPSDP, 2016. Mine closure: Leading practice sustainable development program for the mining industry. *Australian Government, Departments of Industry, Innovation & Science and Foreign Affairs and Trade*. Available from: www.industry.gov.au/resource/Documents/LPSDP/LPSDPMineClosureCompletionHandbook.Pdf

Makivik Corporation, 2014. Parnasimautik consultation report. Available from: https://parnasimautik.com/2014-consultation-report/

MERN, 2017. Guidelines for preparing mine closure plans in Québec. Available from: https://mrnf.gouv.qc.ca/documents/mining/guidelines-mine-closure.pdf

MERN, 2020. *Plan de travail 2020–2021: Restauration des sites miniers abandonnés*. Government of Québec. Available from: https://mern.gouv.qc.ca/en/mines/mining-reclamation/reclamation-of-abandoned-mining-sites/

Miles, S., 2022. Hydrogen and industry heavy weights making moves in CQ [12 April press release from the Minister for State Development, Infrastructure, Local Government and Planning]. Available from: https://statements.qld.gov.au/statements/94941

Monosky, M., and Keeling, A., 2021a. Planning for social and community-engaged closure: A comparison of mine closure plans from Canada's Territorial and Provincial North. *Journal of Environmental Management*, 277, 111324.

Monosky, M., and Keeling, A., 2021b. Social considerations in mine closure: Exploring policy and practice in Nunavik, Quebec. *The Northern Review*, 52, 29–61.

Morrison-Saunders, A., McHenry, A., Sequeira, R., Gorey, P., Mtegha, H., and Doepel, D., 2016. Integrating mine closure planning with environmental impact assessment: Challenges and opportunities drawn from African and Australian practice. *Impact Assessment and Project Appraisal*, 34 (2), 117–28.

Neil, C., Tykkläinen, M., and Bradbury, J., eds., 1992. *Coping with closure: An international comparison of mine town experiences*. London: Routledge.

Northern Territory Government, 2020. Regulation of mining activities—environmental regulatory reform consultation paper. Available from: https://denr.nt.gov.au/__data/assets/pdf_file/0011/956891/regulation-of-mining-activities-consultation-paper-122020.pdf

Northern Territory Government (NTG), 2023. Mining projects [website]. Available from: https://industry.nt.gov.au/publications/mining-and-energy/public-environmental-reports/mining/public-mining-environmental-reports/mines

Nungak, Z., 2017. *Wrestling with colonialism on steroids*. Montreal: Véhicule Press.

O'Faircheallaigh, C., and Lawrence, R., 2019. Mine closure and the Aboriginal estate. *Australian Aboriginal Studies*, 1, 65–81.

Papillon, M., and Rodon, T., 2020. The transformative potential of Indigenous-driven approaches to implementing free, prior and informed consent: Lessons from two Canadian cases. *International Journal on Minority and Group Rights*, 27 (2), 314–335.

Pearson, C., and Daff, S., 2013. Indigenous workforce participation at a mining operation in northern Australia. *The Australian Bulletin of Labor*, 39 (1), 42–63.

Poirier, S., and Brooke, L., 2000. Inuit perceptions of contaminants and environmental knowledge in Salluit, Nunavik. *Arctic Anthropology*, 37 (2), 78–91.

Purtill, J., and Littleboy, A., 2023. More And better mine rehabilitation—Lessons from Queensland. World Mining Congress. Available from: www.qmrc.qld.gov.au/__data/assets/pdf_file/0028/317728/more-and-better-rehabilitation-lessons-from-queensland.pdf

Queensland Coordinator-General, 2018. Social impact assessment guideline. Available from: www.statedevelopment.qld.gov.au/__data/assets/pdf_file/0017/17405/social-impact-assessment-guideline.pdf

Queensland Department of Natural Resources and Mines (QDNRM), 2021. Electricity generation map. Available from: https://maps.dnrm.qld.gov.au/electricity-generation-map/

Queensland Department of Resources (QDR), 2022a. Queensland annual coal export—2016–2021. Available from: www.data.qld.gov.au/dataset/annual-coal-statistics/resource/d22a8d8b-7c00-42d2-884a-c438d51cefc3

Queensland Department of Resources (QDR), 2022b. Queensland annual mineral summary—quantity and value of minerals produced 2019–2020. Available from: www.data.qld.gov.au/dataset/annual-mineral-metal-and-petroleum-statistics

Queensland Government, 2019a. Information sheet: Community consultation for progressive rehabilitation and closure plan. Available from: https://environment.des.qld.gov.au/__data/assets/pdf_file/0035/96569/rs-is-prcp-community-consultation.pdf

Queensland Government, 2019b. Mined land rehabilitation policy. Available from: https://environment.des.qld.gov.au/__data/assets/pdf_file/0035/87659/mined-land-rehabilitation-policy.pdf

Queensland Government, 2022. Queensland resources industry development plan. Available from: www.resources.qld.gov.au/__data/assets/pdf_file/0005/1626647/qridp-web.pdf

Queensland Government, 2023. Progressive rehabilitation and closure plans. Available from: https://environment.des.qld.gov.au/__data/assets/pdf_file/0026/95444/rs-gl-prc-plan.pdf

Queensland Treasury, 2024. About the Queensland economy. Available from: https://www.treasury.qld.gov.au/queenslands-economy/about-the-queensland-economy/

Raglan Mine, 2019. Raglan mine closure plan. Obtained by authors.

Roche, C., Brueckner, M., Walim, N., Sindana, H., John, E., the Venembeli Community, 2021. Understanding why impact assessment fails: A case study of theory and practice from Wafi-Golpu, Papua New Guinea. *Environmental Impact Assessment Review*, 89. https://doi.org/10.1016/j.eiar.2021.106582

Rodon, T., 2014. From Nouveau-Québec to Nunavik and Eeyou Istchee: The political economy of Northern Québec. *Northern Review*, 38, 93–112.

Rodon, T., 2018. Institutional development and resource development: The case of Canada's Indigenous peoples. *Canadian Journal of Development Studies*, 39 (1), 119–136.

Rodon, T., and Lévesque, F., 2015. Understanding the social and economic impacts of mining development in Inuit communities: Experiences with past and present mines in Inuit Nunangat—ProQuest. *Northern Review*, 41, 13–39.

Rodon, T., Keeling, A., and Boutet, J.-S., 2022. Schefferville revisited: The rise and fall (and rise again) of iron mining in Québec-Labrador. *The Extractive Industries and Society*, 12, 101008.

Sandlos, J., and Keeling, A., 2016. Toxic legacies, slow violence, and environmental injustice at Giant Mine, Northwest Territories. *Northern Review*, 42, 7–21.

Smith, H. D., 2009. Strangers in a foreign land—developing cultural closure criteria for mines in Australia's Northern Territory. *In*: A. B. Fourie and M. Tibbett, eds., *Mine closure 2009*. Perth: Australian Centre for Geomechanics.

Squires, C., Reinecke, S., and Meyer, S., 2022. *Socio-economic transition for mine closure in Canada—Investigating standards-based solutions*. Toronto: Canadian Standards Association.

Statistics Canada, 2017. *Région du Nunavik [Health region, December 2017], Quebec and Quebec [Province] (table)* (Catalogue no. 98-316-X2016001). Ottawa: Statistics Canada.

Storey, K., 2016. The evolution of commute work in the resource sectors in Canada and Australia. *Extractive Industries and Society*, 3 (3), 584–593.

Thériault, S., Bourgeois, S., and Boirin-Fargues, Z., 2022. Indigenous Peoples' agency within and beyond rights in the mining context: The case of the Schefferville region. *The Extractive Industries and Society*, 12 (3), 100979.

Unger, C. J., Lechner, A. M., Kenway, J., Glenn, V., and Walton, A., 2015. A jurisdictional maturity model for risk management, accountability and continual improvement of abandoned mine remediation programs. *Resources Policy*, 43, 1–10.

Urzedo, D., Pedrini S., Hearps, C., Dixon. K., and van Leeuwen, S., 2022. Indigenous environmental justice through coproduction of mining restoration supply chains in Australia: Advances on mining restoration policy article. *Restoration Ecology*, 30 (S1), e13748.

Vivoda, V., Kemp, D., and Owen, J., 2019. Regulating the social aspects of mine closure in three Australian states. *Journal of Energy & Natural Resources Law*, 37 (4), 405–424.

Werner, T. T., Bach, P. M., Yellishetty, M., Amirpoorsaeed, F., Walsh, S., Miller, A., Roach, M., Schnapp, A., Solly, P., Tan, Y., Lewis, C., Hudson, E., Heberling, K., Richards, T., Chung Chia, H., Truong, M., Gupta, T., and Wu, X., 2020. A geospatial database for effective mine rehabilitation in Australia. *Minerals*, 10, 1–21.

Xavier, A. M., Veiga, M. M., and Zyl, D. V., 2015. Introduction and assessment of a socio-economic mine closure framework. *Journal of Management and Sustainability*, 5, 38–49.

Young, R. E., Manero, A., Miller, B. P., Kragt, M. E., Standish, R. J., Jasper, D. A., and Boggs, G. S., 2019. *A framework for developing mine-site completion criteria in Western Australia: Project report*. Perth, WA: The Western Australian Biodiversity Science Institute.

Part III

Navigating relationships with Indigenous communities

9 Understanding the silent dimensions of social acceptability of a lithium project in the Cree community of Nemaska

Julie Fortin[1]

Introduction

Historically, economic development initiatives, particularly those related to natural resource extraction, have served as pretexts for the perpetuation of colonial relationships among the state, industry, and Indigenous communities. In an effort to rebalance these relationships, the Truth and Reconciliation Commission called on Corporate Canada to adopt the United Nations Declaration on the Rights of Indigenous Peoples (UNDRIP) as a framework for reconciliation and to "commit to meaningful consultation, respectful relationships, and to obtain the free, prior, and informed consent (FPIC) of Indigenous peoples before initiating economic development projects" (2012, pp. 12–13). Although Canada ratified the Declaration in 2010, it was not until 2016 that it fully endorsed it, including the principle of FPIC, which has been included in the Crown's duty to consult.

The debates surrounding the FPIC pits, on the one hand, a procedural approach, which serves primarily to ensure the legality and legitimacy of the Crown's decisions through the establishment of processes to obtain such consent and, on the other hand, a substantive approach, which is an extension of the right to self-determination of Indigenous peoples and ensures their greater participation in the decision-making process. To overcome this impasse, Papillon and Rodon (2017) proposed adopting a relational approach to FPIC, suggesting that the state recognize the rights of Indigenous peoples and consider them as equal partners in decision-making. This collaborative decision-making process should be the result of deliberations held within the communities. This perspective emphasizes the deliberative nature of consent and the need to include all groups in the community. However, a deliberative process implies that all members of the community have equal access to the discussions. Therefore, one of the challenges associated with the implementation of FPIC is to include more marginalized groups, such as youth, women, and elders, in decision-making (Owen and Kemp 2014). Moreover, different perspectives on resource development can generate internal conflicts within communities, which makes deliberations difficult for them (see Horowitz *et al.*'s chapter in this volume). In order to better understand the roots of these conflicts and diverging perspectives, our research focused on the following question:

DOI: 10.4324/9781003406433-13

This chapter has been made available under a CC-BY-NC-ND 4.0 license.

In the context of mining projects, how is the negotiation and deliberation capacity of Indigenous communities influenced by 1) the perception of the project, 2) the emotions generated, and 3) the project's impacts on social cohesion?

These questions will be answered using the case study of the Cree community of Nemaska and the Whabouchi Mine, the first lithium mine in Eeyou Istchee to receive provincial and federal approvals. Located 30 km from the Cree village of Nemaska, the Whabouchi Project is owned by Nemaska Lithium and is devoted to the production of lithium salts for the manufacture of batteries for electric vehicles. The company is notable for the vertical integration of its activities; that is, its project involves both the extraction of spodumene concentrate at the Whabouchi Mine and its processing into lithium hydroxide at a plant to be located in Bécancour, in southern Québec.

Theoretical framework: The two faces of mining

Mining as a trigger of colonialism

A first body of literature denounces the asymmetrical power relations that characterize the interactions between communities and mining companies. MacNeil (2018), for example, uses the term "extractive violence" to describe the impacts of mining activities on communities and the environment, directly threatening their culture: "A mine is never just a hole in the ground. A mine can cause psychological, physical, and spiritual destruction; it can threaten languages and cultures, social systems, and livelihoods. A mine can wipe out a people" (p. 98). Since they are imposed from the outside, disproportionately benefit external interests, involve the dispossession of traditional lands, and generate profound political inequities, mining activities are associated with a form of colonialism and environmental injustice at the expense of local populations (Munshi and Kurian 2007; Willow 2016; Bernauer 2019; Long 2019; Nachet *et al.* 2022). Furthermore, because of the environmental contamination they cause, mining activities are sometimes associated with a form of "slow violence," the toxic legacy of which is perpetuated over decades (Sandlos and Keeling 2016).

Munshi and Kurian (2005) point to the "environmental racism"[2] (CRRF 2024) experienced by Indigenous people, who are often the most affected by the impacts of industrial activities carried out in the name of "sustainable" and "global" development. This body of literature also highlights the strategies employed by mining companies to divide communities, by selecting the actors with whom they speak and marginalizing certain individuals, often those most affected by their activities (Munshi and Kurian 2007; Coronado and Fallon 2010; Viveros 2016; MacNeil 2018; Long 2019).

Moreover, the literature in social psychology shows that the emotions generated by key events within a culture tend to be reactivated each time an event presents similar characteristics to those of the initial situation (Kitayama and Markus 1999; Frijda 2001; Evans-Campbell and Walters 2006). In an Indigenous context, due to their impacts on the territory and on the practice of subsistence activities, mining activities sometimes act as triggers, reminding community members of past injustices, particularly those related to colonial policies (Gordon 2015; Keeling and Sandlos 2015; Gislason and Andersen 2016; Dahlin and Fredriksson 2017; Meade 2017; Harnetty 2017; Bjørst 2020; Jones 2020). Beyond their link to the colonial past, the environmental disruptions caused by development projects can induce a sense of distress called *solastalgia*, which refers to the inability of individuals to find comfort in their environment (Albrecht 2005; Albrecht *et al.* 2007), a situation that can generate worry, anxiety, anger, and a feeling of powerlessness (Norgaard and Reed 2017)

Mining as a control shift

Although it recognizes the power relations discussed above, a second body of literature, mainly from the field of political science, focuses more on the capacity of communities to exercise control over and benefit from extractive activities, whether through negotiation or resistance (Horowitz *et al.* 2017). In Canada, despite the dismal history of mining development on their territories, some Indigenous communities have managed to reconcile their practice of traditional activities while taking part in the wage economy generated by mining activities (Boutet 2014). Others see their inclusion in the mining landscape as a symbol of cohabitation and reconciliation (Desbiens and Sepúlveda 2019). Moreover, this increased participation in mining development also gives some communities the power to be selective and to oppose certain types of development that are at odds with their values (Bernauer and Peyton 2021). Bourgeois and Rodon (2019) highlight, for example, the ability of the Cree of Eeyou Istchee to exercise a right of refusal over mining development in their territory and cite the example of the Matoush Uranium Project. The refusal to negotiate an impact and benefit agreement (IBA) and the lobbying campaigns of the Grand Council of the Crees led to the project being abandoned, despite the fact that it had been approved by the Canadian Nuclear Safety Commission and had received positive recommendations from federal assessment panels (*ibid*). Wanvik and Caine (2017) showed how Métis in Wood Buffalo, Alberta, have demonstrated strategic pragmatism in their relationship with the oil sands industry, by joining forces with other Indigenous groups to gain greater legitimacy in their negotiations, bringing in the expertise of consultants and signing an IBA. The authors also highlight how traditional land-use studies, carried out as part of the negotiations, were an important form of capacity-building within communities. It should be noted, however, that some communities are reluctant to share information on their land use, underscoring the limits of this methodology, which is sometimes reductive and still associated with Western geographical knowledge systems, especially in the context of impact assessments (Joly *et al.* 2018; Westman *et al.* 2020).

Far from being the result of a single factor, the negotiating capacity of communities depends on the political, financial, human, and technical resources available to them; their degree of social cohesion; the strength of their political organizations; their experience with development projects; and the quality of their interactions with promoters (O'Faircheallaigh 2013). However, despite the advances made by certain Indigenous groups, mining projects are often a source of conflict within communities (Fortin 2019; Bernauer and Peyton 2021). Certain groups, particularly women, are often excluded from negotiations of IBAs (Gibson and Kemp 2008; Hall 2019). And, despite the benefits associated with direct and indirect employment, particularly in terms of improved community infrastructure and a greater ability to purchase equipment for hunting and fishing, the increased income generated by mining activities is also associated with an upsurge in social problems in the communities, including drug and alcohol abuse and domestic violence (Rodon *et al.* 2013; Rodon and Lévesque 2015).

Whether Indigenous communities choose to engage or not in mining activities is strongly related to their capacity to implement the principle of FPIC. At present, because of the political uncertainty surrounding this principle, communities have three strategies at their disposal for taking action: Collaboration, contestation, and reappropriation (Papillon and Rodon 2019). Collaboration is achieved through the signing of IBAs. Contestation takes the form of legal action or occupation of the territory. Finally, Indigenous communities can also take ownership of participatory processes, for example, by setting up their own consultation protocols (Leclair, Papillon *et al.* 2019) or by developing their own impact assessment processes. Because they invoke ancestral authority and international principles derived from the UNDRIP, these processes have strong legitimacy and normative value, which makes them difficult to ignore (Papillon and Rodon 2019). In this chapter, we will see that the Cree of Eeyou Istchee are well versed in all three of the above strategies and use them to control resource development in their territory.

Research methodology

This chapter is drawn from a PhD dissertation in public communication. Qualitative data collection methods were employed to develop a theoretical model to analyze different types of relationships between Indigenous communities and the mining industry. This chapter focuses on a case study of the Cree of Eeyou Istchee.

Epistemological posture: Critical pragmatism

Pragmatism places great emphasis on the agency of communities: "The focus of pragmatism is on the human capacity to learn, reason, and make choices in our environments; to respond to, and interact with, our environments; and to adapt to, modify, and shape them in various ways" (Kaushik and Walsh 2019, p. 9). It implies that communities are able to define the issues they consider important and to conduct the research they consider relevant (Morgan 2014; Kaushik and Walsh 2019, p. 9). These issues are then explored further through an investigation, defined as a process to better understand a problematic situation in order to make improvements (Kaushik and Walsh 2019).

In my doctoral research, I have placed particular emphasis on highlighting Indigenous voices and epistemologies, both in the research topic designation, the collection of data, and in the interpretation of results. Not only does my research aim to present the concerns and perspectives of different community members about the mining projects under study, but special effort was also made to ensure that the contextual and historical information for each of the cases came from Indigenous actors, in order to avoid appropriating the history of these nations or projecting my own interpretation of events. Therefore, information on the negotiations surrounding the James Bay and Northern Québec Agreement and other significant political events is recounted by local political actors (see, for instance, Saganash 2011). Lastly, Indigenous perspectives are also taken in my theoretical framework, which presents the work of numerous Indigenous scholars from a variety of disciplines such as psychology (Bombay 2014; Kurtness 2014), social work (Evans-Campbell 2008), environmental science (Basile 2017), anthropology (Koperqualuk 2015), Indigenous studies (Simpson 2008), and cultural geography (Marsh 2010).

Methodological strategy: A case study

The case study lends itself particularly well to the paradigm of critical pragmatism. Indeed, this methodological strategy allows research to focus on complex social issues that are relevant to the actors and that require in-depth explanation. The main strength of case studies lies in the possibility of analyzing multiple sources of data, historical, attitudinal, and situational, in order to develop a holistic understanding of a phenomenon (Roy 2010; Yin 2014). The use of multiple data sources also allows for a triangulation process to validate the data. Because of their contextual scope and the fact that they do not necessarily cover a large sample population, case studies cannot be used for statistical generalization. However, through their attention to the sociohistorical context and holistic dimensions of a phenomenon, case studies lead to an analytical generalization; that is, they help explain a phenomenon by generalizing a set of data to a broader, universal, theoretical proposition (*ibid*). In this chapter, the relationship between the mining industry and Indigenous communities will be analyzed in terms of two ideal types (forced union and control shift), allowing for an analytical generalization. To construct a typology, the researcher must propose an interpretation of reality and make links between the phenomena observed during the survey and the historical and structural characteristics of the environment under study. While its name may be confusing, the ideal type is not a standard or an ideal to be followed, but rather a conceptual model designed to explain a given situation by presenting its main characteristics (Schnapper 2012).

Research conducted in an Indigenous context must be collaborative and responsive to community needs and priorities (APNQL 2014; INQ 2017; ECPT2 2018; CSSSPNQL 2021). The Knowledge Network on Environment Impact Assessment and the Social Impacts of Mining in the Canadian Eastern Arctic and Subarctic,[3] created in 2013 and led by Professor Thierry Rodon, brought together researchers and representatives from different northern political institutions such as the Makivvik Corporation, the Government of Nunavut, the Cree Nation Government, and the Cree Board of Health and Social Services of James Bay. The dialog that took place within this network has helped establish the research needs and priorities of these different actors. My research project, as well as the choice of my two case

Table 9.1 Two faces of mining: Relational ideal types between mining and Indigenous communities

Ideal Types	*Perception of Mining*	*Emotions Generated*	*Impacts on Social Cohesion*
Forced union	- Incompatibility with Indigenous culture - Extractive violence, environmental racism - Threat to identity and health - Seen as an extension of colonial policies	- Anger, cumulative grief, sadness, anxiety, and *solastalgia* - Project acts as a trigger and revives bad memories - Feelings of injustice, powerlessness, and Distrust	- Divisions between the leadership and community members - Exacerbation of community conflicts (internalized racism, traditional authority vs authority imposed by the *Indian Act*) - Threat to the transmission of Indigenous knowledge
Control shift	- Resiliency and experience gained from previous projects - Compatibility with Indigenous culture (negotiation between two worlds) - Community's political weight promotes greater decision-making power	- A sense of clarity about the project (understanding of both positive and negative impacts) - Community is perceived as resilient (sense of control)	- Risk of conflict between the leadership and the rest of the community (if perceived lack of consultations) - Initiatives to increase participation and dialog between community members - Jobs to prevent the exodus of young people - Income to purchase hunting and fishing equipment - Trust in the remediation and restoration processes

studies, is therefore a response to the concerns expressed by this research group, specifically by representatives of the Cree Nation Government in the case of the present chapter. It should be noted that, in parallel to my involvement in the Whabouchi Project, I acted as a consultant to the Department of Environment of the Cree Nation Government between 2015 and 2016. This experience, in addition to contributing to my reflection on the communication and environmental issues of mining projects in an Indigenous context, also facilitated the establishment of contacts for the Nemaska case study.

Data collection

Data collection was conducted in partnership with local organizations. In Nemaska, a collaboration with the Cree Nation Government enabled my first exchanges with the community, and the data were then collected in partnership with the Nemaska Wellness Center. Direct observation of the Whabouchi Project's public hearings was conducted in March 2015. My analysis focused specifically on the position of individuals with regard to the project and the issues raised, and also on the tone of the exchanges, the emotions conveyed by participants, the decoration of the room (presence of petitions on the walls), the applause after certain comments, and other factors. These observations were recorded in a logbook. A transcript of the Whabouchi Project's public hearings was used to supplement the direct observations made at the time of the event. Lastly, a literature review was carried out on all the briefs that were filed during the hearings. The recordings and briefs are available on the COMEX[4] website.

In the fall of 2017, I collaborated on a research project[5] led by Professor Thierry Rodon and Professor Mylène Riva of McGill University, which consisted in assessing the impacts of mining projects on social cohesion and developing indicators to measure community well-being (miyupimaatisiiun) from a Cree perspective. Since I had already worked with the community of Nemaska, and to limit the pressure on the community, I acted as a facilitator to combine the data collection for both projects. With the permission of the community leader, and in collaboration with the Wellness Center, I conducted a focus group with community elders in Nemaska in November 2017. In order to avoid the need for participants to travel, the focus group was held directly at the Multi-Service Day Center, the community's seniors' center. A representative of the Wellness Center was responsible for recruiting participants and providing an interpreter. The community leader was present at the focus group, which was attended by about 15 elders. The focus group was held in Cree, lasted almost 2 hours, and focused on the impacts of residential schools, the importance of traditional food, the impacts of hydroelectric dams (especially on the taste of fish and game), the benefits of mining (purchase of hunting and fishing equipment and construction of houses), the anticipated impacts on children and on social cohesion, and the difficulty of measuring these impacts.

I also conducted three semi-structured interviews with each of the three participants who were recruited either by the Wellness Center or through contacts made during previous visits to the community. These interviews lasted between 30 and 60 minutes, and the questions dealt with the community's vision of development, the perceived degree of influence of community members, their perception of the project proponent and of community leadership, the impacts of mining development, and the emotions aroused by the projects.

The research results were presented to the members of the Nemaska Band Council on 1 March 2019. A hard copy of the presentation was also sent to one participant who was unable to attend and had requested a copy.

The Cree community of Nemaska: The heart of Eeyou Istchee

The Cree community of Nemaska is a community of nearly 800 people located on the shores of Champion Lakes. Its name means "where fish abound." The community of Nemaska serves as the "capital of the Cree Nation" and is home to the headquarters of the Grand Council of the Crees.

The Nemaska Cree have extensive experience with hydroelectric projects. It is worth providing a brief history of the community's relationship to hydroelectric development in order to situate the perceptions of the Whabouchi Project, which will be presented later in the chapter, in a broader historical, social, and political context.

The Nottaway–Broadback–Rupert Project and the relocation of Nemaska Eenouch

> *[W]e had to move on the account of what we were told was progress, and we were in the way of progress. So we were more or less forced to move.*
>
> (Thomas Jolly, former Nemaska Chief)

The community of Nemaska did not always occupy its present location. Until the late 1960s, it sat on the shores of Lake Nemaska. In addition to the residences, the community was also home to a Hudson's Bay Company store. Launched in 1971, the James Bay Project involved the construction of eight hydroelectric power plants in two phases, the first of which was the La Grande Complex. Although it was not the community most affected by the La Grande Project, Nemaska nevertheless suffered indirect impacts from the James Bay Project. In the late 1960s, as the James Bay Project was getting underway, Hydro-Québec was considering two major hydroelectric projects: the Nottaway–Broadback–Rupert (NBR) Project and the La Grande Project. If the NBR Project went ahead, it would flood the community of Nemaska, with the exception of the Hudson's Bay building, the church, and the cemetery. In the summer of 1968, a government representative informed the community that they had to leave. The closing of the Hudson's Bay store in 1970 convinced the Nemaska Cree to abandon their community, a territory they had occupied for centuries. They were relocated to the communities of Mistissini and Waskaganish. Many of the children who attended residential school away from the community were not informed of their family's move and were left with an empty community at the end of their school year (Cree Nation of Nemaska 2021: online). In 1970, although some studies concluded that both projects (NBR and La Grande) were feasible, Hydro-Québec decided to opt for the La Grande Project. The NBR Project was definitively discarded the following year, when new studies showed that the clay soil of the region could lead to technical difficulties (Bolduc 2000, p. 115). Hydro-Québec announced its final decision to build a dam on the La Grande River in 1972. The NBR Project, which had forced the relocation of the Nemaska Cree, was never built. Today, the community's former location has been designated as a historic site (GCC 2013),[6] and each summer the Nemaska Cree hold a traditional gathering at the site called "Old Nemaska." The book entitled *Going Home: The Untold Story of Nemaska Eenouch*, launched in 2022, tells the story of this dark page in Nemaska's history.

The James Bay and Northern Québec Agreement

Work on the James Bay Project began without consultations with the Cree and Inuit populations, which led them to file a court injunction to stop it from going ahead (Savard 2009; Saganash 2011; Feit 2011; Lajoie 2011). On 15 November 1973, Judge Albert Malouf ordered an immediate halt to the work. Even though his ruling was overturned in the Supreme Court a week later, it nevertheless allowed the Cree to enter into negotiations with the Québec government and Hydro-Québec and laid the groundwork for the *James Bay and Northern Québec Agreement* (JBNQA) in 1975. This agreement would lead to the establishment of a new territorial regime that divided the land into three categories.[7] The signing of the JBNQA also enabled the Cree to establish public institutions such as the Cree Regional Authority, the Cree School Board, and the Cree Board of Health and Social Services. The agreement included measures to ensure the involvement of the Cree and Inuit in the protection of the environment on their territory, exclusive hunting and fishing rights, and a program ensuring a basic income for hunters, fishers, and trappers.

However, disagreements over the interpretation of the JBNQA, a lack of administrative latitude given to Cree institutions, and the low level of Cree participation in the economic development of the region, both in terms of employment and natural resource royalties, prompted the Cree to negotiate a new agreement with the government of Québec (Papillon and Sénécal 2011; Oblin 2011). The communities of Nemaska and Waskaganish would also be more directly affected by this new agreement.

The Paix des Braves and the Rupert River Diversion

After several court cases and numerous media actions, the Cree began negotiations that would lead to the adoption of the *Paix des Braves* Agreement in 2002. The agreement was signed on 7 February 2002, in Waskaganish by Grand Chief Ted Moses of the Grand Council of the Crees and the Premier of Québec, Bernard Landry. It provides for the transfer of $3.5 billion to the Cree over a 50-year period. It included the creation of the Cree-Québec Forestry Board, the Cree Mineral Exploration Board, the transfer of Québec's economic and community development bonds to the Cree, jobs and contracts in the forest industry, an annual payment of $70 million to fund the Cree Government, over $800 million in contracts to be set aside for Cree businesses, and the cancellation of the NBR Project (Saganash[8] 2011). In exchange for these measures, the agreement provided for the cessation of legal proceedings against the Québec government and stipulated that the Cree accept the Rupert River Diversion Project.

The Nemiscau Camp and its impacts on the Nemaska community

The Eastmain-1-A Powerhouse and Rupert Diversion Project, valued at $4 billion, called for the flooding of an area of 346 km^2 of land, the construction of four dams and 72 dikes, as well as a partial diversion (71%) of the Rupert River's waters northward via La Grande Rivière, in order to increase the energy produced at Robert-Bourassa, La Grande-2-A, and La Grande-1 generating stations (COMEX 2006). The project also required the construction of a 40-km road, a 101-km power transmission line, and three camps to house around 5,500 workers. Despite the introduction of compensation programs, notably the *Nadoshtin* and *Boumhounan* agreements,[9] such infrastructure has had major impacts on the land and wildlife, cultural identity, food security, and social fabric of the affected communities.

In Nemaska, the social impacts of the Eastmain-1-A Project are mainly attributable to the proximity of the Nemiscau Camp,[10] located 15 km east of the community. Construction began in 2003 on the camp, which is home to nearly 200 workers. Some are from the south, while others are Cree, mostly from Mistissini. Since the building of the camp, Nemaska health and social services workers have noticed an increase in social problems in the community (Torrie *et al.* 2005). This has led to greater insecurity among community members and increased fear of violence and crime. This unstable social climate has caused some elders to leave the community on weekends and take refuge at the former village site of Old Nemaska.

The Cree and the energy transition: *La Grande Alliance*

The energy transition relies on the extraction and processing of critical and strategic minerals. As well as being used in everyday items such as mobile phones and laptops, these minerals are also used in the medical, aerospace, telecommunications, renewable energy, and transport electrification sectors. They are the subject of various economic policies in Québec, including the *Québec Strategy for the Development of the Battery Industry*, the *Québec Plan for the Development of Critical and Strategic Minerals*, and the *2030 Plan for a Green Economy*.

The political legitimacy of the energy transition is also supported by the Cree, as evidenced by the signing of *La Grande Alliance* with the government of Québec in 2020. Initiated by the Cree, this $4.7 billion agreement is designed to ensure long-term economic development in the Eeyou Istchee James Bay region. It includes several projects related to transportation (railway extension and electrification), infrastructure, telecommunications, labor training, land protection, and the extraction and processing of strategic minerals. The agreement is intended to position Québec "at the center of the global mining sector, particularly for lithium" (GNC 2020), an objective reflected in the increasing numbers of lithium projects in the territory of Eeyou Istchee James Bay. In 2015, Nemaska Lithium's Whabouchi Project became the first lithium project in Cree territory to receive provincial and federal environmental approvals. In the following sections, we will discuss the various milestones that have marked the relationship between the promoter and the Nemaska Eenouch.

The Whabouchi Project

Located 30 km from the Cree village of Nemaska, the Whabouchi Project was proposed by Nemaska Lithium and aims to produce lithium salts for the manufacture of batteries for electric vehicles. One factor that sets the company apart is the fact that its activities are vertically integrated; that is, its project involves both the extraction of a spodumene concentrate at the Whabouchi Mine and its processing into lithium hydroxide at a plant to be located in Bécancour, in southern Québec.

In the summer of 2009, Nemaska Lithium entered into discussions with Nemaska Development Corporation, which acquired a 3.6% stake in the company. A few months later, the two parties began talks that would lead to the signing of a memorandum of understanding (MOU) in the fall of 2010. In 2014, the Nemaska Band Council, the Grand Council of the Crees, and Nemaska Lithium signed the *Chinuchi Agreement*, an IBA for the Whabouchi Project.

Creation of the Nemaska Working Group

Nemaska Lithium's Whabouchi Project prompted the Cree Nation government to launch an initiative—the Nemaska Working Group—to prepare community members for the public hearings. The group was formed in December 2014, one month after the *Chinuchi Agreement* was signed. It was composed of representatives from the Cree Nation's Environment Department, community members, and a consultant hired by the Cree government to act as facilitator. The working group set up various activities to compile the concerns of community members and to provide them with assistance if they wished to write a brief, for example, by providing a translation service for elders, or to speak publicly at the hearings. The working group organized home visits, as well as visits to various locations, such as the clinic, to reach out to people with limited mobility. Individual interventions were also possible for people who wished to speak confidentially, whether out of shyness or fear of reprisal. In addition, various information and discussion forums were created, including on local radio and through a Facebook group (Cree Nation of Nemaska 2015). The consultation activities also included focus groups with different segments of the community, such as elders, women, land users, and youth. The focus groups dealt with different themes, including land use, Cree culture, jobs, water and air quality, and safety. It should be noted that a fifth theme—consultation and participation—was added at the request of community members. Indeed, the working group's report (2015) includes several comments regarding the lack of confidence of some community members in the consultation process for the Whabouchi Project.

Impacts of the project on social cohesion

The data presented below are drawn from the Whabouchi Project public hearings, which were held in Nemaska on 30 and 31 March 2015, and in which more than 100 people participated, including more than 30 speakers. Several participants made their presentation accompanied by another person, often a spouse or family member. During this event, a number of community members decried the tensions generated by the project. The project has undeniably led to tensions between tallymen, traditional authority figures in Cree culture, and the community's political leadership, which is derived from a governance structure imposed by the *Indian Act*. Given the sensitive nature of the subjects discussed with participants and the small size of the communities, precautions were taken to ensure confidentiality. As a result, the raw data (recordings and transcriptions) have not been made public, and the extracts presented are accompanied by fictitious names.

In his brief, one tallyman felt that his role had been infringed upon and voiced his frustration with the community's leadership, which he felt should have consulted him first when the deposit was discovered:

> Now, the main concern I have here is on the issue of my status as the TALLYMAN. If they found some type of natural resource or mineral on our family trapline, wouldn't it be more appropriate to advise the tallyman first. Instead, it seems our leadership at the time totally disregarded the importance of the role played by the tallyman. Where do I stand? Nobody consulted with me and even to this date, it's as if the tallyman does not exist. This is where my concern is!
>
> (Translated from Cree, Luke J. 2015)

Another major point of tension concerned the signing of the IBA, for which some community members feel they were not consulted. Almost all the members of the council were, in fact, replaced in the March 2015 elections, which followed the ratification of the agreement. The lack of transparency regarding the content of the agreement was repeatedly highlighted in the submissions to the hearings:

> The process undertaken by the Chief and Council of Nemaska First Nation relating to the Whabouchi Mine is that they did not seek or receive a "mandate from the members of Nemaska First Nation" prior to entering discussions with the Nemaska Lithium which led to the signing of the Agreement known here as Chinuchi Agreement.
>
> (Brief submitted by Clinton F. 2015, p. 1)

> More importantly, we feel that there were no consultations prior to the signing of the Agreement specifically to address concerns regarding the agreement and to obtain the consent of the members. We request that more information be accessible to the members, information that is comprehensive and clear.
>
> (Brief submitted by the women's group 2015, p. 1)

For some individuals, mining activities are incompatible with Indigenous identity. In an interview, one participant made a rather harsh observation about the signing of IBAs, and their impact on the social cohesion of the community and on Indigenous identity: "It's like we're an apple. We're an apple when we signed the agreement: you cut in half, it's split, it's white inside" (Freddy Jolly 2017).

From this perspective, the leaders embody the phenomenon of mimicry, defined by Bhabha (1984), which consists of trying to imitate the characteristics of the members of the dominant culture, to the detriment of one's own culture. Freddy further added: "We're not real Indians. Real Indians fight, protect. Now, they negotiate."

This statement reinforces the idea that individuals in favor of a project may be accused, directly or indirectly, of a form of betrayal of their own culture and thus suffer a certain social ostracism. In this context, the conflicts between those who support a project and those who oppose it take on a much broader dimension, extending in some ways to the conflicts between "traditional" authorities (tallymen and elders) and newer authorities imposed by the *Indian Act* (Morantz 2002). In so doing, the individuals who adhere to the project ensure the (partial) presence of the colonizer and perpetuate the authority of the colonizer over their fellow human beings, thus threatening the traditional way of life (Hammer 2005).

Socio-cumulative impacts of the project

For many community members, the Whabouchi Project acts as a trigger and revives memories associated with the impacts of previous projects and colonial policies, such as residential schools:

> The project that came from Hydro-Québec, that was a big impact, now we have this mining, and going back, you know, with residential schools and all these things. It's like, I wonder sometimes, people are triggered, I think.
>
> (Harry 2017)

Several excerpts from the report of the working group tasked with assessing the impacts of the Whabouchi Project on the community, including the words of a group of young people, are in keeping with this view: "People here have suffered enough through residential schools and hydro development, and the youth feel this will be the same thing" (Cree Nation of Nemaska 2015, p. 36).

In its brief, the women's group agreed:

> The health and safety of our Community as well as maintaining our traditional way of life is very important to us and our people. We have suffered through the negative impacts created by the hydropower development projects that surround our Community and we want this suffering to stop. The suffering will not stop if this project, which will be so close to our homes, is allowed to proceed.
>
> (Women's Group 2015, p. 10)

One participant recalled the mourning process she went through 35 years prior because of the many impacts of the La Grande-1 generating station project on the territory. Although she was in favor of the Whabouchi Project, she asked Nemaska Lithium to provide funds for a workshop to help community members grieve if the mining project goes ahead:

> I remember when Hydro development happened, I was impacted. I was impacted when I seen the La Grande River change. I was impacted when I couldn't cross this river anymore. I was impacted when we were told we cannot walk on the ice anymore. I was impacted when we were told that we

> couldn't drink the water anymore or eat the fish anymore. I was impacted and I was only 5. So I think Nemaska Lithium should also give a little more amount of money for the community, to run a grieving workshop for the community . . . because we are all gonna be grieving. We all are gonna go through a grieving process when this happens and if it does happen. The children are gonna go through it, the youth are gonna go through it.
>
> (Adele 2015)

Beyond the parallels with previous projects, the Whabouchi Project raises concerns about its impacts on water quality:

> Our Group is also worried about the health of the animals, especially the animals that we hunt to feed our families. The water quality will be affected, and the contaminated water will then flow to the Nemaska River. The water will no longer be safe to drink and will affect the fish that we eat.
>
> (Women's Group 2015, p. 6)

The data collected in Nemaska illustrate how, within the ideal type of forced union, the power relationships among industry, the state, and the communities can be interpreted as a form of "environmental racism" or "environmental injustice" (Munshi and Kurian 2005; Powell 2006; Keeling and Sandlos 2009; McGregor *et al.* 2020; Nachet *et al.* 2022). Despite the strong political legitimacy of the energy transition, both among non-Indigenous and Indigenous people, communities affected by the project are the only ones who have to deal with its health risks, risks that, unlike the economic benefits, cannot be shared. The relational ideal type of forced union shows, on the one hand, the need to take into account the colonial history in order to understand the social and emotional impacts of development projects, and, on the other hand, the importance of putting in place resources and accountability mechanisms to mitigate the impacts of mining projects on the physical and mental health of affected communities.

Mining project as a form of resilience

During the public hearings, a totally different perspective was also expressed by some community members. In a brief, the former chief of Nemaska (he was still chief 5 days before the hearings began) emphasized how the Whabouchi Project was part of the community's vision and represented a unique opportunity to become a prosperous, proud, and healthy community without sacrificing its traditional values:

> The Chinuchi Agreement is an opportunity for the Cree Nation of Nemaska to take a major step of progress in the direction of a prosperous and healthy community. This Agreement offers an opportunity for us to realize our common vision, which is to be a proud community, respectful of our individual and collective obligations, demonstrating strong ethics in order to achieve growth to sustainable human and economic development [. . .]. We as leaders

> must recognize these opportunities as they arise and seize them in order to advance our common vision. These opportunities cannot be denied (they may never come around again) [in bold in text]. It is through such partnerships and cooperation that our community shall prosper without sacrificing our traditional values.
>
> (Matthew Wapachee, former chief of Nemaska 2015)

Another participant mentioned that the project will allow the creation of businesses in the community. She pointed out that many Cree are educated and can take on more skilled jobs:

> I do see a lot of opportunities that's gonna come up, a lot of economic opportunities, like I said, Nemaska has an opportunity to create businesses. We're not just laborers. There's a lot of us that are educated.
>
> (Adele 2015)

In the control-shift relational ideal type, the mining project represents an important economic lever while at the same time being compatible with Indigenous culture. In this context, the mining project is envisaged as a tool for strengthening social cohesion and cultural vitality in the community, in particular by preventing the exodus of young people and by generating income to purchase hunting and fishing equipment. Individuals can then seize the opportunities offered by the project and benefit from it.

While the forced union model emphasizes the socio-cumulative impacts of development, the control-shift model draws on the lessons learned from previous projects by making recommendations and submitting requests to the promoter:

> I've experienced working in construction, I've experienced sexual harassment. I understand firsthand what it's like to work in that field and I want the female to be protected. That's why I'm suggesting that there be a union for the workers and at least a support working group for all workers, male and female.
>
> (Adele 2015)

The control-shift model also implies a reliance on the remediation and restoration processes. There is also an emphasis on the notions of cooperation and conciliation, especially with regard to different land uses. In this perspective, the mining project is associated with a form of resilience:

> I have been involved in the discussions of the CQNRA Paix des Braves Agreement. I have been an impacted Tallyman and land-user of the EM1 Projet [. . .]. I survived, I thrived, I lived through these projects; I know I will survive; I will thrive and live through the Whabouchi Project because we are a people strong and vibrant who have been able to adapt to many challenges and obstacles over the years.
>
> (Matthew Wapachee, former chief 2015)

Another participant mentioned that "adapting" has been part of the Cree way of life for the past 30 years: "Cree society can adapt to many changes. That's been our way of life for the last 20–30 years. Whenever happens, we'll adapt and move on" (Bill O. 2015).

Being involved in mining development allows the community to reclaim the economic development tools of the dominant culture, to "negotiate between two worlds," and to ensure cultural continuity (Kirmayer *et al.* 2011; Kurtness 2014; Vanthuyne 2016). In this context, industry is seen as synonymous with empowerment, autonomy, and prosperity, allowing communities to reduce their dependence on government transfers (Fairbrass and Zueva-Owens 2012). However, as we have seen in previous paragraphs, the control-shift model can lead to conflicts between the leadership and the rest of the community. Although this type of relationship is linked to capacity-building and increased negotiating capacity, when community consultations are deemed insufficient, it is still likely to increase inequalities within a community, hence the importance of encouraging the establishment of internal deliberative forums.

Lastly, the control-shift model assumes that a community has significant human and financial resources, a condition that hinges on the strength of local political organizations. In this sense, because of the existence of the *James Bay and Northern Québec Agreement*, the reality of the Cree may differ from other Indigenous communities, particularly in terms of political organization. However, we believe that the dynamics observed in Nemaska are likely to be found in other Indigenous communities, with which they share a common colonial history.

Discussion/conclusion

The data presented in this chapter show two visions of mining development within the same community, a situation that is not unusual. However, the aim of the chapter is to demystify the roots of the conflicts. In Nemaska, several community groups, particularly women, felt excluded from the negotiations surrounding the signing of an IBA, a situation well documented in the literature (Gibson and Kemp 2008; Hall 2019). The presence of the Nemaska Working Group was undoubtedly an interesting initiative to mobilize different groups in the community, but it came too late in the process. Moreover, as shown in the ideal type of forced union, mining activities rekindle old emotions and relational dynamics linked to the impact of colonial policies. Indeed, the testimonies and briefs submitted at the public hearings for the Whabouchi Project, as well as interviews with members of the community, show how, for some individuals, the project acts as a trigger and revives memories of the impacts of other development projects, such as hydroelectric projects and, for some individuals, even residential schools. In addition, the impact of a project on social cohesion is likely to reactivate certain social dynamics reminiscent of those generated by colonial policies. In Nemaska, the project has led to tensions between the tallymen, traditional authority figures in Cree culture, and the community's political leadership, which stems from a governance structure imposed by the *Indian Act*. Well known in the Indigenous world, the analogy of the apple—red on the outside and white on the inside—is used by one individual to describe the impact of signing IBAs and the identity conflict it may represent for Indigenous communities.

Furthermore, the data collected in Nemaska illustrate how, within the ideal type of forced union, the power relationships among industry, the state, and communities can be interpreted as a form of "environmental racism" or "environmental injustice" (Munshi and Kurian 2007; Willow 2016; Bernauer 2019; Long 2019; Nachet *et al.* 2022). Despite the strong political legitimacy of the energy transition, among non-Indigenous and Indigenous people alike, the latter are the only ones who have to deal with the health risks associated with the project, risks that, unlike the economic benefits, cannot be shared. The ideal type of forced union shows, on the one hand, the need to take colonial history into account in order to understand the social and emotional impacts of development projects, and, on the other, the importance of putting in place adequate resources and accountability mechanisms to mitigate the impacts of mining projects on the physical and mental health of the communities affected.

In the control-shift ideal type, the mining project represents an important economic lever that can be reconciled with Indigenous culture. The mining project is seen as a tool for strengthening social cohesion and cultural vitality, notably by preventing the exodus of young people down south and by generating income that can be used to purchase hunting and fishing equipment. This ideal type can take the form of a strategic alliance and result in the signing of an IBA, a petition, or a testimonial in favor of the project. Individuals can then seize the opportunities offered by the project and benefit from it. In this type of relationship, community members draw on the lessons learned and experience gained from previous projects to make recommendations to the proponent. The control-shift ideal type also implies confidence in the remediation and restoration processes. Emphasis is placed on the concepts of cooperation and conciliation, particularly with regard to different land uses. From this perspective, the mining project is associated with a form of resilience. The fact of being involved in mining development thereby enables the community to reappropriate the economic development tools of the dominant culture, to "negotiate between two worlds," and to ensure cultural continuity (Kirmayer *et al.* 2011; Kurtness 2014; Hatala *et al.* 2016; Vanthuyne 2016). In this context, industry is seen as a means to empowerment, autonomy, and prosperity, allowing communities to reduce their dependence on government transfers (Fairbrass and Zueva-Owens 2012).

While it is not uncommon for two visions of mining development to coexist within a community, the Whabouchi Mine is nevertheless a first milestone in the Cree's overall strategy for lithium mining. In fact, since the Whabouchi Project was introduced, two other lithium projects have received federal authorizations in Eeyou Istchee: the Rose Lithium-Tantalum Mining Project and the James Bay Lithium Mine Project. For these two projects, the impact assessment process was carried out jointly by the Cree Nation government and the Impact Assessment Agency of Canada, under the terms of a collaboration agreement signed between the two parties.

Although we are only at the beginning of *La Grande Alliance* and its infrastructure projects that will take place over the next 30 years, the Cree have demonstrated their capacity to manage resource development on their territory. Whether through hydroelectric projects or uranium, lithium, and various other types of mines, they have shown their strong mastery of the three strategies for action: Collaboration, contestation, and reappropriation (Papillon and Rodon 2019). And while there are still challenges ahead, their journey paves the way for other Indigenous nations in this era of energy transition.

Acknowledgments

This research was supported by the Knowledge Network on Mining Encounters and Indigenous Sustainable Livelihoods (MinErAL) and by the Social Sciences and Humanities Research Council of Canada. The author would like to thank the Cree Nation government, the Nemaska Wellness Center, and Nemaska community members for their participation in the study. Any errors are solely the responsibility of the author.

Notes

This chapter presents the results of a research carried out as part of a doctoral thesis in public communication published in 2022. Since then, this author has been employed by the government of Canada. To avoid any misrepresentation, we would like to emphasize that the views and opinions expressed in this article are those of the author and do not reflect the official policy or position of any Canadian federal departments.

1 In its definition of *environmental racism*, the Canadian Race Relations Foundation (CRRF) notes that: "[p]eople of color, Aboriginal groups, working class people, and low-income people are disproportionately exposed to environmental hazards and risks of toxic industrial substances, polluted air, unsafe water, unhealthy working conditions, poor sanitation, and the location of hazardous and toxic facilities, such as incinerators and toxic waste dumps" (2015, para. 1).

2 This research group gave rise to the current MinErAL network, a knowledge network on mining encounters and Indigenous sustainable livelihoods, which brings together partners and researchers from the Canadian North, Fennoscandia, Australia, and New Caledonia.

3 The Environmental and Social Impact Review Committee (Review Committee, or COMEX) is an independent body composed of members appointed by the governments of Quebec and the Cree Nation. It is responsible for the assessment and review of the social and environmental impacts of projects located south of the 55th parallel in the territory governed by the James Bay and Northern Quebec Agreement.

4 The research project is entitled "Mines and communities: Impacts of mining development on social cohesion in Northern Quebec."

5 The Old Nemaska historic site is located within the boundaries of the *Chisesaakahiika* Protected Area Project, which aims to protect it from industrial development, along with Lake Evans and Lake Nemiscau. This proposed area project is part of the Broadback River Watershed Conservation Plan initiated by the Cree (CNG 2013).

6 Category I lands border the villages and are reserved for the exclusive use of the Cree. Category II lands are public lands over which the Cree have exclusive fishing, hunting, and trapping rights, while Category III lands are also public lands over which the Cree have exclusive rights to harvest certain aquatic species and fur-bearing animals.

7 Roméo Saganash, a member of the Cree Nation, was the New Democratic Party (NDP) member of parliament for Abitibi-Baie-James-Nunavik-Eeyou from 2011 to 2019. For over 23 years, he participated in the negotiations of the UNDRIP. In 2020, Roméo Saganash was awarded an honorary doctorate in law by Laval University.

8 These two treaties were signed in 2002 between Hydro-Québec, the James Bay Energy Corporation, the Cree Regional Authority, the Grand Council of the Crees, and four Indigenous bands as part of the *Paix des Braves*. The *Nadoshtin Agreement* sets out measures that integrate and promote Cree labor and businesses and provides for funds for remedial work and compensation for the environmental impacts of development. The *Boumhounan Agreement* involves the creation of a process for the implementation and participation of the Cree in the environmental study leading up to the Eastmain-1-A/Sarcelle/Rupert Project.

9 The Nemiscau workers' camp would be considered by Nemaska Lithium for the construction of the Whabouchi Project, nearly 15 years later.

10 Critical and strategic minerals are in increasing demand, are subject to high supply risk, and have no commercially available substitutes. They include lithium, graphite, cobalt, nickel, and rare earth elements.

References

Albrecht, G., 2005. 'Solastalgia:' A new concept in health and identity. *PAN (Philosophy, Activism, Nature)*, 1 (3), 41–55.

Albrecht, G., Sartore, G.-M., Connor, L., Higginbotham, N., Freeman, S., Kelly, B., Stain, H., Tonna, A., and Pollard, G., 2007. Solastalgia: The distress caused by environmental change. *Australasian Psychiatry*, 15 (1_suppl), S95–S98. https://doi.org/10.1080/10398560701701288

APNQL—Assemblée des Premières Nations Québec-Labrador, 2014. *Protocole de recherche des Premières Nations au Québec et au Labrador. www.bibl.ulaval.ca/doelec/LibreAcces/1262018179.pdf*

Basile, S., 2017. *Le rôle et la place des femmes Atikamekw dans la gouvernance du territoire et des ressources naturelles*. PhD Thesis. Université du Québec en Abitibi-Témiscamingue. Depositum. https://depositum.uqat.ca/id/eprint/703/1/Basile,%20Suzy.pdf

Bernauer, W., 2019. The limits to extraction: mining and colonialism in Nunavut. *Canadian Journal of Development Studies*, 1–19. https://doi.org/10.1080/02255189.2019.1629883

Bernauer, W., and Peyton, J., 2021. Energy extraction, resistance, and political change in Inuit Nunangat. *In*: P. Stern, ed., *The Inuit world*, 1st ed. London: Routledge, 340–358. https://doi.org/10.4324/9780429275470-25

Bhabha, H. K., 1984. Representation and the colonial text: A critical exploration of some forms of mimeticism. *In*: F. Gloversmith, ed., *The theory of reading*, 1st ed. Brighton: Harvester Press.

Bolduc, A., 2000. *Du génie au pouvoir: Robert A. Boyd, à la gouverne d'Hydro-Québec aux années glorieuses*. Montréal: Libre Expression, 259 p.

Bombay, A., 2014. *Origins of lateral violence in Aboriginal communities: A preliminary study of student-to-student abuse in residential schools*. Ottawa: The Aboriginal Healing Foundation.

Bourgeois, S., and Rodon, T., 2019. Mettre en œuvre le consentement préalable, libre et éclairé (CPLE) en Eeyou Istchee: L'exemple du projet Matoush. *Recherches amérindiennes au Québec*, 49 (2), 63–71. https://doi-org.acces.bibl.ulaval.ca/10.7202/107075

Boutet, J.-S., 2014. Opening Ungava to industry: A decentering approach to Indigenous history in Subarctic Québec, 1937–1954. *Cultural Geographies*, 21, 79–97. https://doi.org/10.1177/1474474012469761

Bjørst, L. R., 2020. Stories, emotions, partnerships and the quest for stable relationships in the Greenlandic mining sector. *Polar Record*. https://doi.org/10.1017/S0032247420000261

Canadian Race Relations Foundation (CRRF), 2024. Glossary of terms. *Environmental Racism*, online: Glossary of Terms – Canadian Race Relations Foundation (crrf-fcrr.ca).

COMEX, 2006. *Report Eastmain-1-A and Rupert Diversion hydropower project. Report by the Provincial Review Committee to the Administrator of Chapter 22 of the James Bay and Northern Québec Agreement*. Québec: Gouvernement du Québec.

Commission de vérité et réconciliation du Canada, 2012. *Commission de vérité et réconciliation du Canada: Appels à l'action*. Winnipeg, online: 4-Appels_a_l-Action_French.pdf (nctr.ca).

Coronado, G., and Fallon. W., 2010. Giving with one hand: On the mining sector's treatment of indigenous stakeholders in the name of CSR. *International Journal of Sociology and Social Policy*, 30 (11/12), 666–682. https://doi.org/10.1108/01443331011085259

Cree Nation of Nemaska, 2015. *Report—Impacts of the Whabouchi project*, 30 April. Montréal: Nemaska Work Group.

Cree Nation of Nemaska, 2021. History. Available from: https://nemaska.com/history/

CSSSPNQL Commission de la santé et des services sociaux des Premières Nations du Québec et du Labrador, 2021. *Boîte à outils des principes de la recherche en contexte autochtone*, Wendake, Québec: CSSSPNQL, UQAT, UQO and Réseau DIALOG.

Dahlin, J., and Fredriksson, M., 2017. Extracting the commons. *Cultural Studies*, 31 (2–3), 253–276. https://doi.org/10.1080/09502386.2017.1303428

Desbiens, C., and Sepúlveda, B., 2019. Pekedamkam: Frontierism and the unearthing of Indigenous landscapes in Val-d'Or, Canada. *Journal of Historical Geography*. https://doi.org/10.1016/j.jhg.2019.06.006

ECPT2 Instituts de recherche en santé du Canada, 2018. Énoncé de politique des trois conseils, éthique de la recherche avec des êtres humains. Available from: http://publications.gc.ca/collections/collection_2019/irsc-cihr/RR4-2-2019-fra.pdf

Evans-Campbell, T., and Walters, K. L., 2006. Indigenist practice competencies in child welfare practice: A decolonization framework to address family violence and substance abuse among First Nations peoples. *In:* R. Fong, R. McRoy, and C. Ortiz Hendricks, eds., *Intersecting child welfare, substance abuse, and family violence: Culturally competent approaches*. Washington, DC: CSWE Press, 266–290.

Evans-Campbell, T., 2008. Historical trauma in American Indian/Native Alaska communities: A multilevel framework for exploring impacts on individuals, families, and communities. *Journal of Interpersonal Violence*, 23 (3), 316–338. https://doi.org/10.1177/0886260507312290

Fairbrass, J., and Zueva-Owens, A., 2012. Conceptualising corporate social responsibility: 'Relational Governance' assessed, augmented, and adapted. *Journal of Business Ethics*, 105 (3), 321–335. https://doi.org/10.1007/s10551-011-0968-9

Feit, H., 2011. Le people cri de la Baie-James parle aux gouvernements: développement, gouvernance et co-gouvernance. *In:* J.-G. Petit, Y.-V. Bonnier Viger, P. Aatami, and A. Iserhoff, eds., *Les Inuit et les Cris du nord du Québec: Territoire, gouvernance, société et culture*. Québec: Presses de l'Université du Québec, 119–132.

Fortin, J., 2019. Les dimensions silencieuses de l'acceptabilité sociale au Nunavik: Le cas d'Aupaluk. *Recherches amérindiennes au Québec*, 49 (2), 73–84. https://doi.org/10.7202/1070760

Frijda, N. H., 2001. The laws of emotion. *In:* W. G. Parrot, ed., *Emotions in social psychology*. New York: Psychology Press, 57–69.

GCC – Grand Conseil des Cris, 2013. Plan de conservation du bassin versant de la Broadback. Available from: www.eeyouconservation.com/broadback-conservation-plan/

Gibson, G., and Kemp, D., 2008. Corporate engagement with Indigenous women in the minerals industry: Making space for theory. *In:* C. O'Faircheallaigh and S. Ali, eds., *Earth matters: Indigenous peoples, the extractive industries and corporate social responsibility*. London: Routledge, 104122.

Gislason, M., and Andersen, H., 2016. The interacting axes of environmental, health, and social justice cumulative impacts: A case study of the Blueberry River First Nations. *Healthcare*, 4 (4), 78. https://doi.org/10.3390/healthcare4040078

Gouvernement de la nation crie (GNC), 2020. Signature d'une grande alliance entre le gouvernement du Québec et la nation crie. Available from: www.cngov.ca/fr/la-grand-alliance/

Gordon, S. M., 2015. Narratives unearthed, or, how an abandoned mine doesn't really abandon you. *In:* A. Keeling and J. Sandlos, eds., *Mining and communities in Northern Canada: History, politics, and memory*. Calgary: University of Calgary Press, 59–85.

Hall, R. J., 2019. A feminist political economy of indigenous–state relations in Northern Canada. *In*: M. P. Thomas, L. F. Vosko, C. Fanelli, and O. Lyubchenko, eds., *Change and continuity: Rethinking the new Canadian political economy*. Montreal: McGill-Queen's University Press, 185–202.

Hammer, R., 2005. Postcolonialism. *In*: G. Ritzer, ed., *Encyclopedia of social theory*. https://sk.sagepub.com/reference/socialtheory/n219.xml

Harnetty, B., 2017. Earthquakes and frack-waste: Sounds of extraction-related disaster in Appalachian Ohio. *Cultural Studies*, 31 (2–3), 400–416. https://doi.org/10.1080/09502386.2017.1303434

Hatala, A. R., Desjardins, M., and Bombay, A., 2016. Reframing narratives of Aboriginal health inequity: Exploring Cree elder resilience and well-being in contexts of historical trauma. *Qualitative Health Research*, 26 (14), 1911–1927. https://doi.org/10.1177/1049732315609569

Horowitz, L. S., 2017. 'It shocks me, the place of women': Intersectionality and mining companies' retrogradation of Indigenous women in New Caledonia. *Gender, Place & Culture*, 24 (10), 1419–1440. https://doi.org/10.1080/0966369X.2017.1387103

INQ Institut nordique du Québec, 2017. *Lignes directrices pour la recherche*. Québec: Groupe de travail des Premiers Peuples de l'Institut nordique du Québec.

Joly, T. L., Longley, H., Wells, C., and Gerbrandt, J., 2018. Ethnographic refusal in traditional land use mapping: Consultation, impact assessment, and sovereignty in the Athabasca oil sands region. *The Extractive Industries and Society*, 5 (2), 335–343. https://doi.org/10.1016/j.exis.2018.03.002

Jones, J., 2020. *Confronting settler colonialism when assessing the impact of mining on Indigenous peoples' health and well-being*. PhD Thesis. University of Guelph.

Kaushik, V., and Walsh, C. A., 2019. Pragmatism as a research paradigm and its implications for social work research. *Social Sciences*, 8, 255.

Keeling, A., and Sandlos, J., 2009. Environmental justice goes underground? Historical notes from Canada's northern mining frontier. *Environmental Justice*, 2, 117–125. https://doi.org/10.1089/env.2009.0009

Keeling, A., and Sandlos, J., 2015. *Mining and communities in Northern Canada: History, politics, and memory*. Calgary: University of Calgary Press.

Kirmayer, L. J., Dandeneau, S., Marshall, E., Phillips, M. K., and Williamson, K. J., 2011. Rethinking resilience from Indigenous perspectives. *The Canadian Journal of Psychiatry*, 56 (2), 84–91. https://doi.org/10.1177/070674371105600203

Kitayama, S., and Markus, H. R., 1999. *Emotion and culture: Empirical studies of mutual influence*. Washington, DC: American Psychological Association.

Koperqualuk, L. Q., 2015. *Les traditions liées au droit coutumier au Nunavik*. Montréal: Institut culturel Avataq.

Kurtness, J., 2014. *Tshinanu, nous autres, et moi qui appartiens aux trois Amériques. Entretiens de l'anthropologue Caroline Hervé avec Jacques Kurtness, négociateur innu de Mashteuiatsh*. Québec: Presses de l'Université Laval.

Lajoie, G., 2011. Les défis environnementaux: la contribution de l'évaluation environnementale. *In:* J.-G. Petit, Y.-V. Bonnier Viger, P. Aatami, and A. Iserhoff, eds., *Les Inuit et les Cris du nord du Québec: Territoire, gouvernance, société et culture*. Québec: Presses de l'Université du Québec, 187–200.

Leclair, J., Papillon, M., and Forget, H., 2019. Les protocoles de consultation autochtones au Canada : Un modèle de convergence des systèmes juridiques autochtones et étatique ? *Recherches amérindiennes au Québec*, 49 (2), 25–36. https://doi-org.acces.bibl.ulaval.ca/10.7202/1070756ar

Long, B. S., 2019. CSR and reconciliation with Indigenous peoples in Canada. *Critical Perspectives on International Business*. https://doi.org/10.1108/cpoib-12-2017-0096

MacNeil, K. S., 2018. Let's name it: Identifying cultural, structural and extractive violence in indigenous and extractive industry relations. *Journal of Northern Studies*, 12 (2), 81–103.

Marsh, J. K., 2010. *A critical analysis of decision-making protocols used in approving a commercial mining license for the Beverley Uranium Mine in Adnyamathanha Country: Toward effective Indigenous participation in caring for cultural resources*. PhD Thesis. The University of Adelaide.

McGregor, D., Whitaker, S., and Sritharan, M., 2020. Indigenous environmental justice and sustainability. *Current Opinion in Environmental Sustainability*, 43, 35–40. https://doi.org/10.1016/J.COSUST.2020.01.007

Meade, M. R., 2017. In the shadow of the coal breaker: Cultural extraction and participatory communication in the Anthracite Mining Region. *Cultural Studies*, 31 (2–3), 376–399. https://doi.org/10.1080/09502386.2017.1303433

Morantz, T., 2002. *The white man's gonna getcha: The colonial challenge to the Crees in Quebec*. Montreal: McGill-Queen's University Press.

Morgan, D. L., 2014. Pragmatism as a paradigm for social research. *Qualitative Inquiry*, 20 (8), 1045–1053.

Munshi, D., and Kurian, P., 2005. Imperializing spin cycles: A postcolonial look at public relations, greenwashing, and the separation of publics, *Public Relations Review*, 31, 513–520.

Munshi, D., and Kurian, P., 2007. The case of the subaltern public: A postcolonial investigation of corporate social responsibility's (o)missions. *In*: M. S. Cheney and J. Roper, eds., *The debate over corporate social responsibility*. Oxford: Oxford University Press, 438–448.

Nachet, L., Beckett C., and MacNeil, K. S., 2022. Framing extractive violence as environmental (in)justice: A cross-perspective from Indigenous lands in Canada and Sweden. *The Extractive Industries and Society*, 12, 100949. https://doi.org/10.1016/j.exis.2021.100949

Norgaard, K. M., and Reed, R., 2017. Emotional impacts of environmental decline: What can Native cosmologies teach sociology about emotions and environmental justice? *Theory and Society*, 46 (6), 463–495. https://doi.org/10.1007/s11186-017-9302-6

Oblin, G., 2011. Une analyse des réactions des Cris à l'entente de la Paix des Braves (2002). In J.-G. Petit, Y.-V. Bonnier Viger, P. Aatami, and A. Iserhoff, eds., *Les Inuit et les Cris du nord du Québec: Territoire, gouvernance, société et culture*. Québec: Presses de l'Université du Québec, 87–103.

O'Faircheallaigh, C., 2013. Extractive industries and Indigenous peoples: A changing dynamic? *Journal of Rural Studies*, 30, 20–30.

Owen, J. R., and Kemp, D., 2014. 'Free prior and informed consent', social complexity and the mining industry: Establishing a knowledge base. *Resources Policy*, 41, 91–100. https://doi.org/10.1016/j.resourpol.2014.03.006

Papillon, M., and Sénécal, S., 2011. Traités modernes, qualité de vie et gouvernance des peuples autochtones au Canada: L'expérience des Cris et des Inuit sous la Convention de la Baie-James et du nord québécois. *In*: J.-G. Petit, Y.-V. Bonnier Viger, P. Aatami, and A. Iserhoff, eds., *Les Inuit et les Cris du nord du Québec: Territoire, gouvernance, société et culture*. Québec: Presses de l'Université du Québec, 255–270.

Papillon, M., and Rodon, T., 2017. Proponent–Indigenous agreements and the implementation of the right to free, prior, and informed consent in Canada. *Environmental Impact Assessment Review*, 62, 216–224. https://doi.org/10.1016/j.eiar.2016.06.009

Papillon, M., and Rodon, T., 2019. Le consentement préalable, libre et éclairé : Les défis de la mise en œuvre en contexte canadien. *Recherches amérindiennes au Québec*, 49 (2), 3. https://doi.org/10.7202/1070754ar

Powell, D. E., 2006. Technologies of existence: The Indigenous environmental justice movement. *Development*, 49, 125–132. https://doi.org/10.1057/palgrave.development.1100287

Rodon, T., Lévesque, F., and Blais, J., 2013. De Rankin Inlet à Raglan, le développement minier et les communautés inuit. *Études/Inuit/Studies*, 37, 103. https://doi.org/10.7202/1025712ar

Rodon, T., and Lévesque, F., 2015. Understanding the social and economic impacts of mining development in Inuit communities: Experiences with past and present mines in Inuit Nunangat. *The Northern Review*, 41, 1–27. https://doi.org/10.22584/nr41.2015.002

Roy, S. N., 2010. L'étude de cas. *In*: B. Gauthier, ed., *Recherche sociale: De la problématique à la collecte des données*. Québec: Presses de l'Université du Québec, 199–225.

Saganash, R., 2011. La reconnaissance des droits des Cris de la Baie-James: De la CBJNQ à la Déclaration de l'ONU. *In:* J.-G. Petit, Y.-V. Bonnier Viger, P. Aatami, and A. Iserhoff, eds., *Les Inuit et les Cris du nord du Québec: Territoire, gouvernance, société et culture*. Québec: Presses de l'Université du Québec, 67–85.

Savard, S., 2009. Les communautés autochtones du Québec et le développement hydroélectrique : Un rapport de force avec l'État, de 1944 à aujourd'hui. *Recherches amérindiennes au Québec*, 39 (1–2), 47–60. https://doi.org/10.7202/044996ar

Sandlos, J., and Keeling, A., 2016. Toxic legacies, slow violence, and environmental injustice at Giant Mine, Northwest Territories. *Northern Review*, 42, 7–21.

Schnapper, D., 2012. *La compréhension sociologique: Démarche de l'analyse typologique*. Paris: Presses Universitaires de France.

Simpson, L., 2008. *Lighting the eighth fire: The liberation, resurgence, and protection of Indigenous nations*. Winnipeg: Arbeiter Ring Publishing.

Torrie, J., Bobet, E., Kishchuk, N., and Webster, A., 2005. *The evolution of health status and health determinants in the Cree region (Eeyou Istchee): Eastmain-1-A Powerhouse and Rupert Diversion—sectoral report, volume 2 detailed analysis*. Chisasibi, Québec: Cree Board of Health and Social Services of James Bay.

Vanthuyne, K., 2016. Remembering residential schools, accounting for decolonization through development: Conflicting viewpoints. *In*: B. Capitaine and K. Vanthuyne, eds., *Power through testimony: Reframing residential schools in the age of reconciliation*. Vancouver: UBC Press, 155–176.

Viveros, H., 2016. Examining stakeholders' perceptions of mining impacts and corporate social responsibility. *Corporate Social Responsibility and Environmental Management*, 23 (1), 50–64. https://doi.org/10.1002/csr.1363

Wanvik, T. I., and Caine, K., 2017. Understanding Indigenous strategic pragmatism: Métis engagement with extractive industry developments in the Canadian North. *The Extractive Industries and Society*, 4 (3), 595–605. https://doi.org/10.1016/j.exis.2017.04.002

Westman, C., Joly, T. L., and Gross, L., eds., 2020. *Extracting home in the oil sands: Settler colonialism and environmental change in subarctic Canada*. Oxon and New York: Routledge.

Willow, A., 2016. Indigenous extrACTIVISM in Boreal Canada: Colonial legacies, contemporary struggles and sovereign futures. *Humanities*, 5, 55. https://doi.org/10.3390/h5030055

Yin, R. K., 2014. *Case study research: Design and methods*, 5th ed. Thousand Oaks, CA: Sage.

10 Lateral violence

Effects of external pressures on Indigenous communities

Kristina Sehlin MacNeil

Introduction

This chapter discusses theoretical causes of lateral violence (intragroup conflict) in Indigenous and local communities and draws on previous research on asymmetric conflict and power relations experienced by Indigenous communities in their interactions with extractive industries. The primary focus is on Sámi reindeer-herding communities, Sámi associations, and local communities on the Swedish side of Sápmi, and comparisons with Indigenous communities in Australia.

Sápmi, the homeland of the Indigenous Sámi People, stretches across the northern parts of four nation states: Norway, Sweden, Finland, and the Kola Peninsula in Russia. The Sámi People have lived in this area since long before the creation of the nation states, and their traditional livelihoods include reindeer husbandry, hunting, fishing, and gathering, as well as *duodji* (Sámi craft) and art. It is estimated there are between 80,000 and 100,000 Sámi in the world. Today, some 20,000–40,000 Sámi are estimated to live in the part of Sápmi located in Sweden, most of them in the southern parts of the country; however, these figures are hypothetical (Stoor and San Sebastián 2022). The rather vague figures for Sweden are a result of the *Swedish Data Act*, which prohibits any registration of data that reveal ethnicity or race (Axelsson and Mienna 2021).

The Sámi have been recognized as an Indigenous People in Sweden, as well as a People under the Swedish constitution, since 2011 (Hansen and Olsen 2006; Reimerson 2015, p. 21). In addition, there is a Sámi Parliament, established in 1993, and, since 2000, the Swedish ratification of the Council of Europe Framework Convention for the Protection of National Minorities (in which the Sámi are included) and the European Charter for Regional or Minority Languages (Lantto and Mörkenstam 2015, p. 148; Pietikäinen *et al.* 2010. pp. 16–17). Sweden has ratified the United Nations Declaration on the Rights of Indigenous Peoples but not the Indigenous and Tribal Peoples Convention (ILO 169). Given this, while the situation for Sámi in Sweden might give the semblance of a degree of self-determination, in matters that truly concern the Sámi, they have little say. Evidenced in part by how Sweden has repeatedly been criticized by the UN for neglecting to properly consult Sámi on matters pertaining to land-use and extractive activities, many Sámi communities are finding it increasingly difficult to maintain traditional livelihoods

DOI: 10.4324/9781003406433-14
This chapter has been made available under a CC-BY-NC-ND 4.0 license.

such as reindeer herding. As a result, in March 2022, Sweden adopted a new law on consultation in matters that affect the Sámi People (Sveriges Riksdag 2022). The law is designed to ensure the Sámi People's rights to participate in decision-making processes at all levels of society and to strengthen Sámi influence in matters of special concern to Sámi society (Sveriges Riksdag 2022; Sámidiggi 2022a). Initial reviews of the new law have been both positive, labeling it a step forward, and negative, criticizing it for being too weak and not safeguarding the rights of Sámi reindeer-herding communities (Sámiid Riikkasearvi 2022).

Today, Indigenous and local communities worldwide are affected by a range of pressures, including the destruction of land, sacred sites, and traditional livelihoods due to political oppression and/or extractive activities, and the consequential loss of languages and cultures (Schultz *et al.* 2019; Dahl 2019). We are also witnessing increasing resistance in the many Indigenous and local community protests taking place around the world (Ojong 2020). However, the situations many Indigenous and local communities find themselves in can result in negative consequences that are not always observed. In Australia and Canada, "lateral violence" has been highlighted as one such consequence, where the term refers to displaced violence directed inward toward one's own community rather than toward the oppressor (Clark and Glover 2019).

Lateral violence exists in Swedish and Sámi contexts, too, not least evidenced by the recent outbreak of hatred and racism against the Sámi People in the Kiruna area after the *Girjas* verdict[1] was handed down (Fröberg 2020). There, Sámi reindeer herders have received death threats, and reindeer have been shot and tortured (Sameradion 2020). In Lycksele, the local community experienced intra-community conflict in conjunction with the repatriation of Sámi skeletal remains, which culminated in the repatriation ceremony on 9 August 2019 (Sámidiggi 2022b). And in Härjedalen, Sámi reindeer herders and local community members reported extensive experience of long-term intra-community conflict (Nord 2019).

In light of this, and of the fact that there is no previous research on the causes of lateral violence in Sámi and local communities in Sweden, this chapter asks: What are the underlying causes of lateral violence in Sámi and local communities, and what methods for conflict management and transformation can be used to address it?

Methodology and ethical considerations

This chapter theoretically explores the causes of lateral violence in Indigenous and local communities, the purpose being to find ways to address lateral violence and to formulate methods for conflict management and transformation. To do this, I draw on peace and conflict theory, including theories on lateral violence (Freire 1972; Fanon 1963, 1967), asymmetric conflict (Ramsbotham *et al.* 2016, p. 27), and conflict transformation (Miall 2004). The analysis tools used to explore the causes of lateral violence in Indigenous and local communities included the concepts of cultural and structural violence (Galtung 1969, 1990) and extractive violence (Sehlin MacNeil 2017).

Previous research on which this theoretical discussion rests is qualitative and involved Indigenous communities; therefore, Indigenous methodologies and ethics were and are important. The research was carried out as two single case studies, one in Sápmi with a Sámi reindeer-herding community, where six members of the community participated. I collaborated with Sámiid Riikkasearvi, the interest organization for Sámi reindeer herding in Sweden, which acted as a cultural broker and initiated contact with the Sámi community. The second single case study was conducted with seven members of the Adnyamathanha community in South Australia, where contact was initiated by an Adnyamathanha Elder. In both case studies, the research participants were approached as they had extensive experiences of extractivism on their lands. The research was conducted as part of my PhD thesis, funded by Umeå University in Sweden, and underwent two formal ethics reviews at Umeå University and the University of South Australia. The data were collected through group and individual interviews, conducted as yarning sessions. "Yarning" is an Australian Aboriginal interview method that can "be utilized to form partnerships with Aboriginal communities in order to develop culturally safe and just research" (Dean 2010, p. 6). Yarning is a conversational-style interview method where a researcher can respectfully take part of someone's lived experiences, sharing their own in return and thus forming a relationship. It aims to involve Indigenous people in the research process instead of merely being the researched party. Bessarab and Ng'andu (2010) describe yarning as an interview method that is particularly well suited for sensitive situations, as it accommodates a therapeutic component within a culturally safe setting.

Indigenous methodologies are many and varied (Smith 2021; Kovach 2010); however, they share an emphasis on relationality (Wilson 2001). This chapter's ethical approach rests on the foundation of respect, reciprocity, and relationships (Reid and Taylor 2011), where research—ranging from project design, data collection, and analysis to the publication and dissemination of results—is conducted in collaborative and non-extractive ways, with a firm focus on mutual respect in all relationships between project participants (including the researcher) and "giving back" in terms of community and societal value for the communities involved.

Sámi and Indigenous connections to land: Rights to land

Within the context of people–planet relationships, there are particular ways in which many Sámi people connect with their lands. These are linked to Sámi legend and belief systems, worldviews, livelihoods, culture, language, and identity. The foundation for these connections is the philosophical view that people and planet—or natural environments—are equal and not hierarchically ordered. Thus, humans are not masters of their natural environments but, rather, live in interdependence with them (Nergård 2006; Oskal 2000). A better understanding of these connections to land, sometimes called "spiritual connections," can help us gain deeper knowledge as to why land holds such importance, not only for many Sámi people but also for many Indigenous Peoples globally. While Indigenous Peoples have different languages, cultures, belief systems, histories, and political systems, there are often similarities in terms of connections to the land. Coates (2004) lists Indigenous Peoples across the globe and their creation stories describing how the land was formed and how reciprocal relationships between people and land took shape, namely, that the well-being of the people depended on the well-being of the land. One clarifying example of such connections is First Australian Adnyamathanha People's *Muda*. Muda is described as

> Yura Muda—the Land is like a book, we are taught by our Elders from the Land. They share their knowledge of the journey of our creators and how they shaped and formed the Yarta "Land" physically, spiritually and environmentally. Yura Muda dictates to the Adnyamathanha people the concepts of "Rules for Living," the "Environment" and the "Spiritual World." Through the Muda it dictates to us who we are, where we belong and how we behave.
> (Iga Warta 2022)

Marsh (2010, p. 124) describes Muda as "the unique spiritual link between Adnyamathanha Yuras and Yarta and cultural resources." This gives an insight into the value of land not measured in monetary terms. The land teaches people who they are, where they belong, and how they should behave. These types of connections to land stand in contrast to neoliberal extractivist philosophies, upheld by many extractive industries and the states that support them. The clashes between worldviews have become more common and visible in recent years, and two examples that have gained international attention are the Gállok Mine protest in Sweden and the Dakota Pipeline protest in the United States (Persson, Harnesk, and Islar 2017; Hunt and Gruszczynski 2021). In Sweden, while the connections of the Sámi People to their lands are most often ignored in consultations and negotiations with extractive industries, rights to land are not; however, most Sámi people have no specific rights to their traditional lands.

According to the *Swedish Reindeer Husbandry Act* of 1971, reindeer-herding Sámi, in the part of Sápmi that lies within Sweden's borders, have the right to use land for reindeer grazing (also to a certain extent for hunting and fishing), and this is based on prescription from time immemorial (Allard 2015, pp. 44–45). Despite the fact that reindeer-herding rights are collective rights belonging to the Sámi People, according to the legislation, Sámi must be members of a Sámi reindeer-herding community to be considered reindeer-herders (Allard 2015, pp. 56–59). There are currently 51 Sámi reindeer-herding communities in Sweden. An initial list of these communities was drawn up at the end of the nineteenth century after a proposal by the "Lapp Committee of Inquiry"; however, the demarcation between the communities was ultimately made by the County Administrative Boards, and since then, only minor changes have been made (Lantto 2012, pp. 41–44). Today, Sámi reindeer husbandry is regulated under the *Reindeer Husbandry Act* and handled by the County Administrative Boards in the relevant counties (Brännström 2017, p. 99). In practice, this means that Sámi must own reindeer to be able to use their traditional lands (Löf 2014, p. 47). Consequently, the connections to land are dependent on the reindeer, or at least, the connections based on being able to exist on and live with one's land. Niila Inga, a reindeer-herder in the Laevas čearru community, notes that "the reindeer is the rights holder" (Sehlin MacNeil and Inga 2019, p. 45). When traditional reindeer husbandry is threatened by various forms of extractive violence—including the progress of extractive industries and, in some cases also, tourism—the reindeer-herders' connections to the land are also threatened.

Cultural, structural, and extractive violence

The term "extractivism" is used to describe activities that extract large quantities of natural resources, largely for export. Acosta (2013, p. 62) states that extractivism is also present in other industries than those linked to minerals or oil, for example, farming, fishing, and forestry. Burchardt and Dietz (2014, p. 481) define extractivism as a term "usually used to describe economic models and sectors such as mining that revolve around the extensive extraction of raw materials and their export." I define extractive violence as a form of direct violence linked to extractivism: "a type of direct violence against nature and/or people and animals that is caused by extractivism and that primarily affects peoples closely connected to land" (Sehlin MacNeil 2017, p. 23). The meaning of extractivism is then expanded to an ideology where the driving force is profit-driven extraction, be it of raw materials, resources, culture, knowledge, or experiences. Extractive violence must be considered in a context of cultural and structural violence (Galtung 1969, 1990). The model known as Galtung's violence triangle (Fig. 10.1a) shows how structural, cultural, and direct violence interact. Direct violence consists of physical or psychosocial violence that harms the body; structural violence refers to harmful and discriminatory societal structures; and cultural violence consists of culture, attitudes, and ideas in a society that underpin and legitimize structural and direct violence (Galtung and Fischer 2013). Galtung defines violence as "avoidable insults to basic human needs lowering the real level of needs satisfaction (far) below what is potentially possible" (Galtung and Fischer 2013, p. 43). Galtung's definition of violence is broad and has generated both praise and criticism by fellow scholars (see, e.g., Boulding 1977; Dilts 2012; Walker 2004). Nevertheless, Galtung's violence triangle is an established model, and the concept of structural violence is a well-known and often used tool for analysis (Ramsbotham, Woodhouse, and Miall 2016).

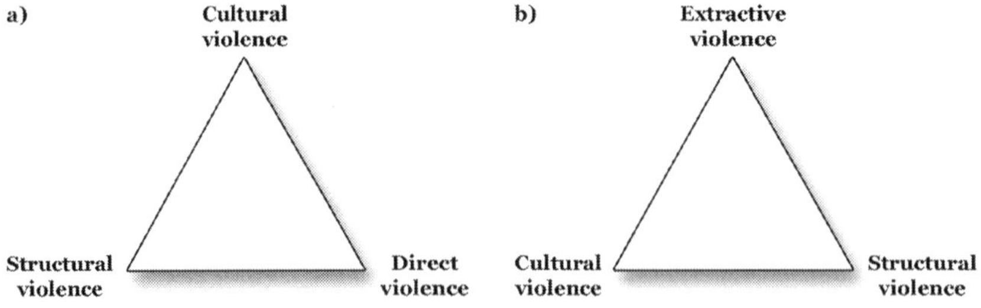

Figure 10. 1a, 1b Galtung's original violence triangle and modification

Fig. 10.1a shows Galtung's original violence triangle and b the extractive violence modification (see Sehlin MacNeil 2017). The different types of violence are always at interplay with each other and are not hierarchically ordered; in fact, they can be placed at any position on the triangle.

In the case of extractive violence, there is a need for the inclusion of Indigenous Peoples' perspectives, such as the above-mentioned Muda or Sámi connections to land. Dewey (1980, p. 246 quoted in Bufacchi 2005, p. 195) describes how in an economically growth-oriented culture, driven by extraction ideology, "energy becomes violence when it defeats or frustrates purposes instead of executing or realizing it. When the dynamite charge blows up human beings instead of rocks, when its outcome is waste instead of production, destruction instead of construction, we call it not energy of power but violence." Violence only occurs when people are harmed. But what if violence against land means that people *are* harmed? John Binda Reid (Reid and Taylor 2011, p. 20) describes how his elderly relatives in the First Australian Kokatha community became physically ill with kidney disease, diabetes, and hypertension after a mining company built a service road that destroyed the Kokatha People's totem, the lizard man that could be seen in the topography. The Kokatha People's worldview and spiritual connection to the land tell them that they will become ill if their sacred sites are destroyed, as this damages the connection between the people and their lands. Can violence against land still be described as productive and constructive if it manifests in people and does them harm?

It is well known that Indigenous Peoples' perspectives must often yield to those of extractive industries (Howlett *et al.* 2011; Lawrence 2014; Nachet *et al.* 2021; O'Faircheallaigh 2016). But how Indigenous Peoples' perspectives are understood by extractive actors, even when they *are* heard, is less known. Previous research shows that Indigenous Peoples' perspectives and multifaceted connections to their lands are often trivialized or ignored in interactions with extractive actors. Instead, Indigenous communities often experience an interplay among cultural, structural, and extractive violence when extractive activities take place on Indigenous lands. The following quotes (Table 10.1) related to experiences of interactions with extractive industries were shared by First Australian and Sámi research participants in two single case studies conducted between 2013 and 2016 (see Sehlin MacNeil 2015,2016).

A significant finding of the research was the intersection of how extractivism was experienced similarly by Sámi and Adnyamathanha peoples, despite existing on opposite sides of the world. The quotes illustrate that, although the Laevas Sámi reindeer-herding community and the Adnyamathanha community in South Australia have different geographical locations, customs, cultures, histories, and current situations—and live within different nation states with different legal and political systems—there are similarities in the experiences of their interactions with extractive industries. Not all interactions were negative, but most were experienced as asymmetric power relations by both Indigenous communities. One example was how the Adnyamathanha community experienced disempowerment when they were not able to use Yura Ngwarla, their language, in consultations

Table 10.1 Direct quotes from Adnyamathanha (Australia) and Laevas čearru (Sápmi) research participants

Theme	*Australia: Adnyamathanha research participants*	*Sápmi: Laevas čearru research participants*
Experiences of asymmetric power relations in interactions with extractive industry actors	"We know that mining proponents are only ever going to be there for themselves, first and foremost. They are there to make a profit and then to get out of there. Governments shouldn't be doing that; they are paid by taxpayer dollars. They shouldn't be taking that attitude, but they are" (Sehlin MacNeil 2016, p. 106) ". . . [O]ur language [Yura Ngawarla] is so central to the way that we are connected to the land but it is also very central to how we are being disempowered, how we're being cut out of the consultation and decision-making processes (Sehlin MacNeil 2016, p. 101)	"You need to be aware that this is the world's largest mine [LKAB, a Swedish government—owned mine] that we are dealing with" (Sehlin MacNeil 2015, p. 80) "We know what happens to a reindeer husbandry area when a mine comes in. And I don't wish that upon other Sámi communities, that the same would happen to them as has happened to us" (*ibid*) "We can't compete and bring out unemployment numbers and financial examples; we don't stand a chance" (Sehlin MacNeil 2015, p. 78)
Experiences of intragroup tensions related to interactions with extractive industry actors	"I've seen the money that mining brings in terms of companies wanting to cut deals with the community. In the past they've tried to do it through individuals . . . It's caused a lot of heartbreak. It's broken families up. It's broken the trust that held the community together" (Sehlin MacNeil 2016, p. 95)	"If we are to coexist, if we are to live together here, you know if we go against this, then we'll get the entire region, the whole municipality against us. What will the consequences for reindeer husbandry be then?" (Sehlin MacNeil 2015, p. 80)

with extractive industry actors because the promised interpretation service was not provided. The result was that many Elders could not take part in the dialogs. As Adnyamathanha knowledge systems are intrinsically linked to both language and land, using Yura Ngwarla can be essential to be able to explain the importance of protecting sacred sites or other specific locations. An example shared by the Laevas Sámi community was how mining impacted reindeer husbandry in profoundly negative ways by destroying grazing lands, threatening to put an end to traditional reindeer husbandry in their area. A further example, given by both communities, was the difficulty of challenging a profit-driven system. Furthermore, both communities shared experiences of interactions with states and governments, as extractive industry proponents, and found particularly high degrees of power inequality in these situations.

In relation to intragroup tensions or conflict, both communities gave different, but similar, examples of how they had experienced the divide-and-conquer tactics of mining companies. In the Adnyamathanha community, individuals had been singled out, and members of the community had received different treatment, causing rifts in the community. The Laevas Sámi reindeer-herding community shared their concerns about how their interactions with a mining company could cause conflicts with the entire region if they were to oppose the opening of a new mine, and how this would be detrimental to their futures as reindeer herders.

These examples are representative of power relations in Indigenous and extractive industry interactions in Sweden and Australia. There are, of course, exceptions where extractive industries and Indigenous communities find ways to collaborate and coexist, but they are not the norm. In situations where asymmetric power relations enable cultural, structural, and extractive violence, there is also great risk of lateral violence.

Lateral violence

"Lateral violence" is a term for displaced violence directed against one's peers rather than one's adversaries. It is one way of explaining minority-on-minority violence as it occurs within marginalized groups where members strike out at each other as a result of oppression. Lateral violence stems from a complex blend of cultural, historical, and social dynamics and includes a range of behaviors such as gossiping, jealousy, bullying, shaming, social exclusion, family feuding, organizational conflict, physical violence, and threats of violence (AHCR 2011).

Prominent scholars associated with the concept of lateral violence include Frantz Fanon (1963, 1967), who observed colonized peoples in Africa, and Paulo Freire (1972), who described lateral violence in Latin America. They both viewed colonialism as oppressive and described how colonized peoples could internalize the colonizers' behaviors, resulting in violent acts toward their own group.

While lateral violence exists in Sámi and local community contexts in Sweden, there is no previous research on this topic. In fact, lateral violence in Indigenous communities is an emerging research topic that has generated a smaller number of studies, predominantly in Canada and Australia, from diverse research fields (see, e.g., Bailey 2020; Clark and Glover 2019; Bombay 2014; Clark *et al.* 2016). Several scholars point out that the term "lateral violence" itself can be problematic, since there is concern that the word "violence" may become a negative label for Indigenous communities (AHCR 2011). However, the existing literature also points to the need for labeling the phenomenon of Indigenous infighting in order to create deeper understandings of it and how to address it (Clark *et al.* 2016). The existing literature on lateral violence in Indigenous communities is focused on either actual acts of lateral violence or the effects it has on Indigenous communities. Although colonialism and oppression are mentioned as underlying factors, there is no focus on, or systematic analysis of, the causes of lateral violence in

Indigenous and local communities. As pointed out by Clark *et al.* (2016), there is very little existing research on lateral violence in Indigenous communities, and, consequently, researchers have relied on overlapping research on oppression, trauma, and racism, for example, which are all relevant to creating a deeper understanding of lateral violence. In Sápmi, studies have also shown that experiencing ethnic discrimination as a Sámi person is common among young adult Sámi in Sweden (Omma 2013).

Since the Swedish side of Sápmi covers northern Sweden, there are a number of smaller towns with a higher presence of Sámi people, many of whom are in some way involved with reindeer herding or other traditional Sámi livelihoods. However, most inhabitants in northern Sweden are not Sámi. This means that small local communities can experience several varieties of lateral violence. For the sake of this chapter, the focus is on two types of lateral violence: That within a Sámi community or association and that within an entire local community. Lateral violence that occurs within an entire local community is more likely to have more dimensions of discrimination linked with prejudice and racism.

Examples of lateral violence in local communities on the Swedish side of Sápmi

The Kiruna area in Norrbotten County is home to several Sámi reindeer-herding communities, namely, Girjas, Gabna, and Laevas čearru. Kiruna is an administrative municipality for Sámi language[2] and has a population of just under 30,000. Kiruna is also home to the world's largest underground iron ore mine, owned and operated by the mining company LKAB, a large employer in the area. In this local community, Sámi reindeer herders and other community members meet on a daily basis, their children go to school together and play on the same sports teams, and they share societal services and facilities. Therefore, the high levels of lateral violence have devastating effects on everyday lives. Sámi reindeer-herding communities in the area have long reported high levels of conflict within the local community (Sehlin MacNeil 2015). There has been some evidence of conflict transformation, notably when local community and Sámi reindeer herders have joined forces to protest mines; however, in recent years, there has also been increasing evidence of racism, discrimination, and conflict in this area (Sehlin MacNeil *et al.* 2018). A recent winning verdict for the Girjas Sámi reindeer-herding community in the *Girjas vs. the Swedish government* case for the rights to hunt and fish on Girjas lands resulted in a hate storm in the Kiruna area, with Sámi reindeer herders receiving death threats and reindeer being tortured and killed (Sameradion 2020).

Lycksele, a municipality in Västerbotten County with a population of just over 12,000, is an administrative municipality for Sámi language. The Lycksele area is also home to five Sámi reindeer-herding communities that have winter grazing lands there: Vapstens sameby, Ubmeje tjeälldie, Rans sameby, Grans sameby, and Malå skogssameby. On 9 August 2019, International Day of the World's Indigenous Peoples, Lycksele hosted a repatriation ceremony, where the remains of some 25 individuals were returned to their original burial site, the old cemetery at

Gammplatsen in Lycksele. This is the largest repatriation in Sweden to date, and it shone a spotlight on Sweden's racist history (Sámidiggi 2022b). Although the process—a collaboration between Líksjuon Sámiensiäbrrie (Lycksele Sámi Association) and the Swedish Church, assisted by the Municipality of Lycksele—was guided by the concept of reconciliation and resulted in a successful repatriation, it uncovered a number of underlying conflict issues that resulted in lateral violence.

The municipality of Härjedalen lies in Jämtland County in the South Sámi area of Sweden. The municipality has a population of just over 10,000. Three Sámi reindeer-herding communities have their traditional lands within Härjedalen: Mittådalen, Ruvthen sijte, and Handölsdalen. In the early 1990s, Härjedalen was the scene of a major legal case over Sámi customary rights to reindeer grazing in the area, where five Sámi reindeer-herding communities were sued by 700 private landowners, as well as a number of forestry companies. In 1996, the verdict was delivered, ruling in favor of the landowners. The entire process caused conflict, and the level of lateral violence in the community has been high since then. Furthermore, in Härjedalen, tourism is an important industry, and Sámi reindeer herders have long raised issues related to the adverse effects of tourism on the well-being of the reindeer and reindeer herding (Nord 2019).

These are all examples of conflicts in local communities within the Swedish side of Sápmi. All the conflicts have connections to land rights, land use, and spiritual connections to the land, suggesting that lateral violence in Sámi and local communities is a real problem with dire consequences for small rural societies.

Lateral violence: Causes and effects?

Previous studies have shown that the combination of cultural, structural, and extractive violence places enormous pressure on Indigenous and local communities. The connections to particular places and pieces of land are also central in this context. When these lands are dug up and destroyed by mining companies; when forests are cut down and the landscape, flora, and fauna altered by forestry companies; when rivers are dammed, lands flooded, and waters and aquatic environments altered by hydropower companies; and when lands are riddled by service roads created to support wind farms, to mention a few extractive activities that impact Indigenous Peoples and, more specifically, Sámi people in Sweden, it impacts Sámi livelihoods like reindeer herding, hunting, and fishing. In addition, Sámi sacred places come under threat or are destroyed. These livelihoods and sacred sites are integral to the revitalization and survival of languages and cultural expressions such as *duodji* (Sámi handicraft) and *yoik* (Sámi song). But when Sámi reindeer-herding communities or Sámi associations voice concerns or protest extractive developments on those grounds, they run the risk of being branded as "backwards" and "anti-growth" or "anti-development," with a consequential increase in discriminatory attitudes toward Sámi, in other words, a reinforced cultural violence. As previously mentioned, Sámi people in Sweden are governed by the Swedish legal system, and limited rights to land are held only by those who have reindeer. Sámi people are thus subjected to asymmetric power relations, where extractive industries and actors or stakeholders have more power and Sámi people, less. Sámi associations and Sámi reindeer-herding communities often find themselves in asymmetric conflict situations when extractive activities are proposed on Sámi lands.

When these pressures, asymmetric conflicts, and power relations—and the consequential cultural, structural, and extractive violence—are placed on a community, it is a common reaction to strike out at one's peers, rather than at the parties responsible for creating the pressures. Colonization deprived groups of people of their power, autonomy, and lands. In response to the injustices, anger and frustration have manifested in violence, not "vertically" toward those responsible for the oppression but "laterally" toward one's peers. There may be several reasons for this. One reason might be that it can be difficult to identify who is responsible when the other party is a state or a multinational company. These can become faceless adversaries, big powers instead of people, creating a feeling of powerlessness. Lateral violence can then be a way to "feel powerful in a powerless situation" (AHCR 2011). In my experience working with Indigenous communities, lateral violence is a symptom of built-up pressure, caused by racism and discrimination, unjust social systems, and extractive activities of various kinds taking place on Indigenous lands. Lateral violence in Indigenous communities, therefore, needs to be addressed by exploring the reasons underlying it and by transforming unjust societal structures and asymmetric power relations. If the causes of lateral violence in Indigenous and local communities are not identified and elucidated, there is a risk of using the wrong tools when applying measures for conflict transformation. As Indigenous communities are already experiencing cultural violence in the form of racism and discrimination, the existence of lateral violence can further fuel cultural violence. This lateral violence may be labeled as an "Indigenous issue" or "Indigenous problem," implying that Indigenous communities are prone to conflict, thus feeding an already existing stereotype. This can, in turn, reinforce the cultural violence that underpins and interplays with structural and extractive violence. As the pressure on the community increases, so does the lateral violence, creating a perpetual cycle of conflict and violence (see Fig. 10.2).

Conclusion

The effects of lateral violence on Indigenous and local communities are often severe. I suggest that conflict transformation could be attempted by addressing the causes of lateral violence. The theory of conflict transformation (see, e.g., Miall 2004 and Lederach 2015), where the contextual and relational aspects of conflicts are emphasized, is of particular interest. Miall (2004, Chapter 4) states that conflict transformation is a process "of engaging with and transforming the relationships, interests, discourses, and, if necessary, the very constitution of society that supports the continuation of violent conflict." The theory of conflict transformation, with its social emphasis, is conducive to research projects with Indigenous communities, as it does not contradict but, rather, complements Indigenous methodologies and Indigenous research ethics. Failing to gain a better understanding of the causes of lateral violence in Indigenous and local communities could potentially lead to increased conflict and violence.

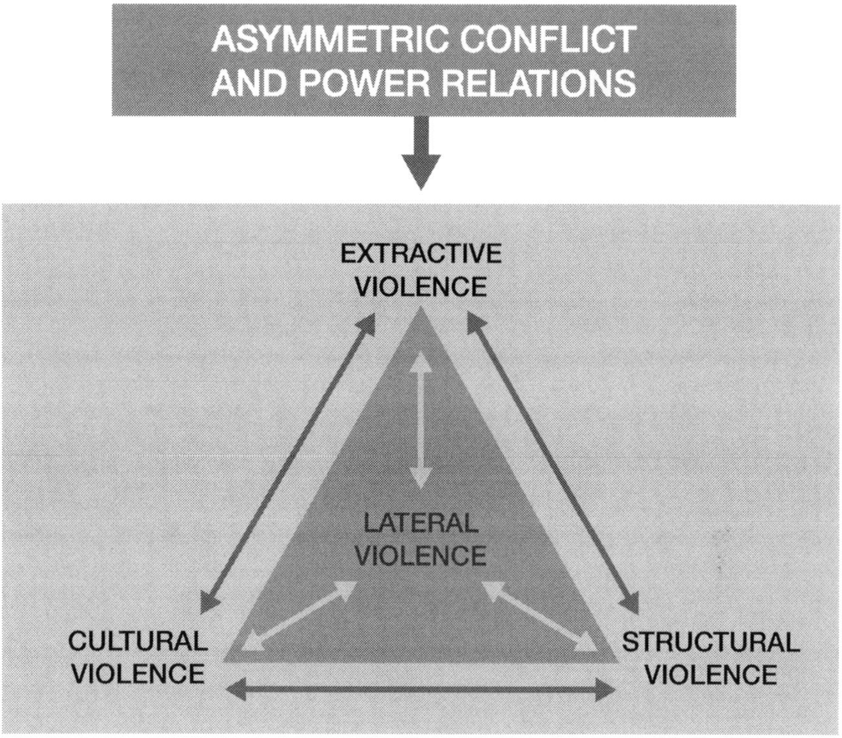

Figure 10.2 Asymmetric conflict and power relations cause extractive, cultural, and structural violence, which in turn causes lateral violence. Lateral violence reinforces cultural violence, which underpins structural and extractive violence. The pressure increases, causing more lateral violence and a perpetual cycle of conflict and violence

Acknowledgments

This chapter was written as part of the research project: *Lateral violence: Effects of external pressures on Indigenous and local communities*, funded by FORMAS, the Swedish research council for sustainable development.

Notes

1 The "*Girjas* case," a decades-long dispute between the Girjas Sámi reindeer-herding community and the Swedish state over the rights to license small-game hunting and fishing on the land of the Girjas reindeer-herding community ended with a positive outcome for the Girjas in 2020. It is seen as a significant legal development in the area of Sámi law in Sweden (see Allard and Brännström 2021).

2 There are 25 Sámi administrative municipalities in Sweden, established in accordance with the *Swedish Act on National and Minority Languages*. These municipalities are required to offer elder care and preschool activities in Sámi languages and aim to promote Sámi culture (Bjärstig *et al.* 2020, p. 6).

References

Acosta, A., 2013. Extractivism and neoextractivism: Two sides of the same curse. *In*: Permanent Working Group on Alternatives to Development, ed., *Beyond development: Alternative visions from Latin America*, 1st ed. Quito, Ecuador: Transnational Institute/Rosa Luxemburg Foundation, 61–86.

Allard, C., 2015. *Renskötselrätt I nordisk belysning*. Göteborg and Stockholm: Makadam Förlag.

Allard, C., and Brännström, M., 2021. Girjas reindeer herding community v. Sweden: Analysing the merits of the Girjas case. *Arctic Review*, (12), 56–79. https://doi.org/10.23865/arctic.v12.2678

Australian Human Rights Commission (AHCR), 2011. *Social Justice Report 2011*. Canberra: Aboriginal and Torres Strait Islander Social Justice Commissioner.

Axelsson, P., and Mienna, C. S., 2021. The challenge of Indigenous data in Sweden. *In:* M. Walter, T. Kukutai, S. Russo Carroll, and D. Rodriguez-Lonebear, eds., *Indigenous data sovereignty and policy*. New York and Abingdon: Routledge, 99–111.

Bailey, K. A., 2020. Indigenous students: Resilient and empowered in the midst of racism and lateral violence. *Ethnic and Racial Studies*, 43 (6), 1032–1051.

Bessarab, D., and Ng'andu, B., 2010. Yarning about yarning as a legitimate method in Indigenous research. *International Journal of Critical Indigenous Studies*, 3 (1), 37–50.

Bjärstig, T., Nygaard, V., Riseth, J Å., Sandström, C., 2020. The institutionalisation of Sami interest in municipal comprehensive planning: A comparison between Norway and Sweden. *International Indigenous Policy Journal*, 11 (2), 1–24. https://doi.org/10.18584/iipj.2020.11.2.10574

Bombay, A., 2014. *Origins of lateral violence in Aboriginal communities: A preliminary study of student-to-student abuse in residential schools*. Ottawa: Aboriginal Healing Foundation.

Boulding, K. E., 1977. Twelve friendly quarrels with Johan Galtung. *Journal of Peace Research*, 14 (1), 75–86.

Brännström, M., 2017. *Skogsbruk och renskötsel på samma mark: En rättsvetenskaplig studie av äganderätten och renskötselrätten*. Umeå: Doktorsavhandling, Skrifter från Juridiska institutionen vid Umeå universitet, 35.

Bufacchi, V., 2005. Two concepts of violence. *Political Studies Review*, 3, 193–204.

Burchardt, H. J., and Dietz, K., 2014. (Neo-)extractivism: A new challenge for development theory from Latin America. *Third World Quarterly*, 35 (3), 468–486.

Clark, Y., and Glover, K., 2019. Lateral violence in Aboriginal communities: From awareness to transformations. *In:* B. Shelley, K. te Riele, N. Browne, and T. Crellin, eds., *Harnessing the transformative power of education*. Leiden: Brill, 16–35.

Clark, Y., Augoustinos, M., and Malin, M., 2016. Lateral violence within the Aboriginal community in Adelaide: "It affects our identity and wellbeing." *Journal of Indigenous Wellbeing*, 1 (1), 43–52.

Coates, K. S., 2004. *A global history of Indigenous* Peo*ples: Struggle and survival*. New York: Palgrave Macmillan.

Dahl, J., 2019. Study on Indigenous Peoples' autonomies: Experiences and perspectives. *United Nations Permanent Forum on Indigenous Issues*. Available from: https://undocs.org/E/C.19/2020/5 [Accessed 26 March 2020].

Dean, C., 2010. A yarning place in narrative histories. *History of Education Review*, 39 (2), 6–13.

Dewey, J., 1980. 'Force, violence and law' and 'force and coercion', *In:* J. A. Boydston, ed., *John Dewey, the middle works, 1899–1924, Volume 10: 1916–1917*. Carbondale IL: Southern Illinois University Press, 211–215 and 244–251.

Dilts, A., 2012. Revisiting Johan Galtung's concept of structural violence. *New Political Science* 34 (2), 191–194.

Fanon, F., 1963. *The wretched of the earth*. New York: Gove Press.

Fanon, F., 1967. *Black skin, white masks*. New York: Gove Press.

Freire, P., 1972. *Pedagogy of the oppressed.* Hormondworth: Penguin.

Fröberg, J., 2020. Januari 23. DN Nyheter Sverige. *Demokratins misslyckande bäddar för oroligheter i Norrland.* Stockholm: Dagens Nyheter.

Galtung, J., 1969. Violence, peace, and peace research. *Journal of Peace Research*, 6 (3), 167–191.

Galtung, J., 1990. Cultural violence. *Journal of Peace Research*, 27 (3), 291–305.

Galtung, J., and Fischer, D., 2013. *Johan Galtung, pioneer of peace research.* Heidelberg; New York; Dordrecht and London: Springer.

Hansen, L. I., and Olsen, B., 2006. *Samernas historia fram till 1750.* Stockholm: Liber.

Howlett, C., Seini, M., McCallum, D., and Osborne, N., 2011. Neoliberalism, mineral development and Indigenous people: A framework for analysis. *Australian Geographer*, 42 (3), 309–323.

Hunt, K., and Gruszczynski, M., 2021. The influence of new and traditional media coverage on public attention to social movements: The case of the Dakota Access Pipeline protests. *Information, Communication & Society*, 24 (7), 1024–1040.

Iga Warta, 2022. Cultural awareness. Available from: www.igawarta.com/documents/Iga%20Warta%20cultural%20awareness%20package%20info.pdf [Accessed 10 November 2022].

Kovach, M., 2010. *Indigenous methodologies: Characteristics, conversations and contexts.* Toronto: University of Toronto Press.

Lantto, P., 2012. *Lappväsendet: tillämpningen av svensk samepolitik 1885–1971.* Umeå: Centrum för Samisk forskning, Umeå universitet.

Lantto, P., and Mörkenstam, U., 2015. Action, organisation and confrontation: Strategies of the Sámi movement in Sweden during the twentieth century. *In*: M. Berg-Nordlie, J. Saglie, and A. Sullivan, eds., *Indigenous politics: Institutions, representation, mobilisation.* Colchester, UK: ECPR Press, 135–163.

Lawrence, R., 2014. Internal colonisation and Indigenous resource sovereignty: Wind power developments on traditional Saami lands. *Environment and Planning D: Society and Space*, 2, 1036–1053.

Lederach, J. P., 2015. *Little book of conflict transformation: Clear articulation of the guiding principles by a pioneer in the field.* New York: Good Books.

Löf, A., 2014. *Challenging adaptability. Analysing the governance of reindeer husbandry in Sweden.* PhD Thesis. Umeå University, Department of Political Science.

Marsh, J., 2010. *A critical analysis of decision-making protocols used in approving a commercial mining license for the Beverley Uranium Mine in Adnyamathanha Country: Toward effective Indigenous participation in caring for cultural resources.* PhD Thesis. Department of Geographical and Environmental Studies, University of Adelaide.

Miall, H., 2004. Conflict transformation: A multi-dimensional task. *In:* A. Austin, M. Fischer, and N. Ropers, eds., *Transforming ethnopolitical conflict: The Berghof handbook.* VS Verlag für Sozialwissenschaften. Wiesbaden: Springer, 67–89.

Nachet, L., Beckett, C., and MacNeil, K. S., 2021. Framing extractive violence as environmental (in)justice: A cross-perspective from Indigenous lands in Canada and Sweden. *The Extractive Industries and Society*, 100949.

Nergård, J.-I., 2006. *Den levende erfaring: en studie i samisk kunnskapstradisjon.* Oslo: Cappelen.

Nord, M., 2019. Rennäring och fjällturism, en konflikt mellan motstridiga intressen. Available from: www.turist.se/rennaring-och-fjallturism/ [Accessed 26 March 2020].

O'Faircheallaigh, C., 2016. *Negotiations in the Indigenous world: Aboriginal Peoples and the extractive industry in Australia and Canada.* New York: Routledge.

Ojong, N., 2020. Indigenous land rights: Where are we today and where should the research go in the future? *Settler Colonial Studies*, 10 (2). https://doi.org/10.1080/2201473X.2020.1726149

Omma, L., 2013. *Ung same i Sverige: livsvillkor, självvärdering och hälsa [Young Sami in Sweden: life circumstances, self-evaluation and health].* PhD Thesis. Umeå universitet.

Oskal, N., 2000. On nature and reindeer luck. *Rangifer*, 20 (2–3), 175–180. https://doi.org/10.7557/2.20.2-3.1511.

Persson, S., Harnesk, D., and Islar, M., 2017. What local people? Examining the Gállok mining conflict and the rights of the Sámi population in terms of justice and power. *Geoforum*, 86, 20–29. https://doi.org/10.1016/j.geoforum.2017.08.009.

Pietikäinen, S., Huss, L., Laihiala-Kankainen, S., Aikio-Puoskari, U., and Lane, P., 2010. Regulating multilingualism in the North Calotte: The case of Kven, Meänkieli and Sámi languages. *Acta Borealia*, 27 (1), 1–23.

Ramsbotham, O., Woodhouse, T., and Miall, H., 2016. *Contemporary conflict resolution*, 4th ed. Cambridge: Polity Press.

Reid, J. B., and Taylor, K., 2011. Indigenous mind: A framework for culturally safe Indigenous health research and practice. *Aboriginal and Islander Health Worker Journal*, 35 (4), 19–21.

Reimerson, E., 2015. *Nature, culture, rights: Exploring space for Indigenous agency in protected area discourses*. PhD Thesis. Umeå University.

Sameradion, 2020. Sara Skum: Förvånad blev jag inte men jag blev chockad av människans grymhet. *Sveriges Radio*. Available from: https://sverigesradio.se/artikel/7417295 [Accessed 10 November 2022].

Sámidiggi, 2022a. Konsultation med Sametinget. Available from: www.sametinget.se/konsultationsordning. [Accessed 10 November 2022].

Sámidiggi, 2022b. Repatrieringen i Lyckselse den 9 agusti 2019. Available from: www.sametinget.se/131941 [Accessed 10 November 2022].

Sámiid Riikkasearvi, 2022. Konsultationsordningen är beslutad. Available from: www.sapmi.se/konsultationsordningen-ar-beslutad/ [Accessed 10 November 2022].

Schultz, R., Abbott, T., Yamaguchi, J., and Cairney, S., 2019. Australian Indigenous land management, ecological knowledge and languages for conservation. *EcoHealth*, 16 (1), 171–176.

Sehlin MacNeil, K., 2015. Shafted: A case of cultural and structural violence in the power relations between a Sámi community and a mining company in northern Sweden. *Ethnologia Scandinavica: A Journal for Nordic Ethnology*, 45, 73–88.

Sehlin MacNeil, K., 2016. On equal terms? Traditional Owners' views regarding nuclear waste dumps on Adnyamathanha Country. *Journal of Australian Indigenous Issues*, 19 (3), 95–111.

Sehlin MacNeil, K., 2017. *Extractive violence on Indigenous Country: Sami and Aboriginal views on conflicts and power relations with extractive industries*. PhD Thesis. Umeå University.

Sehlin MacNeil, K., and Inga, N., 2019. Extraktivt våld och urfolks koppling till mark. *Kulturella perspektiv-Svensk etnologisk tidskrift*, 28 (1–2), 42–51.

Sehlin MacNeil, K., Svonni, C., and Össbo, Å., 2018. *Nyansera debatten om makten*. Luleå: NSD.

Smith, L. T., 2021. *Decolonizing methodologies. Research and Indigenous Peoples*, 3rd ed. London: Zed Books.

Stoor, J. P. A., and San Sebastián, M., 2022. A population-based study on health and living conditions among Sámi in Sweden: The SámiHET study. *International Journal of Circumpolar Health*, 81 (1), 2076383. https://doi.org/10.1080/22423982.2022.2076383.

Sveriges Riksdag, 2022. Lag (2022:66) om konsultation i frågor som rör det samiska folket Available from: www.riksdagen.se/sv/dokument-lagar/dokument/svensk-forfattningssamling/lag-202266-om-konsultation-i-fragor-som-ror_sfs-2022–66 [Accessed 10 November 2022].

Walker, P. O., 2004. Decolonizing conflict resolution: Addressing the ontological violence of westernization. *The American Indian Quarterly*, 28 (3&4), 527–549.

Wilson, S., 2001. What is an Indigenous research methodology? *Canadian Journal of Native Education*, 25 (2), 175–179.

Part IV

Indigenous women and resource development

11 Employment trends for Indigenous women working in the Northern Territory's large-scale mining industry

Real employment opportunities or empty corporate promises?

Jodi Cowdery and Andrew Taylor

Overview

Most large-scale mining in Australia occurs on Indigenous lands and near to remote Indigenous communities (Centre for Aboriginal and Economic Policy Research 2013). The investment and approval processes, together with the operational activities of mining, often create disproportionate and negative consequences for Indigenous peoples living nearby (O'Faircheallaigh 2015, 2018; Kerins 2018). Numerous studies have shown the impacts of large-scale mining (LSM) to be gendered, with women, in particular, experiencing adverse effects more so than men (Carrington *et al.* 2010; Pauktuutit Inuit Women of Canada and University of British Columbia 2016). This suggests that Indigenous women residing in communities adjacent to LSM sites may be disproportionately affected by the negative impacts of mining operations (Parmenter 2011; LaBelle 2015; Parmenter and Drummond 2022).

Large-scale mining projects can result in big increases in employment, with jobs being created during construction of the mine and becoming available during the operational phase as well (O'Faircheallaigh 2015). Employment through jobs created by mining projects has been highlighted as an important benefit of mining (Blackwell and Dollery 2014; Zhang and Moffat 2015), possibly helping offset some of the negative affects experienced by nearby communities. Employment opportunities provided by LSM companies have been used to justify the inception and impacts of mining projects (Paredes and Fleming-Muñoz 2021), yet jobs within the industry are predominantly occupied by non-Indigenous men from outside the region (Bryant and Tedmanson 2005; Gibson and Kemp 2017; Helbert 2018; Denniss 2020).

Although there have been various policies aimed at increasing the employment rates of both Indigenous peoples and women (as distinct groups) in the Australian mining industry, Indigenous women have been—and continue to be—significantly underrepresented in the sector (Parmenter 2011; Parmenter and Drummond 2022). While there have been some increases for both groups, employees who are both Indigenous and women experience what Gibson and Kemp (2017) call a "double blind," where the connection between race and gender is overlooked, as research

DOI: 10.4324/9781003406433-16

This chapter has been made available under a CC-BY-NC-ND 4.0 license.

focusing solely on Indigenous women employed in the mining industry has, to date, been limited (Parmenter 2011; Parmenter and Drummond 2022).

In this chapter, we examine and analyze available data about Indigenous women working at six LSMs in the Northern Territory (NT or Territory) of Australia, where mining remains a major industry. The aim of this chapter is to improve understanding about the opportunities and challenges facing Indigenous women working at LSMs or living local to these mines, and evaluate progress towards greater inclusion, representation, and understanding about the impacts felt by Indigenous women. We profile employment numbers, age, education level, occupation, basis of employment, social marital status, children, and provision of unpaid care, and contrast these with other groups based on indigeneity and gender. We also identify the place of residence of the workforce of the six mines, to demonstrate local versus non-resident employment, and discuss the implications for local Indigenous people and, in particular, Indigenous women.

Introduction

Australia is one of the world's leading suppliers of mineral resources (Mayes and Pini 2014; Australia Minerals 2021), and the sector has experienced significant growth in recent decades, including during the unprecedented mining boom in the decade up to 2013 (Reserve Bank of Australia 2014). The boom was driven by sharp rises in prices for commodities used to produce steel and generate energy, especially iron ore (Australia's biggest export), coal, and natural gas (Reserve Bank of Australia n.d.). The minerals industry is a key contributor to the Australian economy and is considered a major driver in the creation of jobs, revenue, and opportunities for growth and development (World Bank 2009; Pearson and Daff 2013a; Helbert 2018). There are estimated to be a quarter of a million people directly employed by mining in Australia (Department of Jobs and Small Business 2019). In 2018–2019, the resources sector paid approximately AU$25 billion in wages, and the minerals sector paid around AU$40 billion in company taxes and royalties (Australian Government Productivity Commission 2020b).

The beneficial and detrimental effects from mining are well documented globally. Benefits include investments in infrastructure (such as roads and housing) and social services (such as health and education) and possible job "spill-over" effects into other industries in regional and remote areas. Through fly-in/fly-out (FIFO) arrangements for employees, mining creates opportunities for financial wealth to flow into areas outside the remote location of the mine site, for instance non-mining areas and cities where FIFO employees live (Zhang and Moffat 2015).

Despite these economic benefits, LSMs also create detrimental social and environmental impacts. For example, mining activities can draw skilled workers away from other industries such as manufacturing, agriculture, and tourism and create labor shortages and skills pressures due to the higher wages available in mining. The communities and towns close to the mine site may experience a large influx of workers, creating housing shortages and increases in the cost of living for local residents. Mining operations also create dust, noise, pollution, and contamination, as well as destruction of the natural environment, resources, and sites of cultural significance (Zhang and Moffat 2015).

A key attribute of mining is that ore and other deposits are geographically confined, meaning the impacts from extracting minerals are borne largely by the host and/or neighboring communities (Stevens *et al.* 2017; Suopajärvi *et al.* 2017). Many LSMs in Australia are in very remote and sparsely populated areas in proximity to small communities or towns. In the NT, for example, where six LSMs were operating at the time of this research, the Northern Land Council (2022) noted that 80% of the value of minerals extracted in the NT came from mining undertaken on Indigenous lands under the *Aboriginal Land Rights (Northern Territory) Act* 1976 (Cth) and *Native Title Act* 1993 (Cth), with substantial and almost entirely Indigenous populations living nearby (Langton 2013; O'Faircheallaigh and Ali 2017). As a result, LSM operations extensively and disproportionately impact Indigenous landholders, communities, and populations (Langton 2013; Hunter *et al.* 2015; O'Faircheallaigh 2015; Helbert 2018; Sincovich *et al.* 2018).

Past studies have identified a range of negative impacts endured by Indigenous peoples due to mining (O'Faircheallaigh 2015). These include pollution of or disruption to water sources used for drinking, fishing, or hunting (Kerins 2018); disruption or damage of sites of cultural and ceremonial significance (Marsh 2013; O'Faircheallaigh 2018); and potential problems arising from inflows of significant amounts of money to Indigenous communities through mining employment or royalties. The latter includes issues related to drug and alcohol consumption (Resources and Sustainable Development in the Arctic Network 2014; Southcott *et al.* 2018) and internal community conflicts from the uneven distribution of royalties (Brasche 2015). Collectively, these impacts can manifest over time as higher incidences of chronic disease, injury, disability, deaths, mental illness, violence and behavioral issues in the community, and increased pressure on health and other social services (Bohanna and Clough 2012; Muhunthan *et al.* 2017). Furthermore, many impacts from mining are gendered, with women experiencing adverse effects such as family violence and sexual abuse (LaBelle 2015; Sincovich *et al.* 2018) as well as detrimental effects on their capacity to fulfill gendered familial caring responsibilities due to food and water insecurity (Lahiri-Dutt 2012).

Despite this, negotiated land-use agreements in remote areas offer the opportunity for Indigenous communities to negotiate with mining companies for better access to mining benefits. This potential is significant as there are limited alternative work opportunities in very remote settings where LSMs are established, such that remote-living Indigenous people are significantly underrepresented in the labor force (Centre for Aboriginal Economic Policy Research 2016). Negotiated agreements can play a critical role in setting the conditions under which mining occurs on Indigenous land, and many of the positive economic impacts generated in local communities have come about from Indigenous communities successfully negotiating with mining companies for employment and business development opportunities for their local community members (O'Faircheallaigh 2011, 2015, 2018). Conversely, in Australia almost all negotiated land-use agreements contain legally binding confidentiality clauses, making it virtually impossible to publicly review whether LSMs are meeting their obligations to Traditional Owners and Indigenous communities where mining takes place. This lack of transparency also makes it difficult to assess Indigenous communities' perspectives on mining employment and other outcomes (Australian Government Productivity Commission 2020a), requiring the voices of Indigenous peoples whose lands and communities are subjected to LSM operations be tabled through other forms of research.

Indigenous mining employment in Australia

There is an underlying and continuing assumption that mining provides employment opportunities for Indigenous people living in remote communities near mine sites (Pearson and Daff 2013b). This is partly because mining employment may be one of only a limited number of avenues for participation in the cash wage economy (O'Faircheallaigh 2015; Centre for Social Responsibility in Mining, Sustainable Minerals Institute 2018). Furthermore, greater Indigenous participation in the mining workforce has been encouraged by governments and the minerals sector for many decades (Brereton and Parmenter 2008; Taylor 2009).

However, while many members of Indigenous communities would like to participate in the mining industry (Caron *et al.* 2019), proximity alone does not guarantee tangible Indigenous employment outcomes (Blake *et al.* 2014). Several studies have shown that, in general, mining companies source employees from outside the remote locality of the mine, either through FIFO practices or by relocation from outside the region to purpose-built mining facilities close to the mine itself (Bryant and Tedmanson 2005; Centre for Social Responsibility in Mining Sustainable Minerals Institute 2007; Bailey-Kruger 1995; Denniss 2020). Blake and colleagues (2014) point out that the mining industry's practice of importing its workforce through FIFO arrangements is a specific barrier to increased local Indigenous employment, as it allows mining companies to import "highly educated, metropolitan-based labor to work at regional mines," (p. 48) without having to invest in long-term training and planning to develop similar qualifications within the local community and potential local workforce.

Previous studies on employment rates for Indigenous peoples in the Australian mining industry have highlighted they have historically been, and continue to be, significantly underrepresented in the mining workforce (Langton 2013; Pearson and Daff 2013a; Blackham and Temple 2020; Tulele 2020). This is despite substantial increases in the number of Indigenous peoples employed by the mining industry in Australia due to job creation during the national mining boom which spanned 10 years, from 2003 to 2013, and mining company commitments to increase their social licenses to operate near remote Indigenous communities by promising to engage local people in the workforce (Hunter *et al.* 2015). In simple terms, a social license to operate is the societal acceptance of an operation, based on the relationship between the company and the community where it is operating (Moffat *et al.* 2016).

In recent decades, mining companies have come under greater public scrutiny in relation to their global corporate behavior, with growing pressure for them to consider how their operations affect local communities and to exercise better corporate social responsibility. Consequently, mining companies need to go beyond local legal requirements to ensure they operate in a way that respects human rights, engaging in an honest and transparent manner beneficial to all parties throughout all stages of the project (O'Faircheallaigh 2015, 2018; Sincovich *et al.* 2018; Laurence 2021). Employment opportunities, particularly for those living near mine sites who endure most of the negative impacts of mining, are a key component of corporate social responsibility, contributing to the social license of large-scale mining companies to operate near remote Indigenous communities and on Indigenous lands. Therefore, it is reasonable to suggest that failures by large-scale mining companies to facilitate increased and substantial employment opportunities for local Indigenous peoples could constitute a breach of their social license to operate.

Australian Indigenous employment rates in mining increased during the mining boom (Parmenter and Barnes 2021), and it is acknowledged that the Australian mining industry employs a higher proportion of Indigenous people than some other sectors (Tulele 2020). However, Indigenous workers continue to be significantly underrepresented in mining. For example, in 2016, Indigenous peoples of working-age residing in very remote Australia, where most mining takes place, made up 43% of the working-age population. Yet across the entire Australian mining industry, less than 4% of the workforce was Indigenous (Australian Bureau of Statistics 2016b), slightly higher than in 2011, at just over 3% (Australian Bureau of Statistics 2011).

Furthermore, studies by Bryant and Tedmanson (2005), Helbert (2018), and Denniss (2020) have demonstrated that many Indigenous people employed in mining are sourced from a pool of regional- or urban-based Indigenous employees who have already attained experience working for a mining company elsewhere. This potentially negates the benefit of employment for local remote Indigenous communities impacted by mining and may create conflicts between local community members and an "imported" Indigenous workforce. Given the proximity of local communities to mines, these practices repudiate the potential for local residents to be employed locally at the mine and for mining companies to reduce costs and improve localized economic and social outcomes by sourcing some labor from nearby communities.

Turning to women's employment in mining, in Australia specifically, it is estimated that women accounted for about 16% of the mining workforce in 2016 (Australian Bureau of Statistics 2016b). While there have been some recent, small increases in the percentage of women employed in the mining industry, women remain overrepresented in "traditional", gendered, and relatively semi-skilled occupations in mining (such as cleaning, cooking, laundry, ancillary, and administration duties) and underrepresented in decision-making, managerial, operational, and technical expert positions (Lord *et al.* 2012; Jenkins 2014; Deonandan *et al.* 2016; Helbert 2018).

There have been suggestions that these trends exist because mining industry operations are better suited to men (Lozeva and Marinova 2010; Lord *et al.* 2012; Deonandan *et al.* 2016) with long hours, block shifts, and FIFO practices. These suit employees who have minimal caring responsibilities and are available to travel.

Social roles in the private sphere are gendered (such as caring for and raising children), which enables men to participate in the full-time wage economy (such as the mining industry) while restricting women's ability to do so (Mayes and Pini 2014; Helbert 2018). Moreover, mining companies have a patriarchal organizational culture which permits, and even promotes, discrimination in favor of men, instilling preferential progress for men in the industry through training, promotions, and leadership roles, which may be adding to the issue of low representation of women (Kljajevic 2015). Nevertheless, these issues further emphasize the potential for locally based Indigenous women to be employed at remote mine sites since travel, which is quite significant in some cases, may be feasible and because the care and upbringing of children is often shared by wider family members (Charles Darwin University Northern Institute 2015).

Research in Australia and Canada also suggests the lack of engagement with Indigenous women in mining agreement negotiations is a reason for low employment outcomes, given the exclusion of their views at the negotiation stage between Indigenous communities and mining companies (O'Faircheallaigh 2011, 2015; Gibson and Kemp 2017). Importantly, this stage establishes legally binding commitments and conditions for local Indigenous employment from the mining company. This was supported by LaBelle's (2015) study on the role and contributions of Indigenous women in mining negotiations and project development in Canada, which found that decisions guiding mining projects and the impacts on Indigenous communities frequently fail to include contributions from Indigenous women in the community.

Given this context, there is a clear need to grow the engagement and representation of Indigenous women in the mining sector with respect to both their employment and the processes and engagement programs that take place well before a mine becomes operational. Indeed, numerous reports have recommended increased Indigenous employment and women's employment (as separate groups) in the mining industry (Australian Government Office for Women 2007; Brereton and Parmenter 2008; Lord *et al.* 2012; Pearson and Daff 2013a; Centre for Social Responsibility in Mining, Sustainable Minerals Institute 2018, Parmenter and Barnes 2021). However, underrepresentation in both aspects continues, and it is within this setting that we examine Indigenous women's employment in large-scale mines in the NT of Australia, where approximately one in three people are Indigenous.

The NT and its mining industry

The NT is a geographically large jurisdiction in the central north of Australia but has a relatively small population of 252,000 (Australian Bureau of Statistics 2024) and a population density of just 0.2 people per square kilometer (Australian Bureau of Statistics 2018b). One-third of the population identify as Indigenous, much higher than the 3% national figure (Australian Bureau of Statistics 2018a). The capital, Darwin, in the tropical north, is home to over 60% of the NT's total population. The rest of the NT's landmass is defined as remote or very remote by the Australian Bureau of Statistics (2016c), and this is where the remaining residents live (Dyrting *et al.* 2020). Indigenous women make up around 12% of the NT's overall population and 12% of all working-age people, significantly greater than the national figure of 1%. However, in the NT just 28% of Indigenous women of working-age are in the labor force, while nationally the figure is much higher for Indigenous women, at 43%, and for women in general, at 48% (Australian Bureau of Statistics 2016a, 2016b).

For more than a century, mining has been one of the NT's leading industries and an important element in its social, economic, and political emergence as an independent jurisdiction from when it was granted self-government in 1978 (Carment 2003). In 2020, mining accounted for 28% of the NT's gross state product (Minerals Council of Australia 2020), with the combined production value of mines operating in the NT at AU$4.4 billion in 2019–2020. At the time of writing, there were six LSMs operating and producing in the NT (Resourcing the Territory 2021), and according to the Minerals Council of Australia (2020), the industry employed 13,500 people (including those in mining equipment, technology, and service roles). In 2020–2021, the NT government received AU$379 million in royalties from the mining industry, approximately 6% of its total revenue (Department of Treasury and Finance 2020).

Given the significance of the mining sector, together with the NT's distinct population characteristics (where Indigenous women account for a larger portion of the working-age population than across the rest of Australia), and the numerous LSMs operating throughout the region, there is a unique opportunity to examine Indigenous women's employment trends and characteristics. In doing so, we are particularly interested in the extent to which local women (and men) are employed by LSMs that are considered local to where they live.

Data and methods

To analyze the data about Indigenous women working at LSMs in the NT, we produced a customized Australian census dataset enabling comparisons of Indigenous women to Indigenous men, non-Indigenous women, and non-Indigenous men also working in the industry. In addition, we extracted data to ascertain where the workforce for each of the LSMs resided geographically in 2016 and 2011, with a focus on the employment and participation characteristics of Indigenous women from surrounding Indigenous communities, so as to assess the extent of locally sourced employment at the LSMs. While these data are relatively old, they provide a baseline overview of the industry's employment profile and the extent of local sourcing of labor.

In conducting this research, there were two methodological challenges: First, deciding on the range of occupations that should be included under the definition of mining occupations (and therefore employees) and, second, distinguishing between locally sourced employees versus non-local workforce.

Defining the scope of "mining employees"

Previous studies focusing on Indigenous employment in mining found Indigenous employees worked largely in semi-skilled entry-level jobs, and not in trade or professional occupations (Centre for Social Responsibility in Mining, Sustainable Minerals Institute 2007; Parmenter and Barnes 2021). This is also the case for women working in the industry (Perks and Schulz 2020). Although the most common occupations in mining directly relate to extraction and associated machinery, including drillers, miners, shot firers, and metal fitters, Canadian research examining Inuit and Innu employment outcomes affirmed that mine employees who were women were most likely to be employed in culinary, housekeeping, administration, and corporate services jobs (Cox and Mills 2015). Consequently, we took a broad approach to defining mining occupations to ensure the targeted cohort, which we anticipated to be small in numbers, was captured in the data. Thus, drawing on Parmenter's (2008) work examining the experiences of Indigenous women working at Century Mine in northwest Queensland, and Pugliese (2021), who recently investigated mining and gender policies in Congo, we took a similarly broad approach in defining "mining occupations" as any occupation that employed people at a LSM in the NT.

Distinguishing local labor from non-local labor sources

To determine the extent of local labor being employed at each LSM and the potential local labor pool living in nearby communities, we began by identifying the LSMs in operation across the NT at the time of this research. Following is a brief description of the six LSMs.

Rio Tinto Gove Operations

Developed on Yolngu lands (Aboriginal Areas Protection Authority n.d.), the bauxite mine, alumina refinery, and deepwater port established in the late 1960s have been owned by Rio Tinto Ltd. since 2007, and currently employ approximately 300 workers (Rio Tinto 2021a). Prior to the closure of the refinery in 2014, operations employed an additional 1,000 people (Saxinger *et al.* 2016). In 1972, the special-purpose town of Nhulunbuy (also known as Gove) was created, amidst great controversy and resistance, to support mining operations. The township is run by the Nhulunbuy Corporation on behalf of Rio Tinto Ltd. (Nhulunbuy Corporation 2021).

South32 GEMCO

Built on Warnindilyakwa Country (Aboriginal Areas Protection Authority n.d), the manganese mine and port have been operating since 1965 and are owned by South32 (60%) and Anglo America PLC (40%), employing almost 1,000 people (South32 2021). The special-purpose town of Alyangula was created to support the mining operations in the late 1960s.

Nathan River Iron Ore Project

Situated on the lands of the Mara, Alawa, and Yangman peoples (Aboriginal Areas Protection Authority n.d), the Western Desert Resources iron ore mine was established and commenced operating in 2013, but was placed into care and maintenance when the original operator went into liquidation in 2014. Nathan River Resources, a subsidiary of British Marine Group (Britmar), took ownership in 2017 and recommenced production in 2020. It currently employs about 250 people (NS Energy n.d.).

McArthur River Mine

This zinc and lead mine is situated on the lands of the Binbinga Garawa, Gunindiri, Ngandji, Waanyi, and Wambaya peoples (Aboriginal Areas Protection Authority n.d.). It began as an underground operation in 1995 and converted to an open-pit mine in 2006. It is owned by Glencore and employs about 1,000 people (ICN Gateway n.d.).

Bootu Creek Mine

Bootu Creek Mine is located on the lands of the Jingili, Mudburra, and Walmanpa peoples (Aboriginal Areas Protection Authority n.d.). It produces manganese and commenced operations in 2005 as a joint venture between OM (Manganese) Ltd. and GEMCO. Since 2007, it has been wholly owned by OM (Manganese) Ltd. and employs approximately 150 people (OM [Manganese] 2020).

Newmont Tanami Operations

This gold mine is situated in the region of the Walpiri, Pintupi, and Anmatyerre peoples (Aboriginal Areas Protection Authority n.d.). It commenced production in 1983 and has been fully owned and operated by Newmont since 2002. It provides employment for approximately 1,160 workers (Newmont 2020).

We subsequently mapped the location of the LSMs using Google Maps and NT government maps to create Fig. 11.1, which depicts the geographical location of each of the LSMs. Next, surrounding communities (including homelands and outstations) were identified and mapped using the Australian Bureau of Statistics' Indigenous Geographic Structure and Indigenous Regions maps, as they consider language groups and local knowledge about communities (Australian Bureau of Statistics 2021a). The structure has three levels (from largest to smallest): Indigenous regions (IREGs), Indigenous areas (IAREs), and Indigenous locations (ILOCs).

As geography, topography, meteorology, infrastructure, and operational transportation modes in the surrounding region are considered highly diverse across different mines (Kemp and Parmenter 2007), the most meaningful approach was to assess, on a case-by-case basis, the likelihood that a resident living in a community

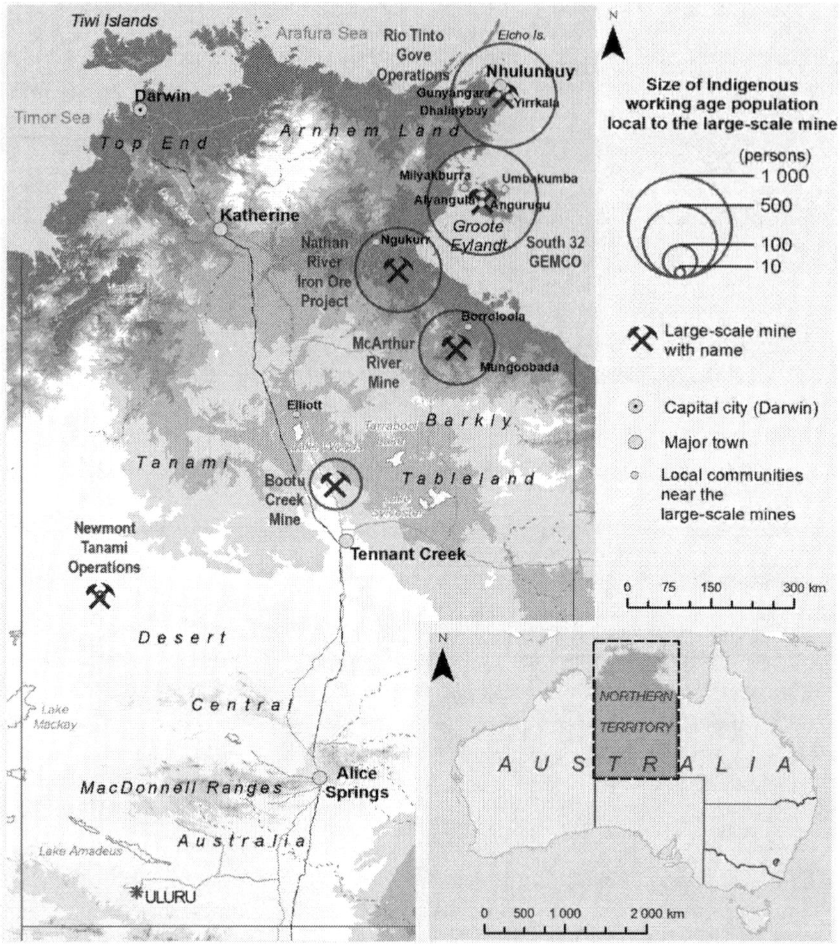

Figure 11.1 Location of the six LSMs in the NT and size of Indigenous working-age populations local to the LSM sites

Source: Fig. 11.1 is based on the authors' analysis of data extracted from the Australian Bureau of Statistics Table Builder, 2016 and 2011 census data

within reasonable proximity to a specific mine, and with adequate and accessible transport infrastructure, could attend the mine site daily for work. This is important in the context of the NT, where all LSMs are in areas classified by the Australian Bureau of Statistics as "very remote" and are often accessible only by small aircraft or unsealed roads that are subject to seasonal flooding. As a result, an arbitrary definition, such as "within a one-hundred-kilometer radius of the mine," would not be suitable. This method is consistent with that used in other studies; for example, see Queensland Local Content Leaders Network (2019).

Desktop research was then undertaken to identify any existing infrastructure and modes of transport available for surrounding community members to access the LSMs. For the communities where infrastructure and transport existed, we analyzed the adequacy of the infrastructure and transport modes to ascertain whether a realistic option existed for local community members to travel to and from the LSM daily to attend work. The communities that met these criteria were deemed local to the mine, and those that did not were omitted.

For example, in Yolngu Country (northeast Arnhem Land), the community of Dhalinybuy is approximately 75 km from Rio Tinto Gove Operations, with road access for community members to travel to the mine most days. Therefore, for this research Dhalinybuy is considered a local community and a potential source of local employees. However, Dhambalinya is approximately 60 km from the same mine and, while closer, is islanded, requiring either a charter flight or a private boat service to connect community members with the mainland, meaning they could not access the mine daily for work. Based on this, Dhambalinya was not considered a local community or a source of local labor for the mine and was omitted.

This process was repeated for each LSM in the NT and is presented in Table 11.1, which shows all communities deemed potential local labor sources and proximity to the LSMs. Fig. 11.1 provides a visual depiction of the same data, noting that some local communities are not pictured due to their small size and geographical location within the boundaries of other larger-sized communities already presented on the map. For example, Yugul Mangi is within Ngukurr; Mabunji, Mara, and Yanyula are inside Borroloola; and Julalikari is within Tennant Creek.

We then used the Australian Bureau of Statistics' Indigenous Areas and Indigenous Locations and Census dataset to establish whether, and to what extent, the LSMs in the NT employ people who live in Indigenous communities (with an emphasis on Indigenous women) considered as being in local proximity to each of the six LSMs.

There are two main types of census data used in the analysis—data based on where people said they usually lived at the time of the census (usual resident data) and data based on the geographical region where people said they worked (place of work data). The latter is applicable to people of working-age, between 15 and 69 years only.

Table 11.1 List of LSMs and surrounding communities deemed a potential local labor source

LSM	*Surrounding Communities Deemed a Potential Local Labor Source*
Rio Tinto Gove Operations	Dhalinybuy, Gumatj and surrounds, Gunyangara, Nhulunbuy, and Yirrkala
South32 GEMCO	Alyangula, Angurugu, Milyakburra, Umbakumba, and some outstations connected to these communities
Nathan River Iron Ore Project	Ngukurr and some of the Yugul Mangi outstations
McArthur River Mine	Borroloola, some of the Mabunji–Mungoobada outstations, Mara, and Yanyula
Bootu Creek Mine	Outstations from the Barkly Tablelands, Elliot surrounds, and Julalikari
Newmont Tanami Operations	Some of the Tanami outstations

It should be noted that population data from the census underreport numbers since 11% of the total population in the NT did not state a response to the question on whether they were of Aboriginal or Torres Strait Islander origin in the 2016 census. To compensate, we allocated the "not stated" responses to this question proportionally, based on the stated responses, a technique commonly used in analyzing Indigenous data from the census. In addition, the net undercount rate (the difference between those missed and those accounted for more than once) for Indigenous people in the NT was 21% in 2016 (Australian Bureau of Statistics 2016a), and around 4% of people in Australia of working-age did not state where they worked. The census is also a snapshot in time and may not reflect peaks in employment, which, in Indigenous communities, can fluctuate dramatically with seasonal changes or ceremonial activities (Regional Development Australia Northern Territory n.d.). These biases in census data cloud the true picture when examining data for small areas. Nevertheless, the data are sufficient to reveal overall patterns and trends in terms of the sourcing of workers for the LSMs as it stood in 2016 and 2011, and while there are some limitations, the results derived from the census provide the best data source available (Academy of the Social Sciences in Australia 2008; Centre for Aboriginal Economic Policy Research 2018).

Collectively, we calculated the potential local labor pool from communities surrounding LSMs to be 1,896 Indigenous women and 1,854 Indigenous men (Australian Bureau of Statistics 2016b, 2016c). The circles around each LSM in Fig. 11.1 demonstrate the size of the Indigenous working-age population considered local to that LSM, from which local employees could be sourced.

We were then able to create a profile of the actual workforce according to the census data to assess the extent of employment of local Indigenous people, Indigenous women, and Indigenous people more generally and to examine the geographic sourcing of the workforce at each mine.

Results

Indigenous people employed in large-scale mining in the NT

In 2016, large-scale mining employed 1,409 people who resided in the NT, up from 1,083 in 2011. This is significantly lower than the approximately 4,500 suggested in the public documents of the mining companies themselves, and the difference is down to workers sourced from outside the NT. Those in the NT represented less than 1% of all Australian mining industry employees in 2016. Emphasizing the very low numbers of Indigenous women employed in mining occupations in the NT, in 2016 fewer than 20 were employed at LSMs, constituting less than 1% of the workforce. This compares to approximately 100 Indigenous men (7% of the workforce), a ratio of one woman for every ten men. With a population in the NT that is one-third Indigenous, it is clear that Indigenous women employed at LSMs in the NT are highly underrepresented in the workforce. However, this also applies to both Indigenous peoples and women in general (as distinct groups), and not just Indigenous women, since, in 2016, only 8% of the NT's local LSM workforce were

Indigenous, and 10% of the non-Indigenous workforce were women. As such, 82% of the workforce in the six LSMs in the NT in 2016 were non-Indigenous men compared to 77% in 2011, indicating that the proportion of men in the workforce increased.

Given the small numbers of Indigenous women working at LSMs in the NT, a separate analysis of their age profile is not feasible. However, when combined with Indigenous men, the profile can be compared to non-Indigenous employees. This shows the Indigenous workforce to be much younger, with 30% under 30 years of age in 2016 compared to 15% for the non-Indigenous LSM workforce, and for those aged 50 years and over, the respective figures were inversed, at 28% in the older age group for non-Indigenous and 17% for Indigenous employees (Fig. 11.2). Nevertheless, there was a higher proportion (31%) of Indigenous employees in their 40s compared to non-Indigenous employees (8%). These differences reflect the younger age profile of the Indigenous population in the NT compared to the non-Indigenous population.

In terms of mainstream education levels for all employees of LSMs in the NT, it was most common to have completed either certificate-level or senior high school, which aligns with the requirements of the most common occupations held, that is, machine operators, drivers, technicians, and tradespeople. For Indigenous women employed at LSMs in the NT, half had Certificate III or IV as their highest education level, and half had year 10 or below secondary education. This is reflected in the data for the occupations of Indigenous women who were all either machinery operators or drivers, or were clerical or administrative workers. The LSM industry in the NT did not employ Indigenous women in any managerial, professional, or technical

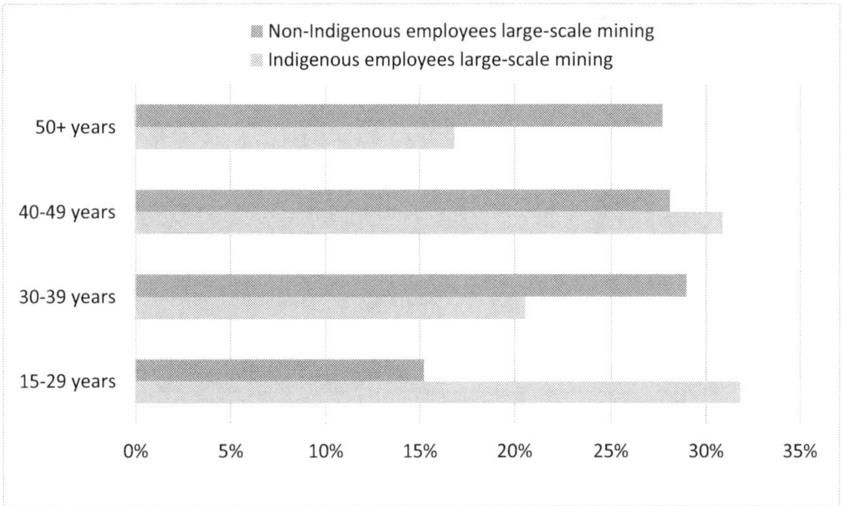

Figure 11.2 Age profile for LSM employees in the NT, 2016

Source: Extracted from Australian Bureau of Statistics Table Builder by the authors

positions (of which there were 738 in 2016 and 571 in 2011). Data on incomes are too sparse to separately analyze for Indigenous women at LSMs in the NT.

In Australia, mining employment continues to be made up of mostly full-time employees, despite a significant increase in part-time employment nationally over the last 40 years (National Skills Commission 2021). Most LSMs operate 24 hours a day, 7 days a week, and work rosters are generally based on two crews working back-to-back 12-hour shifts to keep the mine running continuously. Employees are required to work "time on/time off," for example, two weeks on, one week off; however, the number of days or weeks varies across different companies, locations, and roles (see South32 n.d.; Glencore 2023). Accordingly, we found that the LSM industry in the NT is also primarily a full-time workforce.

In 2016, part-time employees (anyone working fewer than 35 hours per week) only accounted for 6%, up from 4% in 2011; however, no Indigenous women were employed on a part-time basis in either year. In both 2016 and 2011, the group with the largest percentage of part-time employees at the NT's LSMs was Indigenous men (7%, down from 13%). In Australia, part-time workers make up 32% of the overall workforce, but it is more common for women to be employed part-time than men, with women constituting 68% of the national part-time workforce (National Skills Commission 2021). While the literature is unclear on why Indigenous men account for the largest percentage of part-time employees at LSMs across the NT during these years, it may be due to an increase in young Indigenous men undertaking mining traineeships, where part-time, paid, on-the-job training is combined with studying.

Turning to non-Indigenous women working part-time, there was an increase from 4% in 2011 to 6% in 2016. For part-time non-Indigenous men, there was an increase from 2% in 2011 to 5% in 2016. This indicates some improvement in more flexible employment for non-Indigenous employees during the five years between censuses.

Looking at the total combined LSM workforce in the NT, 81% were married (either de facto or registered) in 2016, a decrease of about 5% since 2011. However, of the small number of Indigenous women employed by LSMs in the NT in 2016 and 2011, none reported being married. At the other extreme, for non-Indigenous men, who make up most of the workforce, over 70% were married. Non-Indigenous women share a similar profile to non-Indigenous men, with 62% married, while 59% of employees who were Indigenous men were married, an increase from 29% in 2011.

Very few workers in the NT's LSM industry had children (fewer than 3%). Three-quarters of those who did were non-Indigenous men (76% in 2016 and 89% in 2011), while no Indigenous women working at LSMs in 2016 or 2011 had children (Fig. 11.3).

For the total LSM workforce in the NT, about 30% of all employees had cared for children in the 2 weeks leading up to the census. In 2016, Indigenous men (36%), followed closely by non-Indigenous men (35%) and Indigenous women (31%), reported providing some amount of care for a child. This was different from 2011, where 54% of employees who were Indigenous women provided care, followed by employees who were Indigenous men (40%), both higher than the following census.

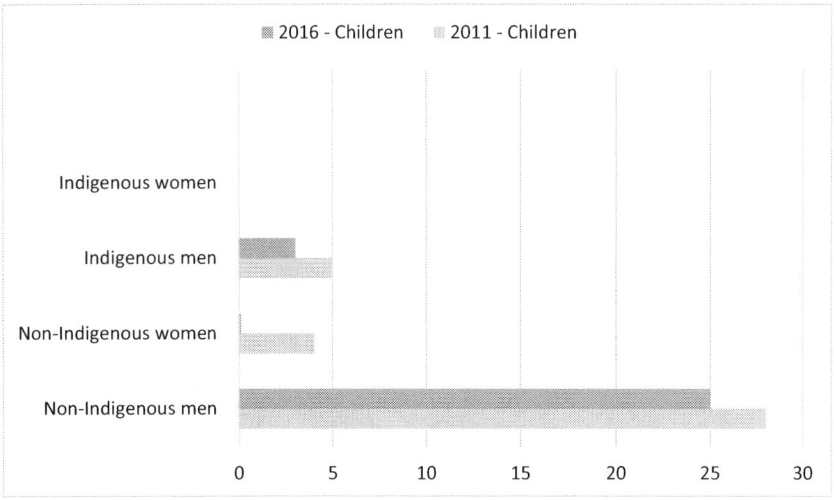

Figure 11.3 LSM employees in the NT and number of children

Source: Extracted from Australian Bureau of Statistics Table Builder by the authors

Care provided by non-Indigenous men was previously lower (29%), but stable for non-Indigenous women.

A smaller portion (around 5%) of the total LSM workforce in the NT had provided some amount of unpaid care, assistance, or help to someone with a disability, long-term health condition, or age-related health issue. Of Indigenous women employed at LSMs in the NT in 2016 and 2011, none reported providing this type of care. Of the small number of LSM employees who reported having provided unpaid care, assistance, or help, the large portion of non-Indigenous men who provided this type of care (85% in 2016, up from 79% in 2011) stood out. This is much higher than the portion of Indigenous men (less than 10% for both years) and non-Indigenous women (6% and 15%, respectively).

Overall, these results provide a profile of the primary characteristics of the NT's LSM workforce as being predominantly non-Indigenous, mainly men, certificate-level or senior high school educated, and employed full-time mostly as machinery operators, drivers, technicians, or tradespeople. Most of the workforce was married (de facto or registered), but only a small percentage had children or provided unpaid care to someone with a disability, chronic illness, or age-related issue.

Geographical sourcing of workers by LSMs in the NT

About 26% of workers at LSMs in the NT live in communities we considered as local to the six mines. However, there were no Indigenous women in this position in 2016, and only 20 or so Indigenous men. For non-Indigenous LSM employees, 43% of women and 38% of men lived local to the mine in 2016, although mostly in purpose-built accommodation for employees of the mine. It is reasonable to conclude from the data that LSMs in the NT employ a very small segment of the working-age Indigenous population who are resident anywhere in the NT and an even smaller segment of Indigenous people living locally to the six LSMs.

Despite these summary results, all 17 Indigenous women employees of LSMs in the NT resided somewhere else in the NT itself, as opposed to other jurisdictions in Australia. All but five lived in either Darwin or the NT's second largest city, Alice Springs. Similarly, 80% of Indigenous men employed by LSMs resided in the NT rather than in other states or territories, but a much higher proportion of non-Indigenous women (41%) and men (52%) were sourced from jurisdictions outside the NT. The Indigenous workforce at LSMs can therefore be described as small, unlikely to live local to the mines, but with a good portion living elsewhere in the NT. From this, and in line with other studies, it can be reasonably hypothesized that many Indigenous LSM workers were not likely to reside in the NT but migrated there for work or other reasons (Charles Darwin University Northern Institute 2019).

These data highlight that out of the 1,409 employees at the six LSMs in the NT in 2016, no Indigenous women and only a handful of Indigenous men (24, or just over 1% of the local potential workforce) were sourced from local communities. Furthermore, locally sourced, Indigenous men were only employed at two of the six mine sites, one of these being the special-purpose town of Nhulunbuy. This is despite 1,896 Indigenous women and 1,854 Indigenous men of working-age living local to LSMs in the NT at the time. For non-Indigenous locals, around 3% of women, and 21% of men who make up the local potential workforce were employed at LSMs. However, 81% of non-Indigenous local employees working at LSMs in the NT resided at Nhulunbuy or Alyangula—both special-purpose towns developed in the 1960s by the initial mine owners to specifically support mining operations and employees and their families who moved to the area to work at the mines. We can add that the picture was very similar in 2011, such that there has been no improvement in the 5 years to 2016 (Australian Bureau of Statistics 2011, 2016b).

Discussion

While Indigenous peoples and women (as distinct groups) are underrepresented in mining employment in Australia, research focusing solely on Indigenous women employees in the industry has been scarce to date. The research in this chapter is a step toward closing the gap in knowledge by focusing on the employment trends and demographics for Indigenous women working at LSMs in the NT. The results are intended to provide a basis for further research and discussion to improve what can be described as very minimal employment of local Indigenous women at LSMs in the NT and to increase awareness of the need for better understanding about why Indigenous women are underrepresented, so as to identify opportunities for change.

As well as overall underrepresentation, Indigenous women were unrepresented in the higher-skilled professional, managerial, and technical roles, emphasized by lower levels of educational qualifications compared to non-Indigenous workers.

Indigenous employees were younger, and Indigenous women were employed only on a full-time basis, tended to be unmarried, were without children, and were not providing unpaid care to someone with a disability, chronic illness, or age-related issue. In line with Parmenter's (2008, 2011) research, we found Indigenous women were not employed in supervisory roles but mainly semi-skilled positions (e.g., truck driving) or roles stereotypically gendered as women's such as administration work.

The LSM industry in the NT sourced a very small portion of its workforce from the working-age Indigenous population living in the NT, and an even smaller portion residing locally to the individual mine sites. LSMs in the NT did not source any Indigenous women from local communities neighboring the LSM sites to be part of its workforce in 2011 or 2016, despite these residents being the most impacted by LSM operations. Most workers were sourced from interstate or were residing in special-purpose towns in the NT specifically established to support a non-local mining workforce and their families. These findings confirm that proximity to the mine alone does not bolster employment opportunities for locals, and especially not for Indigenous residents proximate to mines. This challenges the rhetoric that mining creates employment for local Indigenous communities, which is often used to encourage, gain support for, and justify mining operations on Indigenous lands. It refutes the so-called benefit of employment for local remote Indigenous communities impacted by mining.

Our findings raise the important question of why Indigenous women, particularly Indigenous women from communities local to the mine, are not being employed by LSMs in the NT, especially considering the high cost and ongoing burden of travel expenses for mining companies sourcing non-local, long-distance, commuting workforces. A range of literature has identified an extensive array of barriers preventing Indigenous peoples and women (as separate groups) from seeking, obtaining, and remaining in employment in the mining industry. For Indigenous peoples, barriers can be summarized under three high-level themes: Cultural and spiritual (Altman 2009; Kwaymullina 2018), a lack of recognition and the deficit discourse (Walter and Suina 2019; Diversity Council of Australia and Jumbunna Institute 2020; Tulele 2020), and racial harassment and discrimination in the workplace (Caron *et al.* 2019; Parmenter and Barnes 2021; Delgado and Stefancic 2023). Literature on the barriers preventing increased employment of women in general in mining can be further categorized according to work environment, conditions, and opportunities (Pauktuutit Inuit Women of Canada and University of British Columbia 2016; Native Women's Association of Canada 2018; Kansake *et al.* 2021); inflexible working arrangements (McDonald *et al.* 2012; Mayes and Pini 2014); and sexual harassment and discrimination in the workplace (Bailey-Kruger 1995; Kljajevic 2015).

These barriers negatively impact both Indigenous peoples' and women's employment success and careers in mining, and it is well established that marginalization, discrimination, and harassment based on race and/or gender is unethical and illegal, causes psychological harm, and damages individual health and well-being (Ziersch *et al.* 2011; Paradies 2018). It is likely these barriers make the mining industry unattractive and unsafe for Indigenous women (who fall into both categories) as a place of employment (Parmenter 2008, 2011) and may be contributing to the acute underrepresentation of Indigenous women in the NT's LSM sector.

This underrepresentation is significant because many destructive impacts from mining are disproportionately felt by Indigenous women from local communities near to where mining takes place. Yet, as demonstrated here, employment as one of the main benefits of mining is enjoyed by predominantly non-Indigenous men who are not from these local communities. This is unjust, with social and economic implications for families and communities, as wages received by Indigenous women have been shown to be more often spent on things that benefit families and community (World Bank 2009; Australian Human Rights Commission 2020). The inequitable distribution of resources (and the lack of opportunity to access those resources) may create tension and conflict within communities, further adding to the impacts of mining for Indigenous women.

Strong governance by women regulating relationships to Country, family, community, culture, and spirituality has always existed and been central in Indigenous societies (Dudgeon and Bray 2019). Indigenous women living in the remote areas where mining operations take place are significant holders of knowledge, and many are Traditional Owners with custodial rights and cultural knowledge of their lands and environment, dating back thousands of generations. Yet, our research shows that they are the least likely group to be employed by LSM companies operating in the NT, despite their inherent and important knowledge about environment and sustainability sciences (Whyte 2018). Equally, Indigenous women, specifically, and Indigenous people, in general, appear to be poorly represented in negotiated agreements between Indigenous peoples and mining companies across Australia's major resource-producing regions. O'Faircheallaigh (2015) found that Indigenous communities have expressed a desire to use mining land-use agreements to increase their participation in a mining project's workforce, and Indigenous employment provisions are now commonly featured in land-use agreements entered into between mining companies and Indigenous groups residing near mines (Parmenter and Trigger 2018). Our findings may assist Indigenous landholders to better understand whether LSM companies are meeting the terms of their agreements.

The underrepresentation of Indigenous women employed in mining is not just a local problem. LSM companies are multinational, with operations spanning the world. Rio Tinto, for example, mines aluminum on Yolngu Country (northeast Arnhem Land) in the NT, as well as operating mines, smelters, and refineries in 35 different countries (Rio Tinto 2021b). While our findings highlight the position in the NT LSM industry, the global nature of large-scale mining (Heininen and Southcott 2010; Stevens *et al.* 2017; Suopajärvi *et al.* 2017) suggests these findings may be applied nationally and internationally to bring about positive change. Mining is not the only industry where Indigenous women are underrepresented or where non-Indigenous men make up the majority of the workforce. Therefore, the findings of our research may be applicable and helpful to other similar industries, such as oil and gas, construction, and transport.

However, census data are a snapshot of a point in time, and they do not explain causal relationships. While the literature provides plenty of evidence of the barriers

Indigenous employees and employees who are women have faced in the mining industry, further research is needed to understand why the large-scale mining industry employs minimal numbers of Indigenous women and does not source its workforce from communities local to its operations. Future research should focus on gathering stories about the experience of Indigenous women at LSMs in the NT, as Indigenous women who have worked—or are currently working—in mining are the experts and have the greatest knowledge about how mine work affects them, their families, and their communities. Through sharing stories, employment pathways can be highlighted and issues such as attraction, recruitment, promotion, and retention, as well as the role of an employee's proximity to the mine, may be better understood. Furthermore, improved understanding of race and gender issues through the voices of Indigenous women, including those working in mining in the NT, may uncover the means to reduce workplace barriers, while helping increase overall representation of Indigenous women in the sector through employment. This research is currently underway by the authors, and publication is forthcoming.

Conclusion

In this chapter, we examined and analyzed available census data to describe the demographic profile of Indigenous women working in the NT's LSM industry. We also analyzed the location of LSMs in the NT and surrounding infrastructure to identify sources for potential local workforces. Our findings demonstrated Indigenous women remain persistently underrepresented in the NT's LSM workforce, especially those from local communities neighboring the mines.

We know there are several barriers to increased Indigenous and women's employment and retention in the mining industry centering around culture, spirituality, race, gender, work environment and conditions, harassment, discrimination, and marginalization in the workplace. However, our understanding of how race and gender intersect, and the unique experience of being a "double minority" in the mining sector is limited. Further research focusing on the experiences and voices of Indigenous women employed at LSMs is crucial to comprehend the complex facets of multidimensional marginalization and why the NT's LSM industry employs Indigenous women in such small numbers.

Barriers to employment and discrimination of minority groups continue to exist in the mining industry, notwithstanding decades of policies and schemes aimed at correcting the problem. Continuing with employment policies omitting the intersectionality of race and gender means mining companies may be mistargeting efforts, and social and economic outcomes may fall short of potential. Low engagement also obviates access to much needed labor and inhibits the development of trust-based relationships with local communities.

If mining companies are to continue to rely on the promise of local Indigenous employment as enticement and justification for their large-scale projects on Indigenous lands, it is vital they accept this is not the reality for local Indigenous peoples, particularly local Indigenous women, who are disproportionately impacted by their operations. These are social justice issues that go to the heart of equitable recognition and inclusion of Indigenous peoples and women and, if not addressed, may erode the public's perception of a mining company's corporate social responsibility and hamper its social license to operate near remote Indigenous communities and on Indigenous lands.

Funding and acknowledgments

This research was funded by the Knowledge Network on Mining Encounters and Indigenous Sustainable Livelihoods (MinErAL) and an Australian Government Research Training Program Fee Offset Scholarship. All results and opinions are those of the authors.

References

Aboriginal Areas Protection Authority, n.d. Regions of sacred sites in the NT—Interactive Map. Available from: www.aapant.org.au/sacred-sites/regions-sacred-sites-nt [Accessed 15 February 2022].

Academy of the Social Sciences in Australia, 2008. *Lives of diversity: Indigenous Australia*. Canberra: Academy of the Social Sciences in Australia. https://bit.ly/3KPzbew

Altman, J., 2009. Contestations over development. *In:* J. Altman and D. Martin, eds., *Power, culture, economy: Indigenous Australians and mining*. Canberra: ANU Press, 1–16. https://bit.ly/48mSXIW

Australia Minerals, 2021. Mineral resources information. Available from: https://australiaminerals.gov.au/mineral-resources-information [Accessed 15 February 2022].

Australian Bureau of Statistics, 2011. Census TableBuilder. Available from: www.abs.gov.au/websitedbs/censushome.nsf/home/tablebuilder [Accessed 9 October 2021].

Australian Bureau of Statistics, 2016a. Technical note 2—the undercount in the census and the PES. Available from: www.abs.gov.au/methodologies/understanding-change-counts-aboriginal-and-torres-strait-islander-australians-census-methodology/2016#technical-note-2-the-undercount-in-the-census-and-the-pes [Accessed 10 October 2021].

Australian Bureau of Statistics, 2016b. Census TableBuilder. Available from: www.abs.gov.au/websitedbs/censushome.nsf/home/tablebuilder [Accessed 5 January 2021].

Australian Bureau of Statistics, 2016c. Map of the 2016 remoteness areas for Australia. Available from: www.abs.gov.au/statistics/standards/australian-statistical-geography-standard-asgs-edition-3/jul2021-jun2026/remoteness-structure [Accessed 22 September 2021].

Australian Bureau of Statistics, 2018a. Estimates of Aboriginal and Torres Strait Islander Australians. Available from: www.abs.gov.au/statistics/people/aboriginal-and-torres-strait-islander-peoples/estimates-aboriginal-and-torres-strait-islander-australians/latest-release [Accessed 22 September 2021].

Australian Bureau of Statistics, 2018b. 3218.0—Regional population growth, Australia, 2016–17. Available from: www.abs.gov.au/AUSSTATS/abs@.nsf/Previousproducts/3218.0Main%20Features702016-17 [Accessed 14 February 2022].

Australian Bureau of Statistics, 2021a. Indigenous Structure—Australian Statistical Geography Standard (ASGS) Edition 3. Available from: www.abs.gov.au/statistics/standards/australian-statistical-geography-standard-asgs-edition-3/jul2021-jun2026/indigenous-structure [Accessed 8 October 2021].

Australian Bureau of Statistics, 2024. National, state, and territory population. Available from: www.abs.gov.au/statistics/people/population/national-state-and-territory-population/latest-release [Accessed 3 July 2024].

Australian Government Office for Women, 2007. *Unearthing new resources: Attracting and retaining women in the Australian minerals industry*. Australian Government Office for Women. https://bit.ly/3htZM3x

Australian Government Productivity Commission, 2020a. *Overcoming Indigenous disadvantage: Key indicators 2020*. Australian Government. https://bit.ly/3uLtG7h

Australian Government Productivity Commission, 2020b. *Resources sector regulation*. Australian Government. https://bit.ly/3uLanLq

Australian Human Rights Commission, 2020. *Wiyi yani u thangani (women's voices): Securing our rights, securing our future—Community guide*. Australian Government. https://bit.ly/3KNBHBR

Bailey-Kruger, A., 1995. *The psychological wellbeing of women operating mining machinery in a fly-in fly-out capacity*. Master's Thesis. Edith Cowan University. https://ro.ecu.edu.au/theses/1682/

Blackham, A., and Temple, J., 2020. Intersectional discrimination in Australia: An empirical critique of the legal framework. *UNSWLJ*, 43, 773.

Blackwell, B. D., and Dollery, B., 2014. The impact of mining expenditure on remote communities in Australia: The Ranger uranium mine and the Tanami gold mine in the Northern Territory. *Australasian Journal of Regional Studies*, 20 (1), 68.

Blake, P., Denny, G., Duncan, H., Gianarakis, H., Swanston, C., Wadley, D., and Wilson, T., 2014. Charting the potential for Indigenous employment in the Australian mining industry. *Journal of Australian Indigenous Issues*, 17 (3), 21.

Bohanna, I., and Clough, A., 2012. Cannabis use in Cape York Indigenous communities: High prevalence, mental health impacts and the desire to quit. *Drug Alcohol Review*, 31 (4), 580–584.

Brasche, I., 2015. The thin black line: Living apartheid on Groote Eylandt. *Social Alternatives*, 34 (1), 19.

Brereton, D., and Parmenter, J., 2008. Indigenous employment in the Australian mining industry. *Journal of Energy and Natural Resources Law*, 26, 66–90.

Bryant, L., and Tedmanson, D., 2005. Drilling down: Diversity in the mining industry. *International Journal of Knowledge, Culture and Change Management*, 5 (3), 157–168.

Carment, D., 2003. Presenting mining's past in the Northern Territory. *Journal of Australasian Mining History*, 1 (1), 24–30.

Caron, J., Asselin, H., and Beaudoin, J. M., 2019. Attitudes and behaviours of mining sector employers towards the Indigenous workforce. *Resources Policy*, 61, 108–117.

Carrington, K., McIntosh, A., and Scott, J., 2010. Globalization, frontier masculinities and violence: Booze, blokes, and brawls. *The British Journal of Criminology*, 50 (3), 393–413.

Centre for Aboriginal Economic Policy Research (CAEPR), 2013. *My country, mine country: Indigenous people, mining and development contestation in remote Australia*. Australian National University, CAEPR No. 33 Research Monograph. https://bit.ly/3gSohWt

Centre for Aboriginal Economic Policy Research (CAEPR), 2016. *The ins and outs of the labour market: Employment and labour force transitions for Indigenous and non-Indigenous Australians*. Australian National University, CAEPR Working Paper No. 104/2016. https://bit.ly/3sFmaeq

Centre for Aboriginal Economic Policy Research (CAEPR), 2018. *Indigenous population change in the 2016 census*. Australian National University, CAEPR Census Paper 1/2018. https://bit.ly/34KyAcN

Centre for Social Responsibility in Mining, Sustainable Minerals Institute, 2007. *Indigenous employment in the Australian minerals industry*. Brisbane: University of Queensland. https://bit.ly/3UP9EXa

Centre for Social Responsibility in Mining, Sustainable Minerals Institute, 2018. *Indigenous employment futures in an automated mining industry: An issues paper and a case for research*. Brisbane: The University of Queensland. https://bit.ly/43HPbrD

Charles Darwin University Northern Institute, 2015. *Who cares?—Profiling careers in the Northern Territory*. Darwin: Charles Darwin University Northern Institute, Research Brief Issue 3. https://bit.ly/46hyNyw

Charles Darwin University Northern Institute, 2019. *Northern Territory contemporary Indigenous migration trends*. Darwin: Charles Darwin University Northern Institute, Research Brief Issue 2. https://bit.ly/48ja8va

Cox, D., and Mills, S., 2015. Gendering environmental assessment: Women's participation and employment outcomes at Voisey's Bay. *Arctic*, 68 (2), 246–260.

Delgado, R., and Stefancic, J., 2023. *Critical race theory: An introduction*, vol. 87. New York University Press. https://bit.ly/43MVt9a

Denniss, R., 2020. Stop believing in fairy tales: Australia's coal industry doesn't employ many people or pay its fair share of tax. *The Guardian*, 23 December. https://bit.ly/2QfZfXJ

Deonandan, R., Deonandan, K., and Field, B., 2016. *Mining the gap: Aboriginal women and the mining industry*. University of Ottawa. https://bit.ly/3hpK7m0

Department of Jobs and Small Business, 2019. *Australian Jobs 2019*. Australian Government. https://apo.org.au/node/239966

Department of Treasury and Finance, 2020. *2019–2020 Treasurer's annual financial report*. Darwin, NT: Northern Territory Government. https://bit.ly/3KPlCvC

Diversity Council Australia and Jumbunna Institute, 2020. *Gari yala (speak the truth): Centering the work experiences of Aboriginal and/or Torres Strait Islander Australians*. Diversity Council Australia and Jumbunna Institute. https://bit.ly/3brvuut

Dudgeon, P., and Bray, A., 2019. Indigenous relationality: Women, kinship, and the law. *Genealogy*, 3 (2), 23.

Dyrting, S., Taylor, A., and Shalley, F., 2020. A life-stage approach for understanding population retention in sparsely populated areas. *Journal of Rural Studies*, 80, 439–451.

Gibson, G., and Kemp, D., 2017. Corporate engagement with Indigenous women in the minerals industry: Making space for theory. *In*: C. O'Faircheallaigh and S. Ali, eds., *Earth matters: Indigenous peoples, the extractive industries and corporate social responsibility*. Routledge, 104–122. https://bit.ly/3mN1Ys2

Glencore, 2023. Glencore careers structural engineer. Available from: https://glencorejobs.nga.net.au [Accessed 8 February 2023].

Heininen, L., and Southcott, C., 2010. *Globalization and the Circumpolar North*. Fairbanks, AK: University of Alaska Press. https://bit.ly/41kPyqm

Helbert, M., 2018. Australian women in mining: Still a harsh reality. *In:* L. Stevens, P. Tait, and D. Varne, eds., *Feminist ecologies: Changing the environments in the Anthropocene*. New York: Palgrave Macmillan, 231–246. https://bit.ly/41m584T

Hunter, B., Howlett, M., and Gray, M., 2015. The economic impact of the mining boom on Indigenous and non-Indigenous Australians. *Asia and the Pacific Policy Studies*, 2 (3), 517–530.

ICN Gateway, n.d. McArthur River Mine. Available from: https://gateway.icn.org.au/project/4569/mcarthur-river-mine [Accessed 9 October 2021].

Jenkins, K., 2014. Women, mining, and development: An emerging research agenda. *The Extractive Industries and Society*, 1 (2), 329–339.

Kansake, B. A., Sakyi-Addo, G. B., and Dumakor-Dupey, N. K., 2021. Creating a gender-inclusive mining industry: Uncovering the challenges of female mining stakeholders. *Resources Policy*, 70, 101962.

Kemp, D., and Parmenter, J., 2007. Indigenous women and mining employment in Australia. *Minerals Council of Australia SD07: A Climate for Change*, 29 October–2 November. Cairns: Minerals Council of Australia. https://bit.ly/3LxZeWY

Kerins, S., 2018. Our babies aren't safe, we got lead in our drinking water. *Land Rights News*, August, p. 16. Available from: www.nlc.org.au/uploads/pdfs/LRN-August-2018_web.pdf

Kljajevic, B., 2015. *An investigation into the underrepresentation of women in the Pilbara mining region of Western Australia*. Master's Thesis. Curtin University. https://espace.curtin.edu.au/handle/20.500.11937/862

Kwaymullina, A., 2018. You are on Indigenous land: Ecofeminism, Indigenous peoples, and land justice. *In:* L. Stevens, P. Tait, and D. Varne, eds., *Feminist ecologies: Changing the environments in the Anthropocene*. New York: Palgrave Macmillan, 193–208. https://bit.ly/3LbYwR5

LaBelle, S., 2015. *Aboriginal women, mining negotiations, and project development: Analyzing the motivations and priorities shaping leadership and participation*. Master's Thesis. University of Manitoba. https://mspace.lib.umanitoba.ca/handle/1993/30351

Lahiri-Dutt, K., 2012. Digging women: Towards a new agenda for feminist critiques of mining. *Gender, Place and Culture*, 19 (2), 193–212.

Langton, M., 2013. *Boyer lectures 2012. The quiet revolution: Indigenous people and the resources boom*. Sydney, New South Wales: HarperCollins.

Laurence, D., 2021. The devolution of the social licence to operate in the Australian mining industry. *The Extractive Industries and Society*, 8 (2).

Lord, L., Jefferson, T., and Eastham, J., 2012. Women's participation in mining: What can we learn from EOWA reports? *Australian Bulletin of Labour*, 38 (1), 68–95.

Lozeva, S., and Marinova, D., 2010. Negotiating gender: Experience from Western Australian mining industry. *Journal of Economic and Social Policy*, 13 (2), 7.

Marsh, J., 2013. Decolonising the interface between Indigenous peoples and mining companies in Australia: Making space for cultural heritage sites. *Asia Pacific Viewpoint*, 54 (2), 171–184.

Mayes, R., and Pini, B., 2014. The Australian mining industry and the ideal mining woman: Mobilizing a public business case for gender equality. *Journal of Industrial Relations*, 56 (4), 527–546.

McDonald, P., Mayes, R., and Pini, B., 2012. Mining work, family, and community: A spatially oriented approach to the impact of the Ravensthorpe nickel mine closure in remote Australia. *Journal of Industrial Relations*, 54 (1), 22–40.

Minerals Council of Australia, 2020. *MCA annual report 2020*. Minerals Council of Australia. https://bit.ly/3MYrdm2

Minerals Council of Australia Northern Territory Division, 2020. *Northern Territory mining: Minerals industry strategy 2020–2024*. Minerals Council of Australia. https://bit.ly/3oO3Ka0

Moffat, K., Lacey, J., Zhang, A., and Leipold, S., 2016. The social licence to operate: A critical review. *Forestry*, 89, 477–488.

Muhunthan, J., Angell, B., Hackett, M., Wilson, A., Latimer, J., Eades, A. M., and Jan, S., 2017. Global systematic review of Indigenous community-led legal interventions to control alcohol. *BMJ Open*, 7 (3).

National Skills Commission, 2021. *State of Australia's Skills 2021: Now and into the future*. Australian Government. http://bit.ly/3YBEKCf

Native Women's Association of Canada, 2018. *Indigenous gender-based analysis for informing the Canadian minerals and metals plan*. Native Women's Association of Canada. https://bit.ly/3mLHv6R

Newmont, 2020. *2019 Economic Impact Report—Newmont Australia—Tanami*. Newmont. https://bit.ly/3LzmpQz

Nhulunbuy Corporation, 2021. Job opportunities. Available from: https://ncl.net.au/node/452 [Accessed 8 October 2021].

Northern Land Council, 2022. Mining exploration and production. Available from: www.nlc.org.au/building-the-bush/mining-and-minerals [Accessed 22 February 2022].

NS Energy, n.d. Nathan river iron ore project Available from: www.nsenergybusiness.com/projects/nathan-river-iron-ore-project/ [Accessed 15 February 2022].

O'Faircheallaigh, C., 2011. Indigenous women and mining agreement negotiations: Australia and Canada. *In:* K. Lahiri-Dutt, ed., *Gendering the field: Towards sustainable livelihoods for mining communities*. Canberra: ANU Press, 87–110. *https://bit.ly/40kNNrV*

O'Faircheallaigh, C., 2015. *Negotiations in the Indigenous world: Aboriginal peoples and the extractive industry in Australia and Canada*. Routledge. https://bit.ly/40AVU3T

O'Faircheallaigh, C., 2018. Mining, development, and Indigenous peoples. *Mining and Sustainable Development*, 1, 124–142.

O'Faircheallaigh, C., and Ali, S., eds., 2017. *Earth matters: Indigenous peoples, the extractive industries and corporate social responsibility*. Routledge. https://bit.ly/41kxPzn

OM (Manganese) Ltd., 2020. *Bootu Creek Manganese Mine—Mining management plan 2020–2030*. Northern Territory Government Department of Industry, Tourism and Trade. https://bit.ly/3BpiMrR

Paradies, Y., 2018. Racism and Indigenous health. *In: Oxford research encyclopedia of global public health*. Oxford: Oxford University Press. https://bit.ly/41gwQ34

Paredes, D., and Fleming-Muñoz, D., 2021. Automation and robotics in mining: Jobs, income, and inequality implications. *The Extractive Industries and Society*, 8 (1), 189–193.

Parmenter, J., 2008. Considering the experience of Indigenous women working in the Australian mining industry. *Gender and Mining Conference*, November 2008. Canberra: Australian National University.

Parmenter, J., 2011. Experiences of Indigenous women in the Australian mining industry. *In:* K. Lahiri-Dutt, ed., *Gendering the field: Towards sustainable livelihoods for mining communities*. Canberra: ANU Press, 67–85. https://bit.ly/40lPntm

Parmenter, J., and Barnes, R., 2021. Factors supporting Indigenous employee retention in the Australian mining industry: A case study of the Pilbara region. *The Extractive Industries and Society*, 8 (1), 423–433.

Parmenter, J., and Drummond, F., 2022. 'What did I get myself into?' Indigenous women and mining employment in Australia. *The Extractive Industries and Society*, 12, 101189.

Parmenter, J., and Trigger, D., 2018. Aboriginal cultural awareness training for mine employees: Good intentions, complicated outcomes. *The Extractive Industries and Society*, 5 (2), 363–370.

Pauktuutit Inuit Women of Canada and University of British Columbia, 2016. *The impact of resource extraction on Inuit women and families in Qamani'tuaq, Nunavut Territory: A report for the Canadian Women's Foundation*. Ottawa: Pauktuutit Inuit Women of Canada and University of British Columbia. https://bit.ly/3hoN6ed

Pearson, C., and Daff, S., 2013a. Indigenous workforce participation at a mining operation in Northern Australia. *Australian Bulletin of Labour*, 39 (1), 42–63.

Pearson, C., and Daff, S., 2013b. Transcending hunter gatherer pursuits while balancing customary cultural ideals with market forces of advanced western societies: Extending the traditional boundaries of Indigenous Yolngu people of the Northern Territory of Australia. *International Journal of Cultural Studies*, 16 (2), 189–208.

Perks, R., and Schulz, K., 2020. Gender in oil, gas, and mining: An overview of the global state-of-play. *The Extractive Industries and Society*, 7 (2), 380–388.

Pugliese, F., 2021. Mining companies and gender(ed) policies: The women of the Congolese Copperbelt, past and present. *The Extractive Industries and Society*, 8 (3).

Queensland Local Content Leaders Network (QLCLN), 2019. *Queensland local procurement model position paper: Defining 'local'—Final report*. Brisbane: Queensland Local Content Leaders Network (QLCLN). https://bit.ly/3HQnAsF

Regional Development Australia Northern Territory, n.d. East Arnhem—LGA—Nhulunbuy—Alyangula. Available from: https://profile.id.com.au/rda-northern-territory/about?WebID=160 [Accessed 8 October 2021].

Reserve Bank of Australia, n.d. *Australia and the global economy—the terms of the trade boom*. Australian Government. Available from: www.rba.gov.au/education/resources/explainers/australia-and-the-global-economy.html [Accessed 8 October 2021].

Reserve Bank of Australia, 2014. *The effect of the mining boom on the Australian economy*. Australian Government. https://bit.ly/3bnSJW8

Resources and Sustainable Development in the Arctic Network (ReSDA), 2014. *Gap analysis: Mining development in Canada*. Resources and Sustainable Development in the Arctic Network (ReSDA), Gap Analysis Report No. 2B. https://bit.ly/3rrZD85

Resourcing the Territory, 2021. *Operating mines*. Northern Territory Government. Available from: https://resourcingtheterritory.nt.gov.au/minerals/mines-and-projects/operational-mines [Accessed 8 October 2021].

Rio Tinto, 2021a. Gove. Available from: www.riotinto.com/en/operations/australia/gove [Accessed 8 October 2021].

Rio Tinto, 2021b. Operations. Available from: www.riotinto.com/operations [Accessed 8 October 2021].

Saxinger, G. *et al*., 2016. Boom back or blow back? Growth strategies in mono-industrial resource towns—'east' and 'west'. *In*: A. Taylor *et al*., eds., *Settlements at the edge*. Cheltenham, UK: Edward Elgar, 49–74. Available from: https://bit.ly/3UL2P9b

Sincovich, A., Gregory, T., Wilson, A., and Brinkman, S., 2018. The social impacts of mining on local communities in Australia. *Rural Society*, 27 (1), 18–34.

Southcott, C., Abele, F., Natcher, D., and Parlee, B., 2018. Beyond the Berger Inquiry: Can extractive resource development help the sustainability of Canada's Arctic communities? *Extractive Resource Development*, 71 (4).

South32, n.d. South32 Group Operations Pty Ltd. Available from: https://careers.south32.net [Accessed 8 February 2023].

South32, 2021. GEMCO. Available from: www.south32.net/what-we-do/our-locations/australia/gemco [Accessed 9 October 2021].

Stevens, L., Tait, P., and Varney, D. eds., 2017. *Feminist ecologies: Changing environments in the Anthropocene*. Springer. https://bit.ly/3GXAEOK

Suopajärvi, L., Ejdemo, T., Klyuchnikova, E., Korchak, E., Nygaard, V., and Poelzer, G., 2017. Social impacts of the 'glocal' mining business: Case studies from Northern Europe. *Mineral Economics*, 30 (1), 31–39.

Taylor, J., 2009. Data mining: Indigenous peoples, applied demography and the resource extraction industry. *In:* J. Altman and D. Martin, eds., *Power, culture, economy: Indigenous Australians and mining*. Canberra: ANU Press, 51–72. https://bit.ly/3Rr1Uv2

Tulele, L., 2020. *Employer attitude/behaviour matters: Impact of employer attitude/behaviour on Indigenous employees' skill acquisition and employment experience in the Australian mining and finance/banking sectors*. PhD Thesis. Griffith University. https://research-repository.griffith.edu.au/handle/10072/390784

Walter, M., and Suina, M., 2019. Indigenous data, Indigenous methodologies, and Indigenous data sovereignty. *International Journal of Social Research Methodology*, 22 (3), 233–243.

Whyte, K., 2018. What do Indigenous knowledges do for Indigenous peoples? *In:* M. Nelson and D. Shilling, eds., *Keepers of the green world: Traditional ecological knowledge and sustainability*. London: Cambridge University Press. https://bit.ly/41jwpVN

World Bank, 2009. *Gender dimensions of the extractive industries: Mining for equity* (Working Paper 51114). Washington, DC: World Bank. https://bit.ly/3eKCnsZ

Zhang, A., and Moffat, K., 2015. A balancing act: The role of benefits, impacts and confidence in governance in predicting acceptance of mining in Australia. *Resources Policy*, 44, 25–34.

Ziersch, A., Gallaher, G., Baum, F., and Bentley, M., 2011. Responding to racism: Insights on how racism can damage health from an urban study of Australian Aboriginal people. *Social Science and Medicine*, 73 (7), 1045–1053.

12 Rhetoric versus reality

Understanding employment inequities for Inuit women in mining

Katie Mazer, Justine Becker, and Suzanne Mills

Introduction

In recent decades, the underrepresentation of Indigenous women in mining has attracted increasing attention from researchers and Indigenous organizations (Tallichet 2000; Parmenter 2011; Lahiri-Dutt 2012; Jenkins 2014; Cox and Mills 2015; Pauktuutit 2021). In the context of impact and benefit agreement (IBA) negotiations, which position employment as a central component of benefit sharing, women's limited participation comes into focus as politically problematic, especially given the negative gendered impacts of mining on communities (NAHO 2008; Kudloo *et al.* 2016; Nightingale *et al.* 2017; Hall 2022). The barriers to Indigenous women's employment are both cultural and structural, including gendered stereotypes, fly-in–fly-out (FIFO) employment structures, gender-based harassment, and the increasing skill requirements for mining jobs paired with women's lack of access to training (Tallichet 2000; Mills 2006; Mercier 2011; Rodon and Lévesque 2015; Hodgkins 2018; Hall 2022). Motivated by these concerns, as well as by a desire to increase Inuit employment to meet IBA commitments, mining companies in Nunavik, Canada, have recently adopted a gendered lens in their hiring efforts. The website of Glencore Canada, which operates Raglan Mine in Nunavik, states that "We strive to attract and retain traditionally underrepresented groups, such as women" (Glencore Canada 2022). In 2014, the Kautaapikkut Roundtable, a partnership of regional Inuit organizations and mining companies, was created to find ways to increase Inuit employment in the Nunavik mines (Rogers 2015). Rob Nixon, chair of the roundtable, said in 2015 that "We're already seeing a commitment by these mines [because] hiring women is a key priority" (as quoted in Rogers 2015).

As evidence that there has been progress in closing the gender gap, mining companies have pointed to success stories of Inuit women in mining and increased employment rates. Raglan Mine advertised in a 2020 employee newsletter that "There are many Inuit women at Raglan Mine! They hold a wide variety of positions in all sectors of the organization" (Raglan Mine 2020). These claims are often difficult to verify because of the absence of high-quality employment data about Inuit women in mining. Much of the research on mining employment treats women as a homogenized group and rarely attends to distribution across job types

DOI: 10.4324/9781003406433-17
This chapter has been made available under a CC-BY-NC-ND 4.0 license.

within the industry (Jenkins 2014; Brain 2017; Manning *et al.* 2018; Lutz-Ley and Buechler 2020). Additionally, the companies' increased attention to hiring women does not guarantee improved employment outcomes. Research indicates that even in cases where companies have codified priority hiring of Inuit women in their IBAs, this has sometimes failed to translate into significant improvements in their employment rates (Cox and Mills 2015).

At the same time, while much literature has focused on the marginalization of Indigenous women in mining work (Cox and Mills 2015; Hall 2017) or the negative impacts of mining on women in Indigenous communities (Kudloo *et al.* 2014; Nightingale *et al.* 2017), relatively little has attended to the complexities of Indigenous women's encounters with mining work and the empowering potential they can have (Pauktuutit 2021; Sinclair 2021). Thus, there is a need for more attention to Indigenous women's first-person accounts of their experiences in mining in order to understand their needs and priorities in employment and the ways that mining is succeeding or failing with regard to meeting these needs.

In this chapter, we explore whether this greater attention to Inuit women's employment has actually resulted in better access to better quality jobs for Inuit women. We begin by presenting employment data that indicate persistent, poor employment outcomes for Inuit women. In order to understand the barriers that lead to these poor employment outcomes and attend to the lived experiences of Inuit women, we subsequently outline the results of qualitative interviews with Inuit women and key informants from the mining industry. We explore the complexities of Inuit women's encounters with mining work and consider how company policies and broader structural inequalities interact to produce disparities in mining employment and broader constraints on Inuit women's well-being.

Background

Nunavik is a region in Northern Québec with a population of 13,000 residents, of whom 90% are Inuit (Statistics Canada 2018). Some 98% of Inuit from Nunavik speak Inuktitut as their first language (Inuit Tapiriit Kanatami 2018), while 82% understand English, and just 29% have knowledge of French (Statistics Canada 2018). Owing to a long history of colonial policies, the region has disproportionately high rates of poverty and unemployment. The median individual income for Nunavik Inuit reported in 2018 was CA$25,627, while the median income for non-Indigenous people in Nunavik was CA$79,328 (Inuit Tapiriit Kanatami 2018). Inuit also have lower rates of high-school graduation and post-secondary education than the provincial average, and despite Nunavik having similar labor market participation to the rest of Québec, Inuit employment outcomes in Nunavik are comparatively poor (Lévesque and Duhaime 2021). The employment rate among Inuit adults aged 25–64 years in Nunavik is 12% lower than the provincial average, and for young people in Nunavik aged 25–34 years, the unemployment rate is three times higher than the provincial average (Lévesque and Duhaime 2021).

There are two nickel mines in Nunavik that operate on a FIFO basis: Raglan Mine, owned by Glencore Canada, which has been in operation since 1997, and the

Nunavik Nickel Mine, owned by Canadian Royalties Inc., which began production in 2013. The mines are located just 24 km apart, 100 km and 120 km, respectively, south of Deception Bay, and almost 100 km from the closest community. Because there is no integrated road network connecting them to local communities, the mines are accessible only by plane, via a shared airstrip. These spatial and infrastructural realities have normalized the use of FIFO at both mines, with a standard rotation of 3 weeks at work and 3 weeks at home.

At the time of research, Raglan Mine employed a workforce of about 950 employees, while the Nunavik Nickel Mine directly employed about 419 people. Employees at the Raglan and Canadian Royalties mines are unionized with the United Steelworkers. Both mines are also signatories of IBAs between the companies and several Inuit governance institutions. The IBAs commit the mines to prioritizing the hiring and promotion of Inuit beneficiaries and stipulate that dedicated training and advancement programs for beneficiaries will be provided. The IBAs include language provisions, mandating that lack of language skills (generally, French) will not be a barrier to work where language is not required for the position and that companies make a reasonable effort to ensure supervisors are bilingual (English and French). The agreements also include cultural provisions such as Inuit kitchens on site for beneficiaries to prepare traditional foods and allowances for cultural leaves of absence. Neither agreement includes any explicit provisions related to gender. For reasons of anonymity, we refer to the Nunavik mines in the remainder of the article as Mine 1 and Mine 2.

Methods

We examined employment dynamics at these two nickel mines in Nunavik, Canada, drawing on employment data and qualitative interviews with Inuit women workers and key informants from the mining industry and other relevant organizations. The employment data include information on gender, Indigeneity, job title, seniority, and department. Data from Mine 1 were collected from the United Steelworkers Union, while data from Mine 2 were received directly from the company. The data represent employment in February 2018 for Mine 1 and June 2020 for Mine 2. We conducted interviews with ten Inuit women who either currently or had previously worked in one or both of the mines. Nine of the women were born in Nunavik, but many had since left and were living in Southern Québec when the interviews were conducted. The semi-structured interviews included questions about their employment pathways, experiences in mining, decisions to stay or leave mining, and overall perceptions of mining employment. We also conducted interviews with key informants, including three past and present mine employees working in Inuit recruitment and training, the mining development manager at Makivik Corporation (the Inuit land claim organization), and an employee at an NGO that works with Inuit women. This research is part of a broader examination of Inuit women's access to and experiences of nickel mining employment (see Mazer *et al.* 2022; Mills *et al.* 2023). The research was done in partnership with Pauktuutit and Makivik Corporation and was conducted as part of the Knowledge Network on Mining Encounters and Indigenous Sustainable Livelihoods (MinErAL), an international research network focused on Indigenous livelihoods and extractive industries.

Results

Industry perspectives

Echoing employers' narratives that report improvements in Inuit women's representation in mining, key informants with ties to the mining industry tended to recount success stories. One key informant who had worked in Inuit recruitment at Mine 2 described the success of a number of initiatives to recruit more Inuit women:

> For a long time, we only had one Inuit woman, and then we publicized, we focused on her a lot when we did presentations to students, and when we went to the community for recruitment. And now we're at five Inuit women working underground.
>
> (6)

The mining development manager at Makivik Corporation shared this sentiment: "We have good, nice success stories, as well, at the two mines. They're not only at the kitchen or janitors, but working directly on mine operation, driving heavy equipment, driving heavy equipment trucks, and so on" (2). Often, discussions of progress were based on anecdotes about individual women's successes. When describing the successes of the industry in bringing more Inuit women into traditionally masculine roles, an employee at Mine 2 in Inuit recruitment said the following:

> But the difference is, right now is, we have more Inuit women occupying more other jobs . . . Before 2013 it was a men's—Inuk male that had that position, and now it's the Inuk female. We also had a security agent a few years back that was an Inuk woman, but switched to underground miner. And we have . . . Inuit women working in the warehouse, the tool crib. We have an Inuk nurse, a woman. So, we have more Inuit women occupying different positions like that.
>
> (6)

When interviewees made reference to demographic trends, they generally did not attend to differences in job type, focusing instead on absolute numbers of Inuit or women employed. The mining development manager at Makivik Corporation noted that "When those IBAs were signed there was an informal target of 20 percent Inuit working at both. Recently, like two years ago, I think, [Mine 2] achieved, reached that target, but we're not there yet for [Mine 1]" (2).

Employment demographics

While the rhetoric from the industry tells a promising story, the employment data indicate that, on the contrary, Inuit (and Inuit women, in particular) remain under-represented across the industry and face barriers to long-term employment. As of February 2018, the workforce of Mine 1, which contracts out catering and house-keeping positions, was just 11.9% Inuit. Mine 2 comes close to the informal target with a workforce of 19.6% Inuit; however, there are high levels of job segregation. Of the Inuit working at Mine 2, 57.5% are employed in low-paid catering and housekeeping positions. Inuit women represent 9.6% of the total workforce of Mine 2, yet 80% of these women are employed in catering and housekeeping. If catering and housekeeping positions are excluded, Inuit women comprise just 2.3% of the total combined workforce of Mine 1 and Mine 2, and Inuit, overall, comprise 11.2%.

Employees in catering and housekeeping positions earn some of the lowest wages at the mines; dishwashers, general aides, and janitors earned $24.40/hr at Mine 2 in 2021, representing the lowest wages in the company besides maintenance staff and Inuit trainees (Glencore and USW 2017). Wages in the kitchen at Mine 2 in 2021 ranged from $24.40/hr up to $37.00/hr for sous chefs, while employees in higher-skilled and traditionally masculine positions like concentrator operators earned between $32.12 and $42.64, heavy equipment operators earned between $33.17 and $39.04, and miners earned between $33.23 and $39.04 (Glencore and USW 2017). At Mine 1, where catering and housekeeping positions are provided by subcontractors, wages in the department are even lower; in 2017, janitors, dishwashers, and kitchen general aides earned $14.52/hr, while their counterparts at Mine 2 earned $22 (O'Reilly 2020). In addition to lower wages, there are also discrepancies in retention and seniority by Indigeneity and gender. The average service time for non-Inuit at Mine 2 is 7.5 years, while the average service time for Inuit men is 4.2 years, and for Inuit women, just 2.9 years. The data show that 57.4% of Inuit women were employed in temporary positions, while the corresponding rates for other groups were 49.6% for non-Inuit women, 28% for Inuit men, and 9.8% for non-Inuit men.

Inuit women's self-described challenges of mining work

Key informant interviews and employment data suggest that both recruitment and poor retention are factors in the overall low employment rates of Inuit women in mining. Inuit women's personal experiences, as recounted in our interviews, shed light on the specific factors that limit the desirability and feasibility of mining work for them, contributing to their low employment rates. The women described many challenges in their experiences of mining employment; however, perhaps the most common and most challenging issue was the incompatibility of the work with parenthood. Structural issues like the spatial organization of mining and company policies create an environment that is inhospitable to workers with care responsibilities at home. The mine sites are distant from the nearest communities and are accessible only by plane. The difficulties presented by the typically long shifts of mining work are further exacerbated in fly-in mines by factors like the need for around-the-clock childcare in the workers' home communities during work rotations.

Company policies also contribute to this difficulty. For example, the organization of work shifts in rotations of 3 weeks on, 3 weeks off. The IBAs provide that Inuit beneficiaries may work in 2-week rotations; however, this remains a long period of time to spend away from home, especially for parents. There is no childcare assistance provided by the mines. One woman interviewed noted that "If kids were allowed up there, and we were allowed to bring out kids and they would have like a daycare there, I'd probably still be doing it today" (21.6). Another woman commented that "I wish I could go back, but my son, I don't want to leave him" (21.4). Those mothers who were able to make mining employment work for them for a length of time were often those who had partners, family, or babysitters at home willing to look after the children. Some became primary earners for their families while a co-parent looked after the children, and others developed reciprocal relationships with family or community members who provided childcare. However, even those who had enough support to manage the logistics of rotational work expressed the difficulty of the extended periods away from home and the fear of burdening family or friends.

Taking pregnancy and parental leave also presented barriers to long-term work. Some women who took breaks from mining while their children were young said that it hindered their career advancement and meant repeating probation upon return. The companies also prohibit employees from being on site while pregnant, which led some to conceal their pregnancies so that they could continue to work. Meanwhile, another woman described a workplace culture hostile toward pregnancy, noting that her supervisor attempted to prevent her from taking parental leave, and described, "My relationship with my . . . supervisor had gone really sour when I announced to them that I was pregnant" (21.6).

A second major challenge that Inuit women discussed in their experiences of mining work was job segregation. The women knew from experience what the employment data examined here confirm, that is, that Inuit women are clustered in the least-skilled and lowest-paid positions. Several women expressed their frustration with this fact, and one described educational barriers to advancement: "Most [Inuit] are janitors . . . although some Inuit are very much qualified, but on paper, they're not . . . because not a lot of them may have graduated high school" (20.2). Another woman described how "It's like they only hire Inuit as a janitor or dishwasher. Like they do want a job, but it seems like you cannot start from high . . ." (21.10). Interviewees described mixed experiences with training programs meant to assist in advancement. When training programs were readily available and effective, women often reported they were an excellent benefit of mining work. However, in some cases, training programs were canceled or otherwise made inaccessible, which some workers described as a let-down, especially as training opportunities had, in some cases, been a significant draw of mining work.

Additional challenges included those regarding the work environment and workplace culture. Many participants described the difficulties of working and living in a male-dominated environment, with some describing an unsafe environment for women, microaggressions, and instances of sexual harassment. One woman described feeling constantly objectified in the living environment and said that when she tried to talk to a supervisor about it, "they answer like 'oh yeah, well stop having [makeup]' or oh well, 'stop it, you should dress more baggy'" (21.8). The workplace is also predominantly francophone, which created cultural conflicts at times. One participant felt that the environment was hostile to her because she looked, spoke, and dressed differently from the other employees, describing an incident with an HR employee:

> [He] wasn't very nice to me at all because he knew that Inuit . . . don't constantly follow what they're supposed to do, so he thought that—without knowing me—he treated me like I'm already a bad person or like a bad worker. (21.10)

Another commented that "friends of mine don't [have good experiences at work] mainly because . . . their supervisors are not that friendly, especially when you're bilingual and not able to speak French" (20.2).

Experiences of the work environment were mixed, however. The mines profess zero tolerance for harassment and discrimination, and some interviewees found these policies effective. One woman described an incident of sexual harassment that her coworkers reported to HR, after which she received letters of apology from the perpetrator, HR, and her supervisors. She commented "I was really impressed and pleased that . . . they actually red-flagged him and he got in a lot of trouble" (20.2). Unfortunately, however, another interviewee described an incident of sexual harassment that went unaddressed because no one had witnessed it: "I had to write it down, but . . . nothing happened" (21.9). Some participants similarly felt that the mines took effective action on workplace racism. One woman, referring to the Inuit hiring and training program at one of the mines, said "I would say Tamatumani is really, really good" at supporting Inuit at work (20.2). However, she also noted that while the company took action on outright discrimination, she felt that they could do more to shift the culture: "These *qallunaat* [non-Inuit] that come up there, they have that lack of knowledge and their comments are based on these lack of knowledge . . . providing a better program of understanding it would change a lot of these *qallunaat* perspectives" (20.2).

As a result of these significant barriers, most of the participants were unable to find long-term employment and financial security through mining work. Many of the women who had since left mining jobs were living in poverty at the time of the interviews, with annual incomes of less than $20,000.

Benefiting from mining work: Inuit women's positive perceptions

Despite the seemingly negative prospects for Inuit women in mining based on the employment data and the challenges described, the women's experiences resisted simplistic, negative narratives. Many women reported having complicated feelings about mining. As one woman put it, "There's days where I'm pro-mining and there are the days I'm against it, even after working there" (21.6). However, in contrast to the body of literature that has sometimes overemphasized women's victimization in relation to mining development, many Inuit women described mining work as having a distinctly positive impact on their lives:

> I was a, you know, young thing, I was not sure what I wanted to do in life, but when I went to [Mine 2] for the first time, I'm like, oh my God I love this place! You know there's such a lot of good people, the money is good, the food is free, you know. So yeah, I really fell in love with [Mine 2]. And it motivates me so much to have a routine. Food is always . . . ready, like the vegetables, or you know the gym is there . . . Really cool.
>
> (21.3)

In addition to the challenges, the women described a host of different benefits that mining employment brought to their lives, the primary one being financial. As one woman noted, since working in mining, "We have everything we need, we don't miss anything. And yeah, I feel lucky to have a really good job like that" (21.3). Women who had worked full time in mining reported incomes of between CA$70,000 and $100,000 per year, representing high earnings relative to the required language skills and educational attainment. Many women described experiences of achieving financial independence, and six of the ten participants reported that they used their incomes from mining to support their families or community. One woman described her desire not only to lift her family out of poverty but also to develop her independence:

> Mainly because, while living in the North, my parents ended up racking up so much debt. And so, I've always helped them out to provide food on the table . . . so obviously my income helped the family a lot, but through the years trying to wean them off me [laughs] so that I can kind of, you know, spread my wings and live at the age of 26.
>
> (20.2)

Mining work was also part of a broader strategy for Inuit women to facilitate their well-being in ways beyond just income. Women described how mining work provided a sense of accomplishment and satisfaction, as well as opportunities for learning and personal development, and they often intentionally pursued positions and environments in which they could thrive. Supporting others and having work relationships based on reciprocity were important to the women, and some described satisfaction in their ability to serve their coworkers through their work. Many of the women also sought out mining work partly as an opportunity to learn professional and language skills in order to improve their career prospects and frequently expressed appreciation for the training opportunities offered. One interviewee described how thrilled she was to advance into a position that she desired and noted that, "If it wasn't for the training, I don't think I would have been able to be confident enough to work alone" (20.2). One participant was able to use her experience from mining work to find another job in her community that better suited her lifestyle, stating that mining "has opened a lot of doors" for her (21.6).

Another benefit that participants repeatedly raised was a sense of connection with community, language, and culture that Inuit women had at the mines. Particularly for the women who had moved south, mining represented a unique opportunity to access employment while living within their home territories, speaking Inuktitut, and engaging with Inuit culture. One woman recalled,

> I really enjoyed getting away from the city. And going up North and being with my second family, whether it be with the same people or getting to know new people, I always really looked forward to the summertime, to be honest. (20.2)

Interviewees explained how much they appreciated certain programs and amenities at the mines that helped facilitate connections between Inuit on site, such as Inuit kitchens and trips out on the land. Many participants described the social and cultural ties that employees fostered through the Inuit kitchen and the access to traditional Inuit foods—one woman described the best part of mining as "having [an] abundance of country food" (20.2).

Similarly, women coming from Southern Québec cherished the opportunity to return to their homelands. This was a somewhat mixed benefit, as many participants brought up discomfort with the environmental impacts of mining. As one woman explained, "I still have my mixed feelings [about mining]. There's moments where I'm, like, what are we doing to our beautiful land? What are they doing?" (21.6). However, the experience of living and working on their homelands remained a distinctly positive aspect of mining employment for many. One participant described how she cherished the opportunity to see wildlife at the mine, noting that the reason that she loved mining was because "I was on the land, so I enjoy all my time up there" (21.4).

Discrepancy between employers' claims and employment data

The results outlined above present certain contradictions. First, the demographic employment data contradict employers' portrayals of Inuit women's inclusion in mining, overall and across diverse job types. Despite the recent attention from companies and the sentiment among those in the industry that the employment situation has improved, the data reveal Inuit women's poor overall employment numbers, segregation into low-skilled positions, and low levels of retention. Certain widely cited claims have misleading connotations: The notion that Mine 2 has reached its Inuit hiring target while Mine 1 still has a way to go obscures the fact that Mine 2 has reached that target with 80% of Inuit women and 36% of Inuit men employed in catering and housekeeping positions, while Mine 1 contracts out these positions. The reality of poor employment outcomes for Inuit women is easily lost in the rhetoric of success stories and simple figures of total employment.

While some individual Inuit women have been able to advance in mining careers, these numbers are not significant, and glaring disparities in employment outcomes by gender and Indigeneity persist. This is likely because, despite public rhetoric, the initiatives adopted by companies have been substantially lacking in efforts to recruit and retain Inuit women. A former employee in Inuit recruitment at one mine commented that, in 2014, when new Inuit recruitment measures were adopted, "We kind of had to change the image of what the population thought of the mine. And a lot of people didn't think that this company was really pro-Inuit" (4). This indicates that the mine was cognizant of promoting an image of greater diversity to the public; however, interviews with former employees in Inuit recruitment suggest that recruitment practices lack meaningful attention to gender. An employee in Inuit recruitment from Mine 2 responded to a question about whether their Inuit recruitment efforts specifically targeted women: "No. But we don't target just men either. We target Inuit" (6). This was echoed by an employee in Inuit recruitment at Mine 1: "We usually take all Inuit women, men applications. When we promote, we try to promote both of them" (5). The mining development manager at Makivik Corporation further commented that despite community-based efforts to promote Inuit employment, he did not see initiatives focusing on women specifically: "There's not enough effort to attract Inuit women . . . Like, the community representatives, they have a big focus on the local employment, but they don't specify for Inuit women . . . Not enough, in my opinion" (2).

Moreover, some of the publicized efforts to address women's barriers have not amounted to substantial changes. A former employee in Inuit recruitment involved with the Kautaapikkut Roundtable described how she had never seen the results of one major research project in the region about Inuit women's employment in mining. "We did presentations at the end to [the regional government], I guess to kind of prove what we did with that money, right? But besides that, I've never seen like a final report, or . . . I never really heard anything after" (4). Some individuals in Inuit recruitment have certainly applied effort toward attracting and retaining Inuit women; however, these efforts sometimes ended up with frustration against the companies. A former employee in Inuit recruitment observed that at one of the mines,

> in a way there was kind of like more talk than action. "Oh yes, this year we're going to aim for this," or . . . "we're going to offer this." But in the end, they never do half the things they promised they would do.
>
> (4)

The companies have also taken few steps to address the biggest barriers that Inuit women face in the workplace. Aside from accommodating Inuit workers for 2-week rotations, as mandated in the IBA, there are no policies in place to assist Inuit women with navigating care responsibilities. Employees in Inuit recruitment at both mines brought up childcare as one of the biggest barriers that women face, yet no company policies address this issue, and no accommodations exist for shorter work rotations than 2 weeks. The issue of segregation of women into low-quality jobs is often addressed through individual and informal methods such as hoping that increased representation will naturally inspire more Inuit women to strive for advancement. Dedicated Inuit training programs do not target women specifically. Companies have tried to address workplace culture issues with policies like zero tolerance for harassment; however, issues persist. Thus, despite persistent employment inequities for Inuit women and companies' purported desire to employ more Inuit women, companies are not solving issues related to recruiting Inuit women or eliminating barriers to improve retention. The focus on representation through individual stories allows companies to appear to be following through on IBA commitments, while brushing over the question of whether Inuit women are receiving material benefits from mining employment on the whole.

Contradiction between employment data and women's positive experiences

The data indicating low employment rates and poor retention also seem to contradict the positive experiences of mining employment described by many of the interviewees. Despite the fact that mining was not able to provide secure, long-term employment for most participants, many recalled their time in mining fondly and described many benefits that they were able to derive from it. Many also expressed a wish that they could return to mining work. How do we make sense of Inuit women's positive experiences in mining employment despite the significant barriers they faced?

There are several factors underlying this discrepancy. The first is the relative benefits that mining offers within the context of other opportunities available to Inuit women, as there are few employment opportunities in Inuit communities. One interview participant described why she believed that, overall, mining provides good opportunities for Inuit women: "Sure, because it's a real job. What are they gonna do up North? . . . Because there's not really work there" (21.4). An employee at an organization that works with Inuit women emphasized that this lack of local work opportunities means that women may stay in mining jobs even in the face of significant problems at work:

> Especially in more smaller and remote communities who sometimes can't leave these jobs because it's their only livelihood, but they encounter a lot, a lot of problems on site, especially when they are in rotation two weeks away from home, away from family members.
>
> (3)

She also discussed how issues like the housing crisis in the North, which leads to situations of precarious and overcrowded housing in most communities, can further push women to maintain positions at fly-in worksites.

While a growing number of Inuit have migrated south to urban centers in recent decades, they remain likely to experience poverty and other social and economic struggles in southern cities (Kishigami 2015). A 2012 study on Inuit women in

Montreal found that while respondents commonly cited negative reasons for leaving their homelands for the city, such as abuse, substance problems, or housing or work shortages, many women were low-income or experiencing homelessness in Montreal (Kishigami 2015). Many of the women interviewed described the difficulties they faced finding work in Southern Québec, particularly without speaking French. Others had a difficult time finding work in southern cities without higher levels of educational attainment: "I so regret, I so regret that I didn't finish my school. But I tried my best before, so it's not easy not finishing my high school, to look for a job" (21.5). Women who had worked in Southern Québec prior to working in mining often recalled low-paid positions, such as warehouse shipping clerk, production line worker, or jobs in hospitality, such as at fast-food restaurants, convenience stores, or cafés. Many were employed part-time or precariously.

The scarcity of opportunities available to Inuit women further contextualizes their positive evaluations of mining work. Given the shortage of options and the social problems often faced in Inuit communities, women may have been more tolerant of some of the serious challenges they faced in the mining industry because it offered a good salary, decent working conditions, and an accessible hiring process. Mining is one of the few industries offering such essential job benefits to Inuit communities. Understandably, the relative quality of this work compared to other opportunities causes many women to emphasize the unique benefits of mining, even if their positions are ultimately of relatively poor quality.

A second important factor in understanding women's positive evaluations of mining work is the resourcefulness and resilience that Inuit women demonstrate in navigating mining employment. The women deliberately sought out positions that would allow them to escape poverty, achieve financial independence, and support their families and community. Additionally, many were intentional in seeking out positions and circumstances that would facilitate their well-being more broadly, such as chances to learn and build skills to further their professional and personal development, and ways to connect with community, homeland, and Inuit culture. The women demonstrated notable resilience in navigating the disadvantages that they faced in mining work in order to pursue these ends. They developed various strategies to navigate barriers, such as cultivating childcare solutions based on reciprocity, building communities within the mines that acted as support networks, and individually pushing for opportunities to advance into more desirable positions. Thus, the women navigated mining employment not simply as victims of poor circumstances but as agents seeking pathways to facilitate their overall well-being. Through the strategies they employed, they were often able to overcome some of the barriers they faced and derive many benefits from mining employment, at least for a period of time.

This highlights a third important factor framing the discrepancy between Inuit women workers' positive accounts and the employment data, namely, that many of the women were able to have positive experiences in mining employment at a given point in their lives, until it became unsustainable. For many of the women, the benefits of mining work outweighed the challenges until life events—most notably, pregnancy—tipped the balance. Once they had children, the strain on the women and their families often became too much, and they were forced to leave their jobs even if it meant a return to poverty. The mining development manager at Makivik Corporation observed that, "Every time I see the new, coming Inuit employments—employees—I see a younger generation. So, young ladies with no families, I believe. They have, let's say, easier access" (2). He indicated that mining employment is accessible for young women with no children but less so for women who are mothers. The spatial and structural barriers to mining work became insurmountable once women had families to care for, effectively excluding a large proportion of Inuit women.

Conclusions

Despite IBAs laying out employment opportunities for Inuit as a primary compensation for extraction within their territories, this benefit has long been unevenly distributed along gendered lines within Inuit communities, even as the harms of mining development are disproportionately shouldered by women. While the industry has sought to improve its reputation and has reported successes employing Inuit women, Inuit women continue to be largely excluded from many of the benefits of mining careers. They are still employed in remarkably low numbers, clustered in low-paid entry-level positions, and experience a high turnover rate. In interviews, women described a number of ways that they were able to have quite positive experiences in mining, but only few of them were able to enjoy secure employment in mining for a long-term period, owing to the formidable barriers and disadvantages they described. It is clear that mining companies are not sufficiently ensuring that Inuit women have equitable access to high-quality, long-term employment. Formal Inuit recruitment initiatives have lacked attention to gender, despite the long-standing inequities, and companies have failed to take transformative action to address the major barriers that inhibit Inuit women from accessing successful mining careers. There is much room for improvement in company policies to make employment more accessible for Inuit women; in particular, we suggest three major strategies for improvement based on our interviews.

The first is to acknowledge and support women's dual roles as not only workers but also caregivers. The current structure of mining work, which has been organized around a traditionally masculine workforce, effectively excludes a large proportion of women, given the reality of their responsibilities as mothers. The women interviewed gave several suggestions that they saw as having potential to alleviate some of the strain on mothers: Shorter work rotations (such as 1 week on, 1 week off), shorter shifts, or childcare assistance. The second recommendation is to develop dedicated advancement strategies for Inuit women in order to channel them into more skilled and highly paid positions, including following through on training programs. Lastly, we suggest taking proactive steps to shift workplace cultures to make them more comfortable for non-male and non-francophone minorities. Zero-tolerance policies for harassment and discrimination have made progress toward checking cases of outright aggression; however, interviewees' descriptions of the workplace atmosphere make clear that there is a long way to go for these spaces to feel safe and welcoming for Inuit women. More comprehensive education programs and evidence-based approaches to dealing with smaller incidents like microaggressions could help improve the work environment.

The women's descriptions of their positive experiences in mining also highlight areas that can be reinforced in order to enhance the desirability and sustainability of mining jobs for Inuit women. Interviewees often spoke very fondly of amenities provided by the mines that facilitated social and cultural connections between Inuit, such as the Inuit kitchen and spaces for traditional crafts. Likewise, they were enthusiastic about excursions such as fishing trips out on the land, with several participants noting that they only wished they had more time to enjoy such benefits amid the tiring 12-hour shifts of mining work. Expanding and enhancing programs that facilitate social connection, leisure, and recreation at the mine sites may go some distance toward improving retention among Inuit women.

However, while such policies may help improve the desirability of mining work, it is also clear that many of the barriers to women's sustainable employment in mining are inextricably linked to broader structural disadvantage. The actions of mining companies alone cannot compensate for Inuit women's lack of options in career paths, lack of educational opportunities, and the serious economic and social problems in the North that push women out of their communities. This requires that we examine the questions we are asking about Inuit women and mining. Is the exclusion of Inuit women by mining companies the main problem?

We cannot understand what prevents Inuit women from being able to access secure, long-term employment in mining without understanding the broader context of systemic disadvantage in which they are situated. The colonial history of the region has left behind a legacy of persistent poverty and social problems that severely limit the opportunities of Inuit women to access a variety of career paths and constrain their avenues for success within the industries available to them. An employee at a nonprofit organization that works with Inuit women discussed her sense that, beyond the actions of mining companies, the actions of governments and the infrastructure in the North need to be called into question:

> Of course, the responsibility of the industry is a question first and foremost, but also the broader obligations of the territorial and provincial governments, and also the federal government . . . The way the . . . land claim agreements are framed, because they are geared toward more self-autonomy and that doesn't take into account, particularly, capacity to monitor these operations, it's sort of something that's had the federal government kind of scot free . . . oftentimes a lot of the issues, across the board, are often rooted in the lack of critical infrastructure in the North.
>
> (3)

These comments point to the inherent limitations of IBAs as the only mechanism for Inuit communities to control development within their territories. What are framed as instruments for greater Inuit autonomy and self-governance can end up acting as justifications for government neglect that allow ongoing injustices in Arctic communities to persist.

These factors highlight the complex and multidimensional nature of women's exclusion from the mining industry. It is essential to apply a critical lens to companies mining within Inuit territories who fail to provide accessible and equitable employment for Inuit women. However, employment disparities must also be understood within the context of broader structural disadvantage and the intergenerational effects of colonialism. In addition to problematizing company policies, it is important to question the effectiveness of tools such as IBAs at creating a context in which Inuit women can succeed and thrive. Finally, it is essential to recognize that Inuit women are not merely victims of mining development but agents striving for well-being as they navigate their complex relationship to the mining industry with resilience and resourcefulness. The stories of the women in this small sample tell us about the types of employment benefits that Inuit women need and value in their quests for well-being. Furthermore, their stories highlight some of the ways in which the mining industry and governments have thus far failed to follow through on the promise that mining employment would provide Inuit communities with pathways to prosperity.

Acknowledgments

We would like to thank Makivik Corporation and Pauktuutit for their partnership in this project; Emily Mina, Melissa Reid, and Cynthia Cookie-Simmard for their guidance on project design and knowledge mobilization; Becky Han and Sharoni Mitra for their contributions to the research; and Séverine Bouard for her collaboration. Deborah Simmons and Karla Williamson also helped with the analysis of the interview data. This work was supported by the MinErAL network through funding from the Social Sciences and Humanities Research Council of Canada (grant number 895-2016-1015), Institut Agronomique néo-Calédonien, and the New Research Collaboration Program of the France–Canada Research Fund.

References

Brain, K. A., 2017. The impacts of mining on livelihoods in the Andes: A critical overview. *The Extractive Industries and Society*, 4 (2), 410–418. https://doi.org/10.1016/j.exis.2017.03.001

Cox, D., and Mills, S., 2015. Gendering environmental assessment: Women's participation and employment outcomes at Voisey's Bay. *Arctic*, 68 (2), 246–260. https://doi.org/10.14430/arctic4478.

Glencore and USW, 2017. Convention collective Glencore Corporation Canada–Mine Raglan et Syndicat des Metallos, Section Locale 9449 Mai 2017-Avril 2022. Available from: https://metallos9449.ca/convention [Accessed June 12 2024].

Glencore Canada, 2022. Raglan mine: Our culture. Available from: www.glencore.ca/en/raglan/careers/our-culture [Accessed 25 October 2022].

Hall, R., 2017. *Diamonds are forever: A decolonizing, feminist approach to diamond mining in Yellowknife, Northwest Territories*. Toronto: York University. http://hdl.handle.net/10315/34474

Hall, R., 2022. *Refracted economies: Diamond mining and social reproduction in the North*. Toronto: University of Toronto Press.

Hodgkins, A., 2018. Lost in translation? Exploring outcomes of Nunavut's resource development training and employment policies for Inuit of Northern Baffin Island. *The Northern Review*, 47, 31–57. https://doi.org/10.22584/nr47.2018.003

Inuit Tapiriit Kanatami, 2018. Inuit statistical profile 2018. Available from: www.itk.ca/2018-inuit-statistical-profile-3/ [Accessed 25 October 2022].

Jenkins, K., 2014. Women, mining and development: An emerging research agenda. *The Extractive Industries and Society*, 1 (2), 329–339. http://dx.doi.org/10.1016/j.exis.2014.08.004

Kishigami, N., 2015. Low-income and homeless Inuit in Montreal, Canada: Report of a 2012 research. *Bulletin of the National Museum of Ethnology*, 39 (4), 575–624.

Kudloo, R., Czyzewski, K., Tester F., Aaruaq N., and Blangy, S., 2014. *The impact of resource extraction on Inuit women and families in Qamani'tuaq, Nunavut Territory: A qualitative assessment*. Ottawa: Pauktuutit Inuit Women of Canada, The University of British Columbia School of Social Work. www.pauktuutit.ca/wp-content/uploads/Report-Final-Jan-2015.pdf

Kudloo, R., Czyzewski, K., Tester, F., and Blangy, S., 2016. *The impact of resource extraction on Inuit women and families in Qamani'tuaq, Nunavut Territory: A quantitative assessment*. Ottawa: Pauktuutit Inuit Women of Canada, The University of British Columbia School of Social Work. www.pauktuutit.ca/wp-content/uploads/Quantitative-Report-Final.pdf

Lahiri-Dutt, K., 2012. Digging women: Towards a new agenda for feminist critiques of mining. *Gender, Place & Culture*, 19 (2), 193–212. https://doi.org/10.1080/0966369X.2011.572433

Lévesque, S., and Duhaime, G., 2021. *Employment in Nunavik: Profile and trends*, ed. G. Duhaime. Quebec: Université Laval.

Lutz-Ley, A., and Buechler, S., 2020. Mining and women in Northwest Mexico: A feminist political ecology approach to impacts on rural livelihoods. *Human Geography*, 13 (1), 74–84. https://doi.org/10.1177/1942778620910901

Manning, S., Nash, P., Levac, L., Stienstra, D., and Stinson, J., 2018. *A literature synthesis report on the impacts of resource extraction for Indigenous women*. Ottawa: On Behalf of the Canadian Research Institute for the Advancement of Women, Prepared for the Canadian Environmental Assessment Agency.

Mazer, K., Mills, S., and Bouard, S., 2022. Pathways to nickel mining employment among Inuit women in Nunavik, Canada and Kanak women in New Caledonia: A comparative study. *The Extractive Industries and Society*, 10, 101088. https://doi.org/10.1016/j.exis.2022.101088

Mercier, L., 2011. Bordering on equality: Women miners in North America. *In* K. Lahiri-Dutt, ed., *Gendering the field: Towards sustainable livelihoods for mining communities*. Canberra: Australian National University Press, 33–47.

Mills, S., 2006. Segregation of women and Aboriginal people within Canada's forest sector by industry and occupation. *Canadian Journal of Native Studies*, 26 (1), 147–171.

Mills, S., Simmons, D., and Williamson, K. J., 2023. In pursuit of the "good life": *Inuit Qaujimajatuqangit* and Inuit women's movements in and out of mining work. *Arctic*, 76 (2), 126–142. https://doi.org/10.14430/arctic77518

NAHO, 2008. *Resource extraction and Aboriginal communities in northern Canada: Gender considerations*. Ottawa: National Aboriginal Health Organization. https://ruor.uottawa.ca/bitstream/10393/30625/1/Gender_EN.pdf

Nightingale, E., Czyzewski, K., Tester, F., and Aaruaq, N., 2017. The effects of resource extraction on Inuit women and their families: Evidence from Canada. *Gender & Development*, 25 (3), 367–385. https://doi.org/10.1080/13552074.2017.1379778

O'Reilly, M., 2020. *Union-Indigenous engagement in the mining sector*. Unpublished Master's Thesis. McMaster University. https://macsphere.mcmaster.ca/handle/11375/25333

Parmenter, J., 2011. Experiences of Indigenous women in the Australian mining industry. *In* K. Lahiri-Dutt, ed., *Gendering the field: Towards sustainable livelihoods for mining communities*. Canberra: Australian National University Press, 67–86.

Pauktuutit, 2021. *Addressing Inuit women's economic security and prosperity in the resource extraction industry*. Ottawa: Pauktuutit Inuit Women of Canada. https://pauktuutit.ca/project/addressing-inuit-womens-economic-security-and-prosperity-in-the-resource-extraction-industry/

Raglan Mine, 2020. Katinniapik: Raglan mine employee newsletter, 14 (3) (May/June). www.glencore.ca/.rest/api/v1/documents/2daca834e49663d3484654c53e4a3100/ENG_Katinniapik_May_June_2020.pdf

Rodon, T., and Lévesque, F., 2015. Understanding the social and economic impacts of mining development in Inuit communities: Experiences with past and present mines in Inuit Nunangat. *The Northern Review*, 41, 13–39. https://doi.org/10.22584/nr41.2015.002

Rogers, S., 2015. *Nunavik women say family demands keep them from jobs at mines. Nunatsiaq News*. Iqaluit: Nortext Publishing Corporation. https://nunatsiaq.com/stories/article/65674nunavik_women_say_family_demands_prevent_them_from_working_in_mini/

Sinclair, L., 2021. Beyond victimisation: Gendered legacies of mining, participation, and resistance. *The Extractive Industries and Society*, 8 (3), 100870. https://doi.org/10.1016/j.exis.2021.01.005

Statistics Canada, 2018. *Nunavik [Inuit region], Quebec (table). Aboriginal population profile. 2016 Census* (Catalogue no. 98-510-X2016001). Ottawa: Statistics Canada. Available from: http://www12.statcan.gc.ca/census-recensement/2016/dp-pd/abpopprof/index.cfm?Lang=E [Accessed 13 May 2022].

Tallichet, S. E., 2000. Barriers to women's advancement in underground coal mining. *Rural Sociology*, 65 (2), 234–252. https://doi.org/10.1111/j.1549-0831.2000.tb00027.x

13 A mine for women? Trajectories of Kanak women in the nickel industry in New Caledonia

Guillaume Vadot, Christine Demmer, Séverine Bouard, and Mathilde Baritaud

A rapid and unique process of feminization

This chapter deals with a subject that has not yet been fully explored in New Caledonia: The wage economy of Kanak women. The material on which it is based consists primarily of detailed biographical interviews with Kanak women employees of Koniambo Nickel SAS (KNS), a multinational nickel-processing mining company that has been fully operational since 2013, near the capital of the North Province, Koné (see map, Fig. 13.1). Nine years ago, Christine Salomon observed a process of feminization at the plant under construction, especially in the areas of catering and cleaning (Salomon 2015). Like Marie Salaün (2014), she noted the hiring of young Indigenous people, an initiative promoted by the local government and the mining company itself, and indirectly supported by various training and work integration schemes. While our survey confirms the prevalence of women employment on the site, it also highlights a profound change in the type of jobs held, which are increasingly in the core industrial business in the concentrator/plant and in the mine (particularly the role of machine operator). By examining a sample of biographical trajectories, this work looks at what it means for Kanak women employees to work in what is still considered a man's job.

In many respects, the process of feminization in mining discussed here can be compared to other industries, for example, the textile industry in Western Europe in the nineteenth century (Canning 1996), factories during World War I (Downs 1995; Mak 2017), the electronics sector in the 1970s (Kergoat 1982), and, more recently, the automobile industry (Beaud and Pialoux 2002), in the wake of the industrial decentralization of France in the 1960s (Gallot 2015). In each case, women have been rapidly projected into occupations and, more broadly, into social roles once deemed to be reserved for men, provoking a series of adjustments in the representations associated with employment and in gender relations in and beyond the workplace. However, it is important to distinguish what is happening in New Caledonia from the European context on several points. On the one hand, while the New Caledonian labor market is classically divided along gender lines (21% of employees in the manufacturing industry are women, compared to 81% in the hotel industry, e.g.), it is within the Kanak community that the employed workforce is most feminized. While the gap between men and women in employment is almost

DOI: 10.4324/9781003406433-18

This chapter has been made available under a CC-BY-NC-ND 4.0 license.

ten points for the population as a whole (61.8% of the men in New Caledonia vs. 52% of the women), it is only 3.8 points (50.2% vs. 46.4%) among Kanak people (ISEE 2019a). This smaller gap should therefore be kept in mind in order to better study the current and future gendered division of labor among the Kanak.

New Caledonia, as a former settlement colony, is nonetheless still marked by considerable segregation between the Indigenous community and the rest of the population, and gendered assignments are highly pronounced and present on all sides. One of the aims of our research was to explore the framing effects of this (post)colonial context, which have been intensely debated and reformulated since the recognition of New Caledonia as an overseas country ("Pays d'Outre-mer") that now has a *sui generis* status and that, since the *Nouméa Agreement*, has seen a gradual transfer of powers to it, with the exception of those of a sovereign nature. The process of feminization is therefore both amplified and, in some respects, undermined. In this context, our work explores the tensions between the promotion and disqualification of the Kanak women's workforce. In so doing, it aims to highlight the consubstantial relationships between a variety of social markers in the composition of workspaces, reminding us that "*travailleuse n'est pas le féminin de travailleur*" (Kergoat 2012), which translates loosely as "it's not enough to merely add the feminine ending to the masculine form of 'worker'" to denote a women worker, and that, even more so in this case, being a Kanak woman employee is not the same as being a non-Kanak woman employee.

A labor market that remains ethnically divided

Although New Caledonia has a GDP per capita comparable to that of Italy, it has an exceptionally low employment rate due to its weak economic development outside the capital, Nouméa. In 2018, the employment rate was only 56.9% compared to 66% in mainland France, and over 73% and 77%, respectively, in Australia and New Zealand. The rate drops to 48% for the Kanaks, the largest community in New Caledonia, with just over 41% of the total population, compared to 24% for Europeans (ISEE 2019). The Southern Province alone accounted for 82% of salaried employment in 2018 (IDC NC 2018).

Deprived for decades of freedom of movement and confined to reserves until the end of the *Indigénat* regime in 1946, the Kanak people entered the market late. This situation did not prevent Kanak men (primarily) from being massively put to work to perform forced labor and services in the first half of the twentieth century (Merle and Muckle 2019). In the mining sector, after the mobilization of convicts from mainland France at the end of the nineteenth century, employers relied on Asian migrant workers until the 1950s (Merle 2012). Significant recruitment of Kanak men began during the nickel boom of the following two decades (Freyss 1995). At the same time, Kanak women entered the workforce, primarily as domestic workers (Grochain 2007).

The proportion of Kanak women in employment has, however, increased steadily over the past 15 years. As studies on ethnic inequalities in access to employment in New Caledonia have shown, this integration is taking place mainly in low-skilled occupations, with three-quarters of male and female Kanak wage earners being employees or workers (Gorohouna and Ris 2015). The contemporary relationship of most Kanak people to the wage-earning sector remains characterized by the fact that part of their food requirements are covered by the practice of a food-producing economy in former cantonment areas that have become "customary lands" (Bouard and Sourisseau 2010). Although efforts are underway to reverse past trends of late and limited schooling (Salaün 2001), the differences in qualifications remain significant (see below) and constitute an obstacle to hiring (Ris 2013).

Who are the Kanak employees of KNS?

The Northern Plant and the rebalancing in New Caledonia

For the Kanak independence fighters who have been waging their battle since the mid-1970s, the control and development of nickel—of which New Caledonia holds 15% of the world's reserves and which represents up to 25% of its GDP, depending on the year—is part of the battle for full sovereignty (Demmer 2015, 2018; Le Meur 2017). The creation of the Northern Plant, an alternate name for KNS, which includes an industrial site (Vavouto) and a mining site (Koniambo), was a condition of the negotiation of the *Nouméa Agreement* signed with France and the loyalists in 1998 and is part of the process of economic rebalancing initiated in 1988, at the time of the Matignon–Oudinot accords. KNS is 51% owned by Société minière du Sud Pacifique (SMSP), a company in which the pro-independence North Province holds the majority of shares, while 49% are held by the multinational Glencore. The industrial mining complex is located next to the capital of the Kanak-majority North Province and is intended to counterbalance the concentration of economic power in the South Province, where the capital Nouméa and the historic SLN plant are located, and which, according to the latest census in 2019, accounted for 74.8% of the country's total population.

From the perspective of the sectoral and hierarchical division of labor, the configuration of KNS's gender mix is relatively conventional. In 2017, all communities combined, 26% of KNS's female workforce was in the mine and 30% in the plant, while support roles, which include positions traditionally considered female, accounted for 44% of the female workforce.[1] Recalling that when she arrived in 2007, she was one of only two women machine operators at KNS, Géraldine[2] adds: "After that, the women were in the offices, of course."

The company's management, while less feminized, is also marked by a very classically gendered division of labor. From the outset, women have mainly held positions in human resources, communications, or the environment division, which led one of them to say that she "does not have the impression that it has become more feminized."[3] In 2015, only 6% of KNS managers were women, compared to 13% of executives, 31% of supervisors, and 54% of rank-and-file workers. Women had thus already become a majority at the bottom of the pyramid.[4]

However, a closer look at the situation of Kanak women reveals the original nature of the feminization process in play. Indeed, it is worth noting that within the Kanak community in KNS, the ratio of men to women in management positions is reversed, with women holding the majority. This result can be explained by the stronger academic results of Kanak girls (see below). In 2018, two of three Kanak heads of department were women, whereas this proportion was only four of thirteen for non-Kanak women.

In fact, whether they are members of the management team, multiskilled operators at the plant, or drivers of mining equipment required to extract nickel ore (carried out in the open air), the new female employees of KNS are generally qualified workers whose professional training is often more advanced than that of the male colleagues who preceded them. This is one of the specificities of the phenomenon observed. However, it is expressed differently, depending on the position and status of the job.

In addition to the divisions identified above is the divide between KNS and its subcontractors. The latter are in charge of site development, the management of the "mine front," or revegetation projects and are often owned by the local Kanak chiefdoms as compensation (Bouard *et al.* 2019). However, among these subcontractors, the distribution of jobs by gender is much more set in stone. The workforce specializing in construction or maintenance is mainly male. Floriane says that she was the only female welder at Somainko from the time she was recruited in 2010 until the end of 2018.[5] Conversely, the workforce of subcontractors in charge of revegetation is mainly female. At Thor, which specializes in machine maintenance and fuel supply, "the boys do the refueling and we did the cleaning [of the drivers' cabs]."[6] Other characteristics distinguish these women employees from the women operators at KNS: They are less qualified and less educated, and half of them come from the local tribes (areas of residence created with the cantonment and still subject to specific land tenure),[7] whereas salaried employment is very largely associated with mobility outside the Indigenous communities for the rest of the women employees.

Two final points bear emphasizing. On the one hand, while there are more and more Kanak women working in production today, they are still mainly supervised by men. On the other hand, the experience of these women is different, depending on whether they work in the plant or in the mine. In the plant, unlike in the mine, they do not have a sense of having entered a "male bastion."

> In the mine, women are very much in the background. Maybe in the mine it's a much older profession than we have in the plant. Let's say that here the men and women have all seen the plant evolve at the same time. That is to say that there were not men who started working there and that the women arrived afterwards; we all followed the growth of the plant, we all learned

at the same time. (. . .) As a result, we don't have the same mentalities in the mine and in the plant.

(Solange, production technician at the KNS plant)

When local employment benefits female graduates

As far as the political framework for hiring at the Northern Plant is concerned, it is not the feminization of the workforce that has been the strongest constraint for management, but rather the challenge of safeguarding local employment at different levels. KNS has committed to prioritizing local recruitment (Fig. 13.1), according to the terms negotiated with the country's elected representatives and the state. The rebalancing approach between provinces, combined with considerations linked to the environmental impact for the Indigenous communities close to the industrial site, intersects with that of the country's autonomy, which is enshrined in the *Nouméa Agreement*.[8]

As far as the female presence is concerned, I don't think it's something that Koniambo Nickel was looking for. I'd like to be able to say 'Yeah, we said

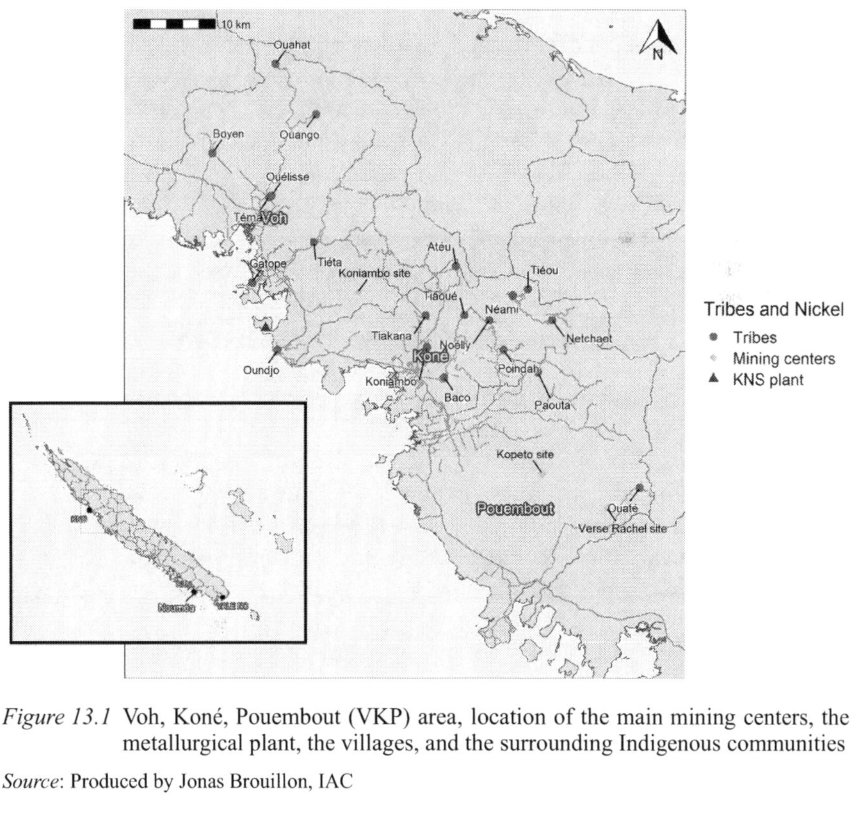

Figure 13.1 Voh, Koné, Pouembout (VKP) area, location of the main mining centers, the metallurgical plant, the villages, and the surrounding Indigenous communities

Source: Produced by Jonas Brouillon, IAC

> let's do it, 30% and all that' . . . No, that's not true. On the other hand, from the start [we] respected and enforced a recruitment structure. And that was something that had been negotiated with the Northern Province and the government. We had to make sure that the priority pools had been targeted before we could go any further. So the first basin was the [ecologically] impacted areas. So basically, the tribes around the site. The second was the VKP area. Voh, Koné, Pouembout. [Then] the North Province. [Then] the Island Province. And then the Southern Province. And if we had simply made sure, or each of our subcontractors had made sure, that these recruitment pools had been exhausted, then we had the right to go and look for expatriates. With the backing of Cap Emploi and therefore of the Northern Province, the backing of the territory, the backing of the state.
>
> (Nicolas, former HRM and then communications director of KNS)

At the same time, the increasing professionalization of the mining industry's occupations, influenced by the standards promoted by the multinationals, has contributed to raising the expectations of recruiters in terms of qualifications. In order to understand how these two developments have benefited Kanak women seeking employment, it is necessary to examine the distribution of educational success in New Caledonia. Research into ethnic inequalities in employment in the archipelago has shown that access to diplomas is the main factor in the development of disparities (Ris 2013), while for all ethnic groups combined, a diploma is more essential for women than for men in order to obtain a job (ISEE 2019a). The latest available data (2019 and 2014 censuses) show that there are still drastic differences between the communities: While 62% of Europeans have at least a *baccalauréat* (and 43.7% at least one higher education diploma), this rate is only 21.7% (and 6.5%) for the Kanak. However, among the latter, women are generally more successful: In 2014, they represented 58% of Kanak graduates with a *baccalauréat* or higher—this proportion being 49% among Europeans by comparison.[9] For those interested in colonial history, this specific commitment to schooling from young Kanak girls is a reminder of the way in which schooling may have been privileged among the subalterns of the Indigenous population, thus constituting a path to social mobility (Bayart 2006; Jézéquel 2003).

Among the Kanak women interviewed who were employed by KNS, those who were in a relationship were almost systematically more qualified or educated than their spouses (six of seven cases at the technician level and above, and eight of nine for operators; the opposite was true for subcontractor employees). The same observation can be made about their brothers. Florence, who is in charge of a strategic department, is an engineer like her older sister, while their first brother has a BTS (senior technologist certificate), and the second has not yet started higher education 2 years after his *baccalauréat*. The same is true of Lynda, a buyer. She and her two sisters each obtained a 5-year degree, while their brother did not finish high school. Rose, a former training manager, and Sonia, an HR advisor, who both accumulated their educational capital through continuing education, before becoming supervisors at KNS, as well as Solange, a technician at the plant who has a DEUST

(French scientific and technical university diploma) in metallurgy, have all helped a spouse with no qualifications gain employment. For the operators, spouses who work generally do so without qualifications and with a more precarious status, in the mines or in the building trade, while others make a living from hunting or fishing in the Kanak communities and from social and family allowances.

Another indication of the way in which this educational success differential among Kanak women contributes to the feminization observed can be found in the overrepresentation of women from the Loyalty Islands, particularly among managers and supervisors. No fewer than six of the ten managers interviewed were from this province, which is home to less than 7% of the New Caledonian population. This is in line with the observation previously made that people from the Loyalty Islands are more academically successful than young people from Grande-Terre (Nicolas 2010), which can be understood in light of the long-standing integration of this population in Nouméa, including at the time of the cantonment and forced labor.

Specific professional training courses taken by women

Women are also overrepresented in the vocational training schemes that prepare them for mining jobs, to the extent that these indirectly contribute to the feminization of KNS, compensating for the underrepresentation of women among spontaneous applications. "Let's say that for every ten CVs I get, one CV is from a woman," says the HR manager at the Koniambo Mine.[10] It is therefore not surprising that, in our sample, more than half of the interviewees had taken part in the schemes mentioned here, a proportion that climbs to 90% for the female operators at KNS.

The main training courses that feed into the process of entry into employment for Kanak men and women differ from vocational training as it is known in mainland France, and are part of the affirmative action measures designed to close the gaps that have opened in the past between Kanak and non-Kanak people. Our survey revealed that technical and managerial staff have benefited from these so-called rebalancing policies, some of which were included in the political agreements of 1988 and 1998 (*Operation 400 cadres* and, later, *Cadres avenir*, which also targeted non-Kanak people). Two of the interviewees, Karine and Sonia, attained their bachelor's degree, thanks to the support of the North Province and the Chamber of Commerce and Industry; Florence is working toward a master's degree at Haute Ecole de Commerce (HEC) with the support of the New Caledonian government; and Lynda acquired her executive status by joining the army, which enabled her to go to France and then take evening classes at National Conservatory of Arts and Crafts (**CNAM**). Conversely, none of the three non-Kanak managers interviewed had taken part in such schemes. Several of the interviewees went through the "juvenile school" system, a scheme that specifically targets "bush" high school students by helping them obtain a scientific or technical *baccalauréat* and through one of the two DEUST (*baccalauréat* + 2 years' higher education) courses that prepare them for jobs in the mines. These latter courses generally have a much higher percentage of women students than the occupations for which they prepare (46.3% of the students in the metallurgy DEUST between 2010 and 2019 were women).

The least qualified women have often taken part in two schemes: The *Mission d'Insertion des Jeunes* (MIJ)[11] and the *Régiment du Service militaire adapté* (RSMA).[12] The former is an association created 30 years ago, after the Matignon Agreements, to promote the integration of the most disadvantaged young people. Meanwhile, with two detachments based in the North, the first of which was created in 1986 at the height of the Kanak political turmoil, the RSMA emerged in the wake of a system created in 1961 specifically for the overseas territories. It is seen as a godsend for many young people aged 16–25 years who have difficulty accessing employment, boasting an integration rate of 72%, with 43% of those positions in long-term employment. Nowadays, employers and elected officials (including independents) value those who have joined its ranks, and the program appears to give renewed focus to young people who are considered to be "losing their bearings" on a cultural level (Salomon 2020). Young Kanak women have responded positively: Of the 619 young people accepted by the RSMA in 2019, 45.2% were women, and this proportion has been rising steadily in recent years. Although few young people go on to work in the mines (around ten in Koné), the RSMA has the advantage of covering the financial costs of the heavy goods vehicle license that machine drivers need in order to work in the mines.

The Centre de formation aux techniques de la mine et des carrières (Technical training center for mining and quarrying; CFTMC), which provides the certificate of aptitude for safe driving (CACES), is also specific to the territory and has become a condition of employment for drivers. In 2006–2007, the New Caledonian government temporarily imposed a quota of 50% for women, 90% of whom found a job, compared to 82.4% for men (Testenoire 2017). Even though the tests conducted by mainland psychologists for entry into training (aptitude and motivational tests based on knowledge of the profession) put women at a disadvantage (*Ibid.*), this appears to be compensated for by gendered differences in relation to addictions and risky behavior, which are systematically sanctioned by all the schemes mentioned. At the Poro training center, tests for cannabis use are carried out. A positive test leads to exclusion, regardless of the stage of progress of the learner in question. Employers such as KNS, keen to limit risky behavior, also carry out such tests.

However, although surveys that have sought to measure the addictive practices of young Kanak men and women have emphasized both their exceptional intensity and the fact that there is little difference between the genders in terms of consumption, they have also shown that boys are more often regular smokers (Hamelin *et al.* 2008; Estivals 2009) and that their consumption is more frequently associated with risky behavior (Baromètre santé jeunes 2019). Among the respondents, several saw cannabis use as a past stage in their lives, clearly different from the one that saw them gain stable employment in the nickel industry ("Here you have many advantages when you don't smoke cannabis," explained one trainee). Rosemarie, the only woman to declare that she smokes (and grows) cannabis, is, on the other hand, particularly unstable in terms of employment. She lives close to the Vavouto plant, in a Kanak community, has no qualifications, and only works on an occasional basis for a subcontractor. The strict attention paid to addictive practices in the context of training for nickel jobs, and at KNS itself, is therefore likely to work in favor of the candidates.

Ultimately, as with studying for diplomas, women appear to be better able to meet the institutions' expectations. For example, one CFTMC trainee explained that it was important to be able to "put on a show" in order to be taken on, in this case by emphasizing her status as a responsible single mother and her determination to get by, in other words, to conform to the self-presentation expected by such schemes.

It should be emphasized that the integrative virtues attributed to specific training courses—especially those judged by the public authorities through a bleak and stigmatizing discourse as being "the last chance"—unequally benefit the people concerned. Although communication campaigns in high schools or BTS (Higher National Diploma (HND) technical courses), or a visit to a Cap Emploi employment agency, encourage these pathways, this type of knowledge is not evenly distributed, as Marie Salaün (2014) previously noted with regard to young Kanak people who left to train in Québec for mining jobs. Membership in kinship networks that are better integrated than others into New Caledonian society (through trade unionism, politics, or even religion) undoubtedly plays a supportive role in encouraging young people to take part in existing schemes.

A career in nickel . . . or more broadly in employment?

At the time the plant was built, Christine Salomon (2015) showed that women did not really anticipate a specific profession but rather sought simply to gain employment. Our survey shows that, even today, women operators, whether employed by KNS or by subcontractors, do not generally see themselves as building a professional career in the nickel sector. Employment at KNS comes at a time when they have had some initial experience of odd jobs or contracts with local institutions (town halls or schools), which familiarize them with the range of vocational training and jobs available. This is the case, for example, for Doriane (age 30), who had already been with the company for 8 years at the time of the survey. She joined RSMA only 18 months after discontinuing her studies (in her first year of vocational training). She then took a job in a laundry and then obtained her driver's license, which opened up a position as a driver in Nouméa, the key to accessing the upper stratum represented by CFTMC and employment in the mines. After joining KNS, Doriane took advantage of her right to training to become certified as a volunteer firefighter and then obtained a soil testing qualification which, after several years of operating machines, has enabled her to join a subcontractor specializing in this field.

For managers, on the other hand, the move to KNS, a high-tech multinational, represents a key step in the accumulation of specialized professional capital. Whether they are in production or support roles, Kanak members of the management team have the characteristic of inscribing their relationship to their job in a professional trajectory that is much more linear than that of subordinate employees, and which—much more often, although not systematically—began with the pathway choices that guided their initial training (6 cases out of 10). The prestige associated with the experience acquired at KNS is a career resource, as emphasized by many of the women interviewed. The political significance of this experience also counts for these managers (whereas a majority of the operators do not attach any importance to it), leading one of them, who is committed to the Loyalist camp, to say that she only applied for the job as a joke, while several others spontaneously insisted on the significance of their commitment to KNS as a Kanak. All in all, the ability (and professional need) to portray one's path as the gradual development of a career is a distinctive feature of the interviewees in management.

The answers given in the interview were often similar to the biographies presented on LinkedIn, on Facebook, or in the profiles produced by the SMSP monthly newsletter. As elsewhere, the use of LinkedIn is a distinctive attribute (12 of the 14 managers or supervisors in the sample had a LinkedIn profile, while none of the other 19 respondents did). These pages, which are sometimes introduced by a short personal pitch, make it possible to highlight one's training and present one's career path. The expectations of employers and the reputational issues specific to the restricted space of private sector executives in New Caledonia, through which they progress, are clear.

Whether or not the interviewees viewed their entry into KNS as part of a professional career move, it was nonetheless part of a wider social trajectory. The Kanak employees of KNS are in fact unique in that they most often belong to families (both extended and nuclear) that have accumulated experience in salaried employment, particularly in the mining industry. Taking all statuses together, the proportion of respondents with at least one employed parent is exceptionally high (21 cases out of 26 Kanak women who answered this question), and in over 70% of cases, the parent is a father or grandfather who is a miner. Consequently, these pioneers in terms of gender are less so if we consider their family experience. This explains why, among female workers, those who are not local are generally from the Indigenous communities affected by mining activity on the east coast of the mainland, between the communes of Canala and Ponerihouen, where employment opportunities are otherwise scarce outside this area.

> As for my father, all his brothers drive a truck to the mine. As for my mother, she's more into . . . fields and all that. In fact, nobody works. Her brothers and sisters live off that, the field . . . hunting . . . the products they sell at the roadside.
>
> (Aline, a welder with a subcontractor and originally from the Canala area)

Male ancestry is also often used to justify the attraction of the mining sector, in explanations that attempt to reduce the exceptional dimension of the individual experience.

And if you didn't work at KNS, would you have stayed in the tribe?

I don't think so . . . it wasn't in my plans to stay at the tribe. In fact, it's because my old man was . . . well, he was someone who used to work in the mines and there are a lot of them in the family . . . So I think it was the fact that I always woke up early in the morning, as a child, with my old man . . . Afterwards, it was this rhythm of working hours that I liked, working in the mines.

(Doriane, driver, originally from Nonhoué, Canala)

Among the managers, this mining socialization is coupled with access to an exceptionally high level of schooling within the families. Three of them (four, if we include one father who is a pastor, which can also be considered a qualified job) have parents who hold higher education diplomas, which is not the case for any of the other respondents, and which was the case of less than 1% of Kanak people before the year 2000 (Gorohouna and Ris 2015).

Working to gain autonomy

By shedding light on the weight of previous socialization in the trajectories of employment at KNS, we are not implying that employment is part of a smooth biographical continuity. On the contrary, it is often experienced as a real fork in the road by the interviewees, who emphasized their desire to escape family constraints or to bounce back after a life event.

Once again, we must begin by underlining a specificity common to the Kanak women employees of KNS. Taking all ethnic groups together, single women were clearly overrepresented in the 2017 workforce. Only 15% of 30- to 34-year-olds were in a relationship, compared to 31% nationally (this proportion rises to 26% among 35- to 39-year-olds, compared to 42%). But the Kanaks, who represent two-thirds of the total, are six points above the average proportion of their colleagues who have at least one child (53% compared to 47%). This discrepancy can be explained by the fact that these Indigenous women employees are faced with a social overvaluation of motherhood, with an ideal that may reach 5 or 6 children (Salomon 2000). Raising children, combined with a refusal to depend on the spouse and his kinship group (residence is virilocal in the tribe) or a desire to escape from the conflicts arising in the relationship, drives the search for salaried work in town (in Koné). Among female workers at KNS, and for subcontractors, less than one-third had seen their relationship survive the wage situation (this was the case for 50% of managers and supervisors). Separation is the life event most often associated with entry into training and employment.

Then the second one was born and from then on, I could no longer live like that in a tribe [with her ex-spouse]. Well, my father didn't teach me to live off other people, I'm not a parasite. And with the children, we need money, well

> the nappies, they're not banana leaves. Their father wasn't too . . . he made little commitments here and there. But I understood that this was the way it was with him. And what motivated me to work was the fact that my mother lived at home, it's not easy every day. I can't . . . it was impossible for me to live like that.
>
> (Gina, driver, 2 children, joined KNS at age 24)

Similarly, it was often childbirth at a young age (one-quarter of the respondents had had their first child before the age of 20), with the obligations that entailed, that motivated the decision to bounce back through employment. Rose, along with others, explained that "[her] will and perseverance" stemmed from the daughter she had at age 16. But her example also attests to what constitutes an enabling condition for these transitions: The availability of resources in the family circle that can be mobilized to guarantee their material means, starting with childcare and, thus, women's work. The interviewees emphasized the decisive role of their mother (or her sisters) in initiating their change of direction, during the training and establishment period. These women family members helped them cope with domestic tasks despite highly restrictive working hours (44 hours over 4 days, followed by 4 days off for the operators). Erika's mother looked after her little girl in the community for three and a half years, and now travels back and forth between the commune of Kouaoua on the east coast and Koné to continue assisting her daughter. Gina and Josiane, a cleaning worker, have received the same kind of support and expressed their willingness to give back, by taking care of her little brothers, in the case of the former, or by doing the shopping for her mother every weekend, for the latter.

For some women, their move into employment also stems from dissatisfaction with life in their Indigenous communities, where they (in their own terms) feel "not well established," materially and symbolically, in the husband's clan. Single mothers find themselves in an uncomfortable situation, devalued because of the uncertain status of the child (which in principle falls under the father's lineage but is often cared for by the mother's clan). The case of Erika, who became a driver at the age of 29 after two major events, is a good example of how the search for a job can be an attempt to bounce back from a weakened, initial social position. While the early death of her father had prevented her from obtaining her vocational *baccalauréat* in hotel management, the humiliating abandonment by her ex-husband, an employee at the southern plant who left her for a colleague, threatened to lock her and her daughter into the Kanak tribe of her birth. After a period of "discouragement," Erika first got her HGV license and did minor contracting work at the mine next to her community in Kouaoua and then enrolled at CFTMC, after applying via Cap Emploi.

This increased autonomy is also reflected in changes in consumer behavior (like Gina, who links the signing of her permanent contract to the purchase of a car, which is her pride and joy) and in place of residence. Access to property outside the Kanak community, in the towns of Koné, Pouembout, or Voh (villages with a greater ethnic mix), is very common among the respondents. It applies to a majority of female operators, half of the employees of subcontractors interviewed, and is virtually systematic among managers. Others also rent accommodation outside their community. For these women, the creation of the industrial hub in the North clearly represents an additional opportunity other than Nouméa to assert themselves in relation to men and to free themselves from certain social obligations. This is also seen in the advanced savings and investment strategies they are putting in place, and the efforts made to ensure their children's success at school, which are even more extraordinary at senior levels (two executives have already opened a savings account for their babies, only a few months old).

For most of these women, salaried employment in the nickel industry therefore appears to be a way out. However, while more or less accepted by their family of origin, it is often contested by the spouse, sometimes even violently.

Spouses who feel threatened by women's work

For the women interviewed who were still in a relationship, or who entered into a new one once they were employed at KNS, their testimonies were consistent regarding the sense of threat that women's work poses to the position held by the man (the case of a woman driving a mining machine that is larger than her husband's is very symbolic in this respect). Arguments and even physical violence form a recurring pattern, characterized in particular by the man's jealousy of his partner's male colleagues and his anxiety about the opening up of the domestic sphere.

> We've been separated for a while now. Because he was too . . . sometimes we argued for hours. He would even wake me up with a start and say "how come you're talking to so-and-so?" I would say, "He's a colleague from work. What's wrong with you?" From then on, I was so upset that I said, "either my work is taking a hit, or my family is." So I said no, ah no, I've struggled to get here, I'm not going to give up my job. Because when I came to work here, it was because we were already having problems as a couple. That's what motivated me to leave home. I was at home with the children and he sometimes went to work, didn't come home in the evening, came back two days later and all that . . . After that, it built up over time. And I said, I'll wait until the children grow up a bit and then I'll start to do it. That was with the little one. The second one I was [secretly] pregnant with him when I was in training in Poro.
>
> (Doriane, driver, 2 children, joined KNS at age 20)

In fact, even when the couple stays together despite them both being employed, problems can still persist. Solange, a technician who joined KNS after completing her DEUST diploma, has been pressured by her husband for 5 years to return to the community, where he has land and the support of his clan. In the end, he gave in to his wife's wishes and came to live with her and their two children when she bought a house in Koné by herself. Sonia, an HRM, had the same experience: Her husband finally gave up trying to prevent her from moving to the village of Koné when she gave him an ultimatum about maintaining their relationship, and the children now move between the apartment she bought and the Kanak community.

The trajectory followed by Marie-Noëlle, a mother of eight children, who joined KNS at the age of 40 via Cap Emploi and CFTMC, corroborates several of the above remarks. Her reason for joining the mine, after escaping to Nouméa for a year and working in a series of precarious jobs, was to put an end to the feeling of confinement she felt in the community and to the humiliation inflicted by an ex-husband for whom she had dropped out of school at the age of 14. The new husband she met in Koné works for a small mining company. He is less qualified than she is, but is also less jealous than his predecessor. After 3 years on the machines, Marie-Noëlle took and passed a competitive examination to become a care assistant.

> I took the exam. And I passed! And my husband fell on his ass. Yes, he did. My children said to me "Hey Mum, Dad said wow, she passed the exam!" I told them "Well, why, did he think I was stupid?"
>
> (Marie-Noëlle, eight children, former machine driver-turned care assistant, January 2019)

While Marie-Noëlle expresses pride in having learned to drive mining machinery and in having escaped from a humiliating situation, and while she is keen to convince her daughters that they, too, can build their independence, she also insists on the fact that "in the end, mining work is not women's work." This assessment, sometimes shared by the women concerned, highlights the ambivalence of Kanak women's contemporary experience toward work.

The workplace: Resource or vector of heteronomy?

Although work often represents a turning point in the lives of Kanak women, it should not be seen as the start of a linear path for their future. The ambiguous place given to female recruits in their new workspace, and the emotional isolation they say they feel, often leads them to take different directions.

Noting the feminization of its workforce, the Koniambo Nickel company has undertaken efforts to promote its women employees. Each year, it invests in the celebration of International Women's Day on 8 March and sponsors the Orange the World Campaign, run by UNESCO to mark International Day for the Elimination of Violence against Women, on 25 November. During our survey, these initiatives were lauded not only by the members of the management team in charge of publicizing them but also by the women employees, who freely praised the greater openness of the multinational KNS to feminization compared to other mining employers (particularly those from local fortunes). This attention to gender violence is nevertheless accompanied by a racializing discourse, also widespread in other multicultural contexts, which consists of criticizing the society of the other under the guise of championing feminism (Fassin 2006). The implicit narrative here is one of a dichotomy between an Indigenous social world described as the site of violence against women and a "modern" world that is egalitarian and respectful of individuality. This framing is present in almost all the interviews conducted with non-Kanak respondents, who return to the same traumatic event—the rape of a female employee by a man from a neighboring Indigenous community. These culturalist formulations, however, overshadow the shared—albeit unequal—dimension of domestic violence in New Caledonian communities (Hamelin and Salomon 2004) and merely reinforce the criticisms of Kanak feminists who have long denounced racism and gender inequalities in the domestic sphere and in (post) colonial relations (Réveil Canaque 1972). In this context, gaining employment at KNS, while initially experienced as a form of emancipation, has nevertheless led several Kanak women employed by the company to feel a conflicting sense of loyalty, affecting both the personal sphere and the feeling of belonging to a collective identity sharpened by Kanak nationalism.

This is all the more true when the employer uses a second type of essentialism, seemingly flattering for women, to justify the feminization process. That is, if women are needed in the nickel industry, it is by virtue of certain dispositions that are feminine by nature and that apparently make them better operators or better drivers of the mining machinery, more conscientious, and careful with the machines.

> The different management levels at Koniambo Nickel praise their female employees for their diligence, their respect for procedures regarding attendance at work without being under the influence of various substances, their care for the equipment, and their productivity.
>
> ("The Koniambo Nickel Woman," KNS press release, 6 March 2015)

> They are more productive. Less gear breakage, so I'd say they're gentler . . . they're more careful too. But I don't want to offend men because it's in their nature to be hard. But women are very gentle and very productive, so it's valuable for the teams.
>
> (Rose, CFTMC consultant, quoted above)

This type of assessment is not specific to the studied field and only rarely refers to representations of Kanak culture. It has even been identified by Michel Gollac and Volkoff (2002) as a constant among rationalizations for the move to hire women in industry, promoting a "putting to work of gender stereotypes." From a subjective point of view, we must consider the ambiguity of the mode of self-legitimization that it proposes to female employees, who are reduced to a series of gendered assignments that the employer does not fail to solicit in return. The premise of increased female productivity is in fact based on allusions to two types of qualities: On the one hand, "gentleness" or "thoroughness," which amounts to cost savings linked to wear and tear on the equipment through increased mobilization at work (concentration and patience when driving), and, on the other hand, "seriousness," "discipline," or "availability," which underline the possibility of a specifically extensive use of female employees.

The professional positioning proposed to Kanak women workers in this way invariably puts them at odds with their male colleagues, whose position is discredited and who react with various forms of denigration. In addition to this "positive" gendered discourse, there is also the much more derogatory discourse, already identified during the plant construction phase, of the condemnation of Kanak women as "easy" women who, by working, deviate from the valued form of female sociability confined to the domestic sphere. This view is held not only by Kanak men but also by others. It also reflects colonial discourses—and, in particular, the familialist discourse of missionaries, who were ambivalent about Kanaks working.[13] In the testimonies collected, places of work (and training) are hyper-sexualized, perceived as places marked by the free and "uncontrolled" movement of women. After describing KNS and its living quarters as a "brothel," the spouse of a female respondent, a welder, stated: "I've seen twelve maternity leaves. They're not the real fathers. And the first six months, well, the next six months I saw about twenty of them." In this context, many women employees describe the code of conduct they have adopted to avoid being labeled as "easy."

> In fact, when you work in the mine, you really have to put up barriers with the men. You have to have a protective threshold. After that, you have to be careful with your clothes, you can't be too forward, you have to be careful. (. . .) Where you go out, who you go out with . . . When you want to go for a drink, sometimes it is interpreted differently. They'll say, "Oh, he's slept with her or she's led him on." You know? It's easy to be labeled, isn't it?
>
> (Lynda, purchasing manager at KNS who previously worked in production, then in HR at SLN)

In the mines, women employees are confronted with symbolic aggression from their male colleagues. "We were constantly hassled," says Géraldine, a former sprinkler driver, recalling, like several of the other women interviewed, the offensive and sexualizing "jokes" made by the operators on the radio used for mine coordination operations. Three of the interviewees recounted direct incidents of sexual harassment. The publicization and repression of such behaviors at KNS remain clearly limited, despite the procedures put in place on paper. While one CFTMC manager only raised the issue to talk about how a victim had "lied" about her consent, two others at KNS redirected the interviewer's focus and wanted to talk about "feelings of harassment" instead. Among them, Florence, a manager, explained that she had been confronted with the "problem" several times but, in the absence of "evidence," had organized amicable mediation within her department.

The process of feminization studied here has therefore led to a power struggle between the sexes in the workplace, similar to that observed by Stéphane Beaud and Pialoux (2002) in the French car industry in the early 2000s, but with the specificities linked to the postcolonial context of New Caledonia. What is at stake is the (de)legitimization of female recruitment, leading to a multitude of trials experienced on a daily basis, which crystallize episodically, such as during the strike carried out at SLN in the commune of Thio, in January 2020, against what was denounced in a leaflet as the "excessive hiring of women."

The respondents' accounts demonstrate that they find the resources to oppose this delegitimization and continue working. More than two-thirds of the women interviewed mentioned they have constructed a personality that keeps certain representations of female socialization at bay in order to put forward others that are a far cry from the ethos of submission (Salomon 2000). Having a character ("a bad temper" or "being ready with a quick reply," "being able to lose it" or "blow a fuse," "being authoritarian," "carrying a big stick," etc.) is key here for many female employees. "I'm often told that I'm a ballsy woman," says Lynda (quoted above) with a touch of pride, a sentiment also shared by Géraldine (a driver), when she recounts that her mother and uncle immediately identified her as a "tomboy" who surpassed her brothers in mechanics.

From this point of view, wage employment also provides resources that enable the transformation of gender relations to be reinforced. However, this form of self-affirmation can place women employees in peril by pushing them to an unusual level of investment in the job. Whether they were with plant technicians like Solange, HR advisors like Sonia, drivers like Géraldine, or plant operators like Nadia or Andy, the interviews we conducted are full of examples showing that the integration of gender norms leads them to make themselves more available in general, and during periods of "all hands on deck," and to take on unpaid tasks more than others. As a result, work becomes ever more intrusive.

The interview with Anna, an employee of a subcontractor firm specializing in cleaning, shows that the "strength" that women put forward in their work capacity can backfire. Her husband, a welder for KNS, intervened vehemently on several occasions to underline the harm done by her "workaholism," accusing her of being "too nice" and recounting how he had once gone to her boss to "tease him" about the overtime she had accepted. Hyper-availability for work is sometimes part of a care-based approach which, when mobilized in the workplace, increases tensions with regard to domestic responsibilities, especially when pointed out by the spouse. The path followed by Graziella, an operator and then technical director in a revegetation company, before becoming a stay-at-home mother at the age of 37, is a good example of this. As the granddaughter of a chief, she wanted to "not live the life of [her] mother" and was burdened by the failure of the sheep farming project she had set up after completing her BTS diploma because of a lack of accessible land ("My father explained to me that, as a girl, you have no right to your own property."). After 5 years of odd jobs, her meeting with Mr. "M" marked a turning point. Hired as an operator in his company, she climbed the ladder, thanks to her unlimited commitment to her work (weekends and holidays):

> That is to say, it even went as far as the snack bar. There were some, they couldn't pay for their snacks, so what do you do [sigh], "Well hang on, yes, I'll help you out." But the thing is that the help lasts all month. So I had my ex-boyfriend who used to say to me, "Oh, put the brakes on, you're not their mum." Sometimes he would say to me, "You take more care of them than of

> your daughter." I said "Yes, that's true." (. . .) So no, but I'd rather look after my kids at home than lose my mind with people who are forty years old and don't know how to manage their lives.
>
> (Graziella, November 2018)

Now at home, Graziella takes care of all the domestic chores, looks after the children of her new partner's son in addition to her youngest child, and puts her professional skills to use to handle the administration of the subcontracting business run by her husband. "Maybe it's our fault too, we're the ones who give them this habit," she says. This doubling of responsibilities, and the burden to reconcile them that it places on the interviewees, also marks the trajectory of Gina, a driver who is hyper-available for work, but also a single mother of two children, and who takes in three of her younger brothers ("I'm kind of responsible for everyone."). For her, these multiple commitments nevertheless appear in a more gratifying form, as a vector of authority in the family, which she feels she shares with her own father.

An unsustainable job? the challenging quest for personal balance

The above observations shed light on the discomfort felt by many Kanak women employees at KNS, notwithstanding the autonomy conferred by their jobs. Whatever their status and hierarchical level, many of the women interviewed described their commitment to mining as untenable over time. They feel they devote too much energy to their job and that it generates too many conflicts, to the extent that it calls into question the place given to motherhood and marriage. Lynda, a single, childless executive who describes herself as impervious to men's remarks, confessed that she was badly hurt when a colleague told her that she had "only work in [her] life." Our survey identified another recurring theme: Several women used the resources they had gained in their jobs to make a second shift in life, motivated by the desire to have more time for themselves or for their children. Like Graziella, Géraldine made a concerted move toward motherhood and the domestic sphere as she entered her forties, after working 10 years in the mine. Leaving her job, she relied on her partner and devoted herself to their baby, whom she did not plan to leave with a nanny as her colleagues did. The interview with Géraldine underscored a polarity between strong professional identification and pride, and this self-discovery as a "very maternal mother."

Three other interviewees, who were single and had no children, expressed their intention to leave their jobs or switch to part-time work in order to consider starting a family. Isabelle, a technician at KNS at the time of the survey, has since done this by taking advantage of her skills to set up her own business in Maré, her hometown. After describing the life she would have had outside of work and her mother's existence in the tribe as so many deterrents, Nadia, age 39, described how, in 2012, she was recruited and then separated from her first spouse, with whom she had two boys, and expressed her joy at being able to accommodate the "little family" she is rebuilding in the house she purchased in Pouembout and to buy everything her "baby doll" (born 4 months ago) needs.

While these reversals may suggest the integration of gender norms, some of them emerge from the loyalty conflicts identified above. For example, investment in "customary obligations" (attending exchanges between kinship groups within the Kanak community) on the occasion of a birth, a wedding, or a bereavement, where the women are in the kitchen—sometimes for several days in a row—raises questions. The cultivation of a field of yams, presented as a condition for taking part in customary ceremonies with dignity, is the subject of divided opinions during the interviews. While Lynda abruptly rejects the interviewer's question ("Me, in a field??"), Josiane asserts that without this investment of oneself in the land, "you are nothing" as a Kanak, and Nadia, along with others, maintains that giving it up "is not good." Some women told the interviewer about the tactics they had developed to reconnect with their community after a period of distancing themselves from it due to their job commitments. Graziella, for example, has restarted her yam field after a long period when, like many of her colleagues, she was content to buy yams for the customary dates. She says that she takes advantage of available opportunities to show her crop to her father, with whom she had been estranged since her move into employment. For Rose, getting closer to her community represents a new stage in her life, following a phase of what she considers "excessively intense" employment. Having become a part-time consultant after starting work at the age of 18, she considers that maintaining her ties is a matter of education for her children, a way of emphasizing a form of Kanak socialization that involves knowledge of kinships. Her first decision after her resignation was to marry according to custom (at age 33 and with 4 children).

Conclusion

While it is important not to infer a one-sided interpretation of the relationship between these Kanak women employed in the nickel industry and "customs" as a moving set of specific norms and practices, we cannot fail to note the insistence of several of them on this identity marker as both social (conferring a place in a clan and a chiefdom) and cultural. After phases of disconnection that were felt to be excessive, these respondents are trying to make up for lost time, revealing the burden placed on them by particularly strong family assignments, given their gender, but from which Kanak men are not exempt either. For Kanak women employed in the nickel industry, gender relations and racialized representations of sexual relations make the alternatives more divisive and brutal. Especially for those from Indigenous communities further away from the industrial site, and who did not grow up in the city, the task of reconciliation is more difficult, and this is reflected in their frequent life changes. On the other hand, the fact that they are more successful in their schooling and training gives them advantages, just as the gradual legitimization of their presence in these men's occupations can encourage a transformation of gender relations and representations and herald more equal relations within their Kanak communities, albeit not without considerable effort on the part of these women.

In this respect, the biographical trajectories presented here shed a historically situated light on Kanak women's employment and suggest further investigation is required from a non-essentialist intersectional perspective. More specifically, we may ask how the social and personal situation of Indigenous women will evolve if their presence in the nickel industry becomes even more commonplace. To what extent will their presence alter the way the New Caledonian working class is viewed, of which the mining sector, as in many other places, constitutes a paradigm?

Acknowledgments

This work was supported by the Knowledge Network on Mining Encounters and Indigenous Sustainable Livelihoods (MinErAL) through funding from the Social Sciences and Humanities Research Council of Canada (grant number 895-2016-1015), Institut Agronomique néo-Calédonien (IAC), and the Centre Norbert Elias (CNRS). We are deeply grateful to the former and current KNS staff and to all interviewees for their time. Any errors are solely the responsibility of the authors.

Notes

1 This analysis of the 2017 KNS workforce was carried out by Séverine Bouard. It is worth recalling that women represented 28.9% of employees in the industry in mainland France in 2018 (INSEE 2020).
2 Fictitious first names were used to mask the identity of the interviewees.
3 Interview with Kelly, performance coordinator, October 2018.
4 www.koniambonickel.nc/actualite/press/communique-de-presse-06d31?ln=En, [accessed 23 June 2020].
5 Interview with Floriane, October 2018.
6 Interview with Josiane, November 2018.
7 In New Caledonia, the term "tribe" refers to the areas of residence created with the colonial cantonment and still subject to a specific land tenure system. At that time, the creation of the legal entity "tribe" on the land of the "reserve" was intended to control the Kanak. This process led to the introduction of a system of indirect rule that made the "chief" responsible for the actions of his "subjects." At the end of the *indigenat*, both the tribe and the particular land tenure system were preserved to enable a different personal status to be maintained. Today, "tribes" constitute an emic category of social and spatial belonging. It is to this entity that we refer when we use the term here.
8 The *Nouméa Agreement* established New Caledonian citizenship based on a restricted electoral body for local elections and on preference for hiring New Caledonian citizens. A text was adopted, albeit with difficulty, in this regard for the private sector in 2010 and another for the public sector in 2016.
9 Proportions calculated from data available at www.isee.nc/population/jeunes-femmes-seniors/femmes.
10 Interview with Sonia, HRM for KNS mine management, December 2018.
11 This association is in charge of supporting the integration of youth into the workforce.
12 Adapted military service program.
13 In 1936, Catholic missionaries obtained a change in the rules for forced recruitment of Kanak workers in the name of defending the family. No more than one-third of adult men were allowed to work outside the tribe, with conditions attached to the hiring of married men. The hiring of women was limited to those without children (Merle and Muckle 2019).

References

Baromètre santé jeunes, 2019. La santé en chiffres. Conduite de véhicule à moteur. Available from: www.santepourtous.nc/la-sante-en-chiffre/barometre-jeunes/tabac-alcool-cannabis-et-autres

Bayart, J.-F., 2006. *L'État en Afrique: La politique du ventre*. Paris: Fayard.

Beaud, S., and Pialoux, M., 2002. Jeunes ouvrier(e)s à l'usine. Notes de recherche sur la concurrence garçons/filles et sur la remise en cause de la masculinité ouvrière, *Travail, genre et sociétés*, 8, 73–103.

Bouard, S. (Coord.), Levacher, C., (Coord.), Bencivengo, Y., Decottigny, L., Demmer, C., Le Meur, P-Y., Blaise, S., Burton, J., Enjuanes, F., and Grochain, S., 2019. *Petites et moyennes entreprises minières en Nouvelle-Calédonie. Programme PME Minières en Nouvelle-Calédonie*. Nouméa: CNRT Nickel & son environnement, 108 pages + appendices.

Bouard, S., and Sourisseau, J.-M., 2010, Stratégies des ménages kanak: hybridations entre logiques marchandes et non marchandes. *Natures Sciences Sociétés*, 18 (3), 266–275.

Canning, K., 1996. *Languages of labor and gender: Female factory work in Germany, 1850–1914*. Ithaca, NJ: Cornell University Press.

Demmer, C., 2015, De la diversité des enjeux du nickel pour les kanaks aujourd'hui. *Ethnies*, 37-38, 134–158.

Demmer, C., 2018. Nationalisme minier, secteur nickel et décolonisation en Nouvelle-Calédonie. *Cahiers Jaurès*, 230 (4), 35–52.

Downs, L. L., 1995. *Manufacturing inequality: Gender division in the French and British metalworking industries, 1914–1939*. Ithaca, NJ: Cornell University Press.

Estivals, M., 2009. *Tabac et cannabis chez l'adolescent scolarisé de Nouvelle-Calédonie*. Mémoire pour le DIU de pneumologie pédiatrique, Marseille: Faculté de médecine de Marseille.

Fassin, E., 2006. Questions sexuelles, questions raciales: parallèles, tensions et articulations. *In:* Didier and Eric Fassin, eds., *De la question sociale à la question raciale*. Paris: La Découverte.

Freyss, J., 1995. *Economie assistée et changement social en Nouvelle-Calédonie*. Paris: PUF.

Gallot, F., 2015. *En découdre*. Paris: La Découverte.

Gollac, M., and Volkoff, S., 2002. La mise au travail des stéréotypes de genre. *Travail, genre et sociétés*, 8, 25–53.

Gorohouna, S., and Ris, C., 2015. Inégalités ethniques sur le marché de l'emploi en Nouvelle-Calédonie. *Ethnies*, 37–38, 126–135.

Grochain, S., 2007. *Les Kanaks et le travail en Province Nord de Nouvelle-Calédonie*. Paris: Thèse de sociologie, École des Hautes Études en Sciences Sociales.

Hamelin, C., and Salomon, C., 2004. Parenté et violences faites aux femmes en Nouvelle-Calédonie. Un éclairage sur l'ethnicité différenciée des violences subies au sein de la famille. *Espaces, populations, sociétés*, 2, 307–323.

Hamelin, C., Salomon, C., Lert, F., Vignier N., Bonenfant S., Kanieswski N., 2008. *Situation sociale et comportements de santé des jeunes en Nouvelle-Calédonie*. Rapport d'étude, Paris: Inserm, 112 p.

Institut national de la statistique et des études économiques (INSEE), 2020, *Tableaux de l'économie française, édition 2020*, Collection Insee Références, Paris: INSEE, 266 p.

Institut de la statistique et des études économiques (ISEE), 2019. Enquête sur les Forces de Travail en Nouvelle-Calédonie: résultats 2018. Synthèse n° 40. Nouméa: ISEE, 5 p. Available from: https://dtenc.gouv.nc/sites/default/files/atoms/files/eft-resultats20182.pdf

Institut de la statistique et des études économiques (ISEE), 2019a. Synthèse enquête force de travail 2019. Approche par genre. Synthèse n°46. Nouméa: ISEE, 5 p. Available from: https://www.google.com/url?sa=t&source=web&rct=j&opi=89978449&url=https://

www.isee.nc/phocadownload/non_theme/publications/toutes_publications/Synthese/Emploi-chomage/eft2019-approchepargenre.pdf&ved=2ahUKEwj1p4GjotWGAxUSjK8BHe_LCboQFnoECBAQAQ&usg=AOvVaw0DMFLjTr9wLjCdo2LFgUl5

Institut pour le développement des compétences en Nouvelle-Calédonie (IDC NC), 2018. *Etude prospective emploi-formation 2017: principaux résultats*. Nouméa: DFPC-NC, 20 p.

Jézéquel, J.-H., 2003. Histoire de bancs, parcours d'élèves. Pour une lecture "configurationnelle" de la scolarisation à l'époque coloniale. *Cahiers d'études africaines*, 43 (169–170), 409–433.

Kergoat, D., 1982. *Les ouvrières*. Paris: Sycomore.

Kergoat, D., 2012. *Se battre, disent-elles . . . .* Paris: La Dispute.

Le Meur, P.-Y., 2017. Le destin commun en Nouvelle-Calédonie : entre projet national, patrimoine minier et désarticulations historiques. *Mouvements*, 91 (3), 35–45.

Mak, A., 2017. Conspicuous consumption in wartime? Welsh mining communities and women in munitions factories. *In* C. Peniston-Bird and E. Vickers, eds., *Gender and the second world war: Lessons of war*. London: Palgrave Macmillan, 55–72.

Merle, I., and Muckle, A., 2019. *L'indigénat. Genèses dans l'empire français. Pratiques en Nouvelle-Calédonie*. Paris: CNRS éditions.

Merle, I., 2012. Autour de la condition ouvrière en Nouvelle-Calédonie. Le recrutement des travailleurs asiatiques entre les deux guerres et aujourd'hui sur les chantiers de mines, Rapport pour le projet Gouvernance minière n°11/12, CNRT Nickel.

Nicolas, H., 2010. "Emporter un diplôme dans son sac." Les transformations de la socialisation sexuée à Lifou (1945–2004). *In*: I. Merle, ed., *La Nouvelle Calédonie, vers un destin commun?* Paris: Karthala, 225–244.

Réveil Canaque, 1972. Sexualité et racisme, 20, 5–7.

Ris, C., 2013. Les inégalités ethniques dans l'accès à l'emploi en Nouvelle-Calédonie. *Économie et Statistique* 464 (1), 59–72.

Salaün, M., 2001. Les "moniteurs" kanak ou l'impossible élite indigène en Nouvelle-Calédonie. *Genèses*, 43 (2), 71–88.

Salaün, M., 2014. Des Kanaks au Québec: l'expérience de jeunes autochtones calédoniens en formation aux métiers de la mine à Rouyn et à Sept-Îles. *Recherches amérindiennes au Québec*, 44 (2–3), 55–62.

Salomon, C., 2000. Hommes et femmes. Harmonie d'ensemble ou antagonisme sourd? *In:* A. Bensa and I. Leblic, eds., *En pays kanak: Ethnologie, linguistique, archéologie, histoire de la Nouvelle Calédonie*. Paris: Éditions de la Maison des sciences de l'homme, 311–338.

Salomon, C., 2015. L'emploi des femmes kanak à Vavouto, *Ethnies*, 37–38, 160–186.

Salomon, C., 2020. Service militaire adapté et nostalgie du service militaire obligatoire en Nouvelle-Calédonie. *Outre-Mers, Revue d'histoire*, 108 (406–407), 49–71.

Testenoire, A., 2017. Dynamique et résistances au principe d'égalité réelle; la loi sur l'emploi local et la formation continue en Nouvelle Calédonie. *In:* Foued Laroussi, ed., *Insularité, langue mémoire, identité*. Paris: l'Harmattan, 245–261.

14 Conclusion

Thierry Rodon, Sophie Thériault, Arn Keeling, Séverine Bouard, and Andrew Taylor

The chapters in this volume offer a multidisciplinary exploration of the interactions between Indigenous communities and mining activities across different regions, highlighting the complex dynamics of legal and policy frameworks, environmental impacts, and socioeconomic considerations. A burgeoning literature on these questions has emerged in the past two decades, coinciding with the rapid global expansion of extractive industries associated with the "commodities supercycle." Like mining itself, much of this research is intensely place-based, reflecting the particular configuration of resources, environments, regions, and communities at extractive sites. The key contribution of the MinErAL project and this volume is their explicit attempt to explore comparative perspectives across these sites, in order to gain insights into the patterns and processes of extractive development in Indigenous contexts.

These studies reveal a consistent theme, namely, that despite growing recognition of Indigenous Peoples' rights and widespread discourses on corporate social responsibility (CSR); environment, social, and governance (ESG); and sustainable development, mining operations often prioritize development over these rights (e.g., chapters 1, 3, 5, and 7). Historically, this has been a persistent theme, and it is one that has resulted in limited participation of Indigenous Peoples in decision-making and negotiation processes (Boirin-Fargues and Thériault, Chapter 1) and significant and long-lasting social, health, and environmental impacts (Horowitz *et al.*, chapter 5; Keeling and Sandlos 2017).

The book's comparative analyses from Canada, Fennoscandia, Australasia, and other regions illustrate the complex challenges that mining encounters pose for Indigenous communities. The different chapters highlight Indigenous engagement in negotiating and implementing impact and benefits agreements (IBAs; O'Faircheallaigh and Rodon, Chapter 7); exercising agency within and beyond state- or industry-led decision-making processes (Boirin-Fargues and Thériault, Chapter 1; Bourgeois and Zema, Chapter 4); planning for closure and mine site transition (Keeling *et al.*, Chapter 8); and dealing with the socioeconomic, cultural, and ecological impacts (Horowitz *et al.*, chapter 5) and internal conflicts (Sehlin MacNeil, Chapter 10) generated by mining activities. However, as Bourgeois and Zema stress in Chapter 4, in spite of the asymmetrical power relationship, Indigenous Peoples are sometimes able to assert their political agency, advance new

DOI: 10.4324/9781003406433-19

This chapter has been made available under a CC-BY-NC-ND 4.0 license.

ways to interpret their participatory rights, challenge how the state defines their rights, and create new opportunities to either resist or capture benefits from extractive developments.

From an extractive justice perspective, the analyses in this volume underscore the need for more effective recognition and protection of Indigenous Peoples' rights in mining regulatory frameworks and practices; enhanced accountability of investors and mining companies (Boirin-Fargues, Chapter 3; O'Faircheallaigh and Rodon, Chapter 7); and a reassessment of employment equity (Cowdery and Taylor, Chapter 11; Mills *et al.*, Chapter 12; and Vadot *et al.*, Chapter 13) and the health and socioeconomic impacts of mining (Horowitz *et al.*, this volume; Myette, this volume). These chapters highlight the fact that supposed benefits and opportunities for Indigenous People and their communities are often not realized, creating tension, suspicion, and resentment between Indigenous Peoples and companies. This collective body of research emphasizes the importance of a multidisciplinary approach, with a view to better understanding and addressing such tensions in order to truly prioritize the rights and well-being of Indigenous communities. Beyond its academic and theoretical insights, this research also advocates for strategies and transformative changes that ensure more equitable and sustainable outcomes. Many of the researchers whose work appears in this book are specifically focused on generating knowledge and contributing to community-defined priorities. One area of concern remains the limited contribution of Indigenous Peoples' voices. This perennial concern is, of course, evident not only in research on extractive industries (Baker and Westman 2018; Joly *et al.* 2018) but also across all the social sciences (e.g., Pollard *et al.* 2020). In part, this shortcoming is a reflection of the adherence to Western definitions of knowledge and long-standing epistemological posture over how it is generated (Antoine 2017). Thankfully, there is growing acknowledgment of this issue across the social sciences in academia, and it is worth mentioning, in the context of this work, that greater insights and opportunities for change would likely emerge from the promotion of Indigenous voices to the forefront of works such as this one. Nevertheless, as the research detailed in Chapters 4, 5, 6, 9, 11, 12, and 13 demonstrates, these researchers and, more generally, the MinErAL network, itself, have attempted to prioritize Indigenous voices through participatory approaches and the return of findings to communities, in culturally appropriate formats.

Achieving these goals and addressing ongoing concerns have become more urgent as the race for critical minerals picks up in many countries. Demand for critical minerals has been growing steadily since the early 2010s, driven by the needs associated with the production of new technologies, including the so-called green technologies required for energy transition and carbon emission reduction. To meet these growing needs amidst supply chain tensions brought on by the COVID-19 pandemic and the Russia–Ukraine conflict, many countries have adopted public policies aimed at ensuring a steady supply of critical minerals (IEA 2022). While the regularity of supply and the matching of required minerals with demand are considered essential for a timely and smooth green energy transition, supply chains are highly dependent on a small number of countries extracting and exporting critical minerals. This means that supply is not only price-sensitive but also highly vulnerable to disruptions caused by, for example, geopolitical turmoil, environmental or social conflicts, and governance upheaval.

Faced with this situation, the United States, Canada, Europe, and other jurisdictions have decided to rethink their metal supply or production strategies in order to reduce their vulnerability to fluctuations in supply chains and position themselves in the global market for critical minerals. Certain regions of the world, such as France and Austria—which curtailed their mining activities at the end of the last century, relocating them to countries with lower labor costs and fewer environmental restrictions—are beginning to review their strategies. Mineral security and global instability have led to the reinvigoration of domestic mining production in Europe (*Wall Street Journal* 2023), even if this production is subject to higher and more closely monitored environmental and social performance standards. Other traditional mining jurisdictions, such as Canada, Australia, and several Latin American countries, have also developed policies to incentivize and speed up the development of their critical minerals, creating or renewing tensions with affected Indigenous and local communities (Kingsbury and Wilkinson 2023; Lorca *et al.* 2022; Owen *et al.* 2022). As Cowdery and Taylor (Chapter 11) explain, in these regions almost all of the population are First Nations people, and in Australia's case, the lands, too, are under the ownership of First Nations people through that country's land rights, legal processes, and successful claims.

While critical minerals are deemed essential to the global energy transition and the achievement of net-zero carbon emission targets, the local environmental, cultural, and socioeconomic impacts of their development are largely obscured. The rush for critical minerals places significant pressure on marginalized populations that are already negatively and disproportionately affected by ongoing trauma from colonialism, industrial development, and the impacts of climate change, exacerbating existing climate and environmental justice dynamics (Deberdt and Le Billon 2024; Owen *et al.* 2023; Sneddon 2023; Forget and Bos 2022; Lorca *et al.* 2022; Zografos and Robbins 2020). For example, some of the most sought-after metals are found in Melanesia, in the Pacific, resulting in increased pressure from extractive activities in one of the areas most affected by global warming, which is manifesting itself in rising water levels, violent cyclones, and other catastrophic events. The risk is that the intensification of mining activities will accentuate the effects of climate change on these small island territories, which will find themselves doubly exposed (Bebbington *et al.* 2015).

Indigenous communities and the mineral-rich territories near or on the lands where they are situated are significantly affected by climate change and by past and present extractive activities. The pressure for governments to meet emissions reductions targets is already focusing attention on the need to ramp up outputs from critical minerals (Finn and Stanton 2022; Lorca *et al.* 2022; Kingsbury and Wilkinson 2023; Schmaus 2023). This will be the case in Canada, Australia, Argentina, and Chile, where the majority of current and potential critical minerals projects are, or will be, located in or near Indigenous Peoples' traditional territories (Owen *et al.* 2023), exacerbating the legacy of past and present industrial activities and the impacts of climate change on their communities, lands, and environments (Fortin, Chapter 9; Deberdt and Le Billon 2024). The acceleration of mining for critical minerals, and the relaxation of the authorization processes that can be observed in many jurisdictions (Owen *et al.* 2022), will undoubtedly become more prevalent in future mining encounters. A primary goal of the MinErAL project was to facilitate knowledge translation to support Indigenous partners during these encounters. This book aims to complement the networking, relationship-building, and knowledge-sharing activities enabled by the MinErAL project. Its purpose is to provide a record of the outcomes and observations of mining encounters, in order to capitalize on and share experiences of the involvement of Indigenous Peoples and communities in mining encounters, for their own benefit. The book simultaneously aims to improve understanding among governments and mining companies of the impacts of mining on Indigenous communities.

References

Antoine, D., 2017. Pushing the academy: The need for decolonizing research. *Canadian Journal of Communication*, 42 (1), 113–120.

Baker, J. M., and Westman, C. N., 2018. Extracting knowledge: Social science, environmental impact assessment, and Indigenous consultation in the oil sands of Alberta, Canada. *The Extractive Industries and Society*, 5 (1), 144–153. https://doi.org/10.1016/j.exis.2017.12.008

Bebbington, A. J., Bury, J., Cuba, N., and Rogan, J., 2015. Mining, risk and climate resilience in the 'other' Pacific: Latin American lessons for the South Pacific. *Asia Pacific Viewpoint*, 56 (2), 189–207.

Deberdt, R., and Le Billon, P., 2024. Green transition's necropolitics: Inequalities, climate extractivism, and carbon classes." *Antipode*. https://doi.org/10.1111/anti.13032.

Forget, M., and Bos, V., 2022. Harvesting lithium and sun in the Andes: Exploring energy justice and the new materialities of energy transitions. *Energy Research & Social Science*, 87, 102477.

Finn, K. R., and Stanton, C. A., 2022. The (un) just use of transition minerals: How efforts to achieve a low-carbon economy continue to violate Indigenous rights. *Colorado Environmental Law Journal*, 33, 341.

IEA, 2022. *Introducing the critical minerals policy tracker*. Paris: IEA. www.iea.org/reports/introducing-the-critical-minerals-policy-tracker License: CC BY 4.0

Joly, T. L., Longley, H., Wells, C., and Gerbrandt, J., 2018. Ethnographic refusal in traditional land use mapping: Consultation, impact assessment, and sovereignty in the Athabasca oil sands region. *The Extractive Industries and Society*, 5 (2), 335–343. https://doi.org/10.1016/j.exis.2018.03.002

Keeling, A., and Sandlos, J., 2017. Ghost towns and zombie mines: The historical dimensions of mine abandonment, reclamation, and redevelopment in the Canadian North. *In*: S. Bocking and B. Martin, eds., *Ice blink: Navigating northern environmental history*. Calgary: University of Calgary Press, 377–420.

Kingsbury, D., and Wilkinson, A., 2023. 'We are a mining region': Lithium frontiers and extractivism in Abitibi-Témiscamingue, Canada. *The Extractive Industries and Society*, 15 (9), 101330. www.sciencedirect.com/science/article/abs/pii/S2214790X2300120X

Lorca, M., Olivera Andrade, M., Escosteguy, M., Köppel, J., Scoville-Simonds, M., and Hufty, M., 2022. Mining Indigenous territories: Consensus, tensions and ambivalences in the Salar de Atacama. *The Extractive Industries and Society*, 9 (3), 101047. https://doi.org/10.1016/j.exis.2022.101047.

Owen, J. R., Kemp, D., Harris, J., Lechner, A. M., and Lèbre, É., 2022. Fast track to failure? Energy transition minerals and the future of consultation and consent. *Energy Research & Social Science*, 89, 102665.

Owen, J. R., Kemp, D., Lechner, A. M., Harris, J., Zhang, R., and Lèbre, É., 2023. Energy transition minerals and their intersection with land-connected peoples. *Nature Sustainability*, 6, 203–211. https://doi.org/10.1038/s41893-022-00994-6

Pollard, K., Smith, C., and Willika, J., 2020. Indigenous views on the future of public archaeology in Australia. A conversation that did not happen. *AP: Online Journal in Public Archaeology*, (10), 31–52.

Schmaus, L., 2023. *Mining and climate change vulnerability: A case study of the Tłı̨chǫ.* Master's Thesis. University of Alberta.

Sneddon, S., 2023. Paradise lost? The red right hand of green technology. *Journal of Human Rights and the Environment*, 14 (2), 169–193.

Wall Street Journal, 2023. Europe is embarking on a mining renaissance. Winning over locals is proving a challenge. Available from: www.wsj.com/articles/europe-is-embarking-on-a-mining-renaissance-winning-over-locals-is-proving-a-challenge-b7d14f5f [Accessed 14 February 2024].

Zografos, C., and Robbins, P., 2020. Green sacrifice zones, or why a green New Deal cannot ignore the cost shifts of just transitions. *One Earth*, 3 (5), 543–546.

Postface

MinErAL partner reflections

A: How MinErAL expose the realities of mining development in Nunavik

Over the past two decades, I've had a privileged vantage point to witness the evolution of mining development in the Nunavik region in Northern Québec, Canada. From my successive positions as project geologist and technical director of the Nunavik Mineral Exploration Fund, then as mining development manager on behalf of Makivik, the land claim organization of the Nunavik Inuit, I have monitored the increase in mining claims designations, mineral exploration activities, and mining operations in the region. Despite its remoteness, limited access and infrastructures, and harsh seasonal conditions, the territory is known to hold enormous, unexplored mineral resource potential, a potential that has attracted the interest of the mining industry. The Inuit communities are well acquainted with mining development and have been directly experiencing its various effects, both positive and negative, since the early 1950s. The provincial plan for the development of critical and strategic minerals is now launching a new chapter of accelerated exploration and evaluation of Nunavik's mineral resources. Over the years, part of my role has been to provide Nunavik leaders and regional stakeholders with relevant information on mining projects, technical recommendations, and assistance to help communities engage with mining developers and government agencies, contributing in the best possible way to enable the Inuit to obtain all the right tools and levers to ensure the best decision-making and participation in the mining development process.

Right from the initial invitation and the first meeting of the MinErAL network, I became aware of the vast geographical and interdisciplinary scope of the proposed research projects and studies underway. Bringing together a wide range of experts and researchers interested in the complex relationships between Indigenous groups and mining development, the network has provided an opportunity to share perspectives and discussions from different backgrounds in a mutual way, helping identify the best (and worst!) practices being implemented and highlighting respective approaches and—every so often—similarities across different jurisdictions and territories. The numerous opportunities to share experiences and strategies through conferences and workshops have enabled researchers and network partners to not only enhance their expertise and knowledge but also to forge lasting links and relationships that translate into ongoing collaborations on specific projects and initiatives.

Being part of the MinErAL network has not only provided access to interdisciplinary results and recognized experts leading projects within our communities, but it has also helped expose and share some of the realities of mining development in the Nunavik region. As a partner in the network, it has been stimulating to relate and obtain comparative international results on similar elements spanning the social, environmental, and economical aspects observed in different contexts. The network's annual meetings, conferences, and webinars have facilitated exchanges with other partners who experience common challenges in their mining encounters. These interactions have spurred the creation of policies and frameworks that have proven to be instrumental for the socioeconomic development of communities where the local subsistence economy is often confronted with the industrial market economy.

Since the network's inception, the Nunavik region has been able to contribute to several research initiatives and assist authors through interviews and fieldwork to explore, among other topics, Inuit women's challenges and roles, inequities, and empowerment within operating mining projects; labor relations affecting Inuit employment; community consent, agreement negotiations, and implementation gaps; regional- and community-level socioeconomic development and economic benefits from mining operations; and the social effects resulting from mine closure and remediation. Finally, the opportunity created by the network to travel and meet communities *in situ* has enabled us to observe first-hand how mining and mining operations directly and indirectly influence the daily lives of community members. These visits and meetings have also generated rich exchanges and a clear understanding of the synergies and conflicts between mining promoters and all the multilayered community stakeholders involved. Listening to the testimonies, stories, expectations, and disappointments—some of them emotionally charged—has given us a better understanding of the stakes involved in mining, and regardless of geographical and cultural differences, it is obvious that Indigenous communities face the same challenges and impacts to varying degrees.

It is with great pride that I have observed this book come to fruition, for while it identifies the challenges faced by Indigenous communities confronted with mining development, it also provides avenues for solutions, perspectives, and ideas to improve the relations of communities with the mining industry and governmental bodies and, ultimately, the living conditions of Indigenous populations.

—Jean-Marc Séguin, P.Geo., M.Sc.
Liaison Agent, Qiniqtiq Landholding Corporation of Kangiqsualujjuaq

B: The impact of the MinErAL from the perspective of a Kanak

Although our senior management was initially somewhat averse to the idea of an industry representative taking part in this forum, the ensuing discussions and participation provided us with an opportunity to step back from our daily actions in our relations not only with the communities but also with the world of research.

As it turned out, the MinErAL conferences in Canada and Australia proved to be highly enriching. With topics that addressed mines and the local communities through various lenses and, especially, from a research perspective, I, personally, found it a valuable opportunity to reflect, and it made me more aware of the many similarities among Australia, Norway, Canada, and New Caledonia. The concerns and claims of Indigenous Peoples with regard to industry are real, and their relations with government and elected officials are often precarious.

Indigenous representatives—whether they be Kanak, Inuit, Sami, Aboriginal, or other—share a common goal of preserving their culture and playing a role in development, while mitigating its negative impacts.

While environmental monitoring and local community involvement through employment or contracts are controlled to some degree, the distribution of dividends or royalties often leads to negative consequences in these populations.

Together with representatives from Inuit communities, we discussed the toll on their people's health (drugs and other substance abuse), their relationship to money, and the struggle for influence among clans, families, and chiefs.

These are some of the challenges these communities face, not only with regard to industry and mining policies (where they exist) but also among their own people.

While the Koniambo Project in New Caledonia is unique due to the involvement of pro-independence political parties and the desire to fund the country's independence, the fact remains that the global nickel sector is highly sensitive to market fluctuations.

MinErAL also provided a forum for discussing best practices with the participating communities and with the various researchers and young doctoral students who are valuable resources, with their systemic approach to relations between local communities and the mining industry in both local and global contexts.

The MinErAL network contributed to comparative research on subjects associated with mining and local communities, from Caledonia and Australia to Canada, including, for example, on women's employment in the mining industry.

These discussions allowed us to desacralize the "masculine world of the mine" and address the role of women within the various communities and the strategies they have developed to strike a balance between professional and family life.

When examined separately, the economy, society, the environment, corporate social responsibility, and the energy transition are just concepts, but within the MinErAL network, these concepts take on meaning not only for those of us who work in the industry but also for the scientists and community representatives.

My sincere gratitude to MinErAL for this initiative that, since its creation, has enabled all the direct and indirect actors in the mining sector to get together at least once a year to discuss a wide range of topical issues in a serene atmosphere. And, of course, my special thanks to Thierry Rodon, the initiator of this project.

Jean-Louis Thydjepache
Community Relations Officer,
Koniambo Nickel SAS

Index

Made in the USA
Las Vegas, NV
19 September 2026

b3752232-f3ac-443c-8a1c-47ce8768554bR02